D1090438

Lavery Library
EX LIBRIS
ST. JOHN FISHER COLLEGE

Reagents for Organic Synthesis

ADVISORS

Fieser and Fieser's

Reagents for Organic Synthesis

VOLUME FIFTEEN

Mary Fieser
Harvard University

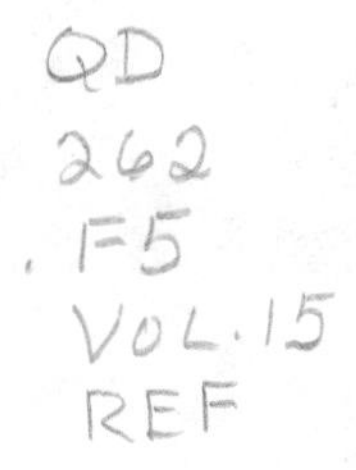

A WILEY-INTERSCIENCE PUBLICATION
John Wiley & Sons, Inc.
NEW YORK / CHICHESTER / BRISBANE / TORONTO / SINGAPORE

ISBN 0-471-52113-2
ISSN 0271-616X
Printed in the United States of America
10 9 8 7 6 5 4 3 2 1

PREFACE

This volume of Reagents includes reports on synthetic uses of reagents published for the most part in 1988 and during the early months of 1989. The manuscript had the benefit of careful review by James R. Gage and Scott Virgil; their advice has been invaluable. In addition, both the galleys and page proofs have been scrutinized carefully by the advisors for this volume. Greg Fu drew the formula for the dustcover. Tarek Sammakia (Dr. T) is the photographer for the picture of some of the advisors.

MARY FIESER

Cambridge, Mass.
February 11, 1990

CONTENTS

Reagents for Organic Synthesis

Acetic anhydride–Pyridine.

Dakin–West reaction.[1] This reaction involves conversion of an α-amino acid to an α-acetylaminoalkyl methyl ketone by reaction with an acid anhydride and a base at 25–135°. When catalyzed by DMAP, the reaction takes place at room

$$H_2N\underset{}{\overset{R}{C}}HCOOH \xrightarrow{Ac_2O,\ Py} AcNH\overset{R}{C}HCOCH_3 + CO_2$$

temperature. Secondary amino acids also can undergo this reaction as well as acids that lack an amino group. The review deals mainly with the preparative scope (64 references).

[1] G. L. Buchanan, *Chem. Soc. Rev.*, **17**, 91 (1988).

Acetonitrile.

Rearrangement of (α-methyldiphenylsilyl)alkyl ketones.[1] These α-silyl ketones rearrange thermally to a mixture of (Z)- and (E)-enol silyl ethers. However, rearrangement in acetonitrile results in only the (Z)-enol silyl ethers (>99:1). These enol silyl ethers are useful precursors to (Z)-lithium enolates.

$$n\text{-}C_8H_{17}\underset{CH_3Si(C_6H_5)_2}{CH}\overset{O}{\overset{\|}{C}}Pr \xrightarrow{CH_3CN} \underset{(Z,\ >99:1)}{n\text{-}C_8H_{17}(H)C{=}C(Pr)OSiCH_3(C_6H_5)_2}$$

$$\xrightarrow{1)\ CH_3Li;\ 2)\ ClSi(CH_3)_3} n\text{-}C_8H_{17}(H)C{=}C(Pr)OSi(CH_3)_3$$

[1] G. L. Larson, R. Berrios, and J. A. Prieto, *Tetrahedron Letters*, **30**, 283 (1989).

2-(Acetoxymethyl)-3-trimethylsilylpropene, $(CH_3)_3SiCH_2\overset{CH_2}{\overset{\|}{C}}CH_2OAc$ **(1), 11,** 259.

Improved preparation:[1]

$$H_2C=C(CH_2Cl)CH_2Cl \xrightarrow[60\%]{Cl_3SiH,\ CuCl,\ N(C_2H_5)_3,\ ether} H_2C=C(CH_2SiCl_3)CH_2Cl \xrightarrow{1)\ CH_3MgBr\ (81\%)\ 2)\ KOAc,\ DMF\ (90\%)} \mathbf{1}$$

[1] B. M. Trost, M. Buch, and M. L. Miller, *J. Org.*, **53**, 4887 (1988).

Acyloxyboranes. Yamamoto *et al.*[1] have used the known reactivity of borane with carboxylic acids to activate acrylic acids for Diels–Alder reactions. Thus addition of $BH_3 \cdot THF$ to acrylic acids at 0° furnishes an acyloxyborane formulated as **1**, which undergoes cycloaddition (equations I and II). The reaction proceeds satisfactorily even when borane is used in catalytic amounts. A chiral acyloxyborane, BL_n^*, prepared from a tartaric acid derivative, can serve as a catalyst for an asymmetric Diels–Alder reaction (equation III).

(I) $CH_2{=}CHCOOH \xrightarrow{BH_3 \cdot THF} \mathrm{>BOC(O)CH{=}CH_2}$ (**1**) $\xrightarrow[72\%]{\text{cyclopentadiene}}$ bicyclo[2.2.1]hept-5-ene-2-carboxylic acid (COOH)

(II) $CH_2{=}C(CH_3)COOH \xrightarrow{BH_3 \cdot THF} \mathrm{>BOC(O)C(CH_3){=}CH_2} \xrightarrow[49\%]{CH_2{=}C(CH_3){-}C(CH_3){=}CH_2}$ 1,3,4-trimethylcyclohex-3-ene-1-carboxylic acid (CH_3, CH_3, COOH, CH_3)

(III) $CH_2{=}CHCOOH$ + cyclopentadiene $\xrightarrow[93\%]{BL_n^*}$ bicyclo[2.2.1]hept-5-ene-2-carboxylic acid (COOH)

(78% ee)

[1] K. Furuta, Y. Miwa, K. Iwanaga, and H. Yamamoto, *Am. Soc.*, **110**, 6254 (1988).

Alkylaluminum halides.

[4+2]Cycloaddition of N-benzylimines and a diene. Diethylaluminum chloride is the best Lewis acid for promoting addition of N-substituted imines (**2**) with the activated diene **1** (**9**,302–303).[1]

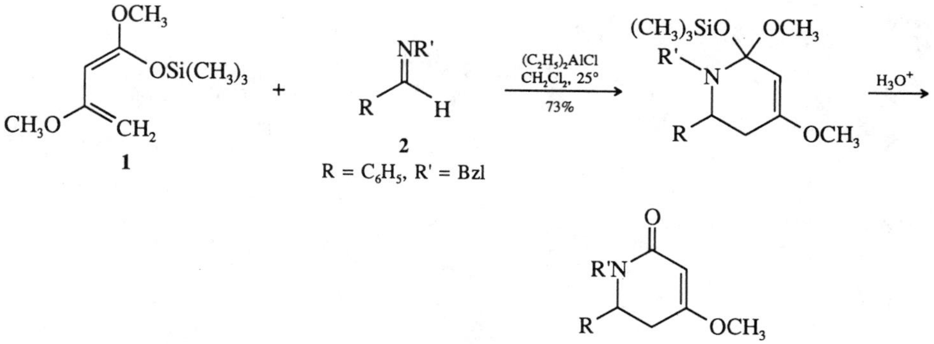

The reaction of the cyclohexylidene ketal **3** of a chiral imine, prepared from L-threonine, gives a single adduct (**4**), which is converted by known reactions into *xylo*-daunosamine (**5**).

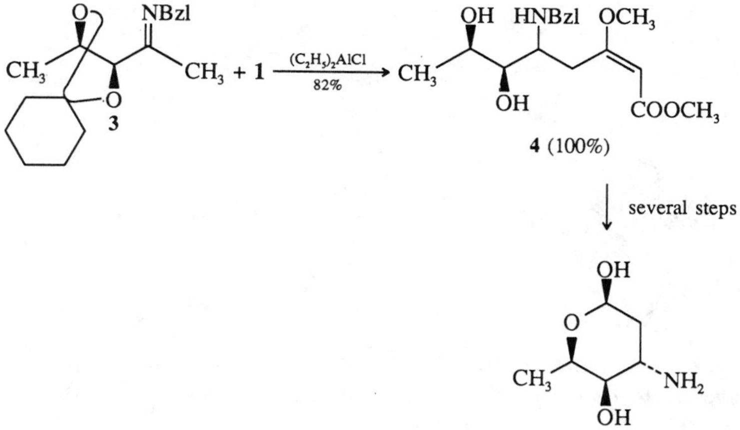

Asymmetric Diels–Alder reaction.[2] The chiral menthyl (S)-3-(2-pyridylsulfinyl)acrylate (**1**) undergoes [4+2]cycloaddition with the furan **2** in the presence of $(C_2H_5)_2AlCl$ at $-20°$ to give the *endo*- and *exo*-adducts **3** in the ratio ~2:1, both in about 93% de. The *endo*-adduct (**3**) was converted by known reactions to **4**, which is opened by lithium hexamethyldisilazide (**12**,257) to the unsaturated acid in 56% yield. Remaining steps to methyl (−)-triacetoxyshikimate (**5**) include debenzylation and acetylation. A similar sequence with *exo*-2 should provide (+)-shikimic acid.

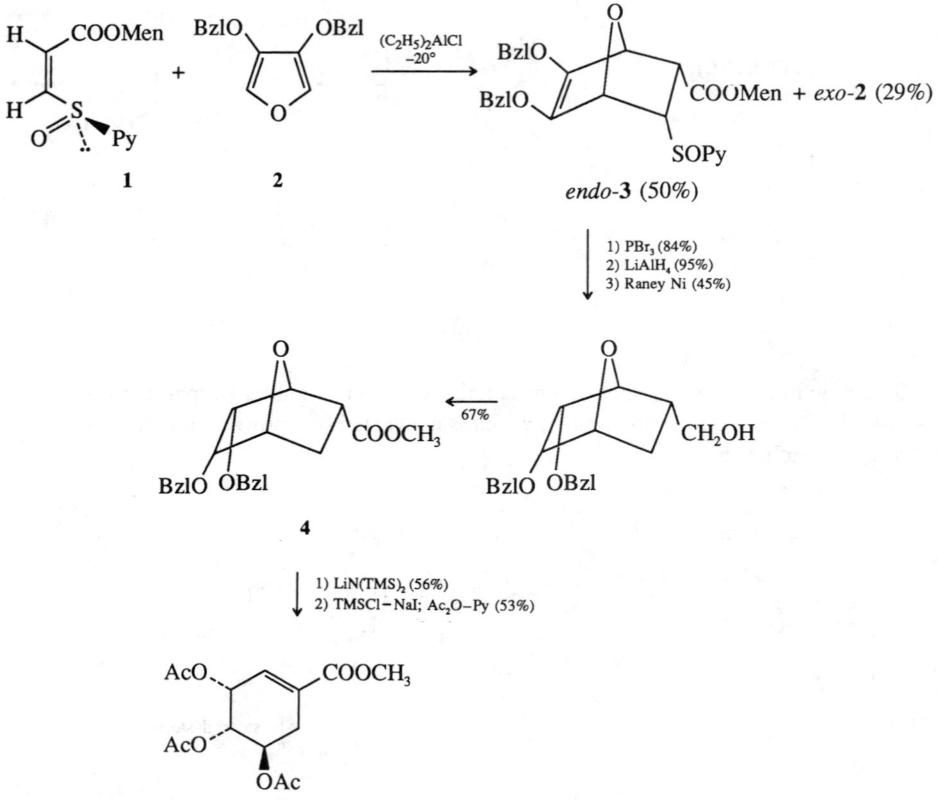

H COOMen BzlO OBzl $(C_2H_5)_2AlCl$ −20° BzlO BzlO COOMen + *exo*-**2** (29%) H S Py O SOPy

1 **2** *endo*-**3** (50%)

1) PBr_3 (84%)
2) $LiAlH_4$ (95%)
3) Raney Ni (45%)

O COOCH$_3$ 67% O CH$_2$OH BzlO OBzl BzlO OBzl

4

1) LiN(TMS)$_2$ (56%)
2) TMSCl−NaI; Ac$_2$O−Py (53%)

AcO COOCH$_3$ AcO OAc

5, α_D−168°

Stereoselective ene reaction.[3] The reaction of the $\Delta^{17(20)}$-steroid alkene **1** with acetylenic aldehydes catalyzed by $(CH_3)_2AlCl$ proceeds with high diastereoselectivity (about 95:5) at both chiral centers.

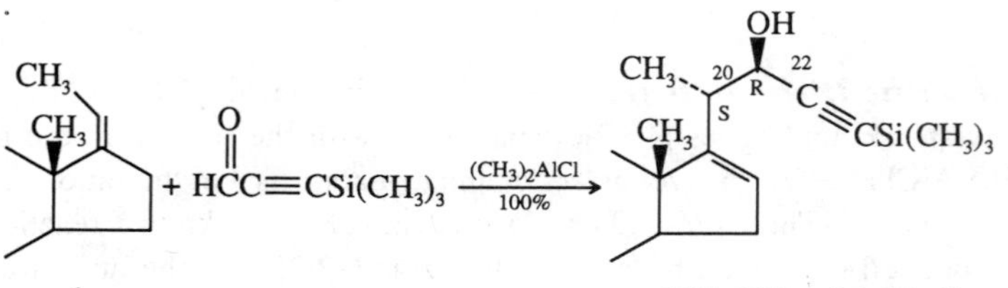

CH$_3$ CH$_3$ O + HCC≡CSi(CH$_3$)$_3$ $(CH_3)_2AlCl$ 100% OH CH$_3$ 20 S R 22 C ≡CSi(CH$_3$)$_3$ CH$_3$

1 (20S, 22R) –**2** (>95:<5)

N-Arylsulfoximines.[4] These imines can be obtained in satisfactory yield by reaction of an arylamine with *p*-toluenesulfinyl chloride followed by oxidation

$$\mathrm{ArNH_2} \xrightarrow[\text{2) } t\text{-BuOCl, } CH_2Cl_2]{\text{1) } p\text{-}CH_3C_6H_4SOCl} \left[C_7H_7-\overset{O}{\underset{N\diagdown Ar}{\overset{\|}{\underset{\|}{S}}}}-Cl \right] \xrightarrow[65\text{-}95\%]{C_2H_5AlCl_2, CH_2Cl_2} C_7H_7-\overset{O}{\underset{N\diagdown Ar}{\overset{\|}{\underset{\|}{S}}}}-C_2H_5$$

(t-butyl hypochlorite) to a sulfoxide. The resulting sulfonimidoyl chloride is then treated with $C_2H_5AlCl_2$ in CH_2Cl_2 at $-78°$ to provide S-ethyl sulfoximines in 65–95% overall yield.

[1] M. M. Midland and J. I. McLoughlin, *Tetrahedron Letters*, **29**, 4653 (1988).
[2] T. Takahashi, T. Namiki, Y. Takeuchi, and T. Koizumi, *Chem. Pharm. Bull.*, **36**, 3213 (1988).
[3] K. Mikami, T.-P. Loh, and T. Nakai, *J.C.S. Chem. Comm.*, 1430 (1988).
[4] M. Harmata, *Tetrahedron Letters*, **30**, 437 (1989).

Alkyl cobalt(III)bis(dimethylglyoxime anion)pyridine, [R-Co$^{(III)}$(dmgH)$_2$py,**1**]

1

These cobaloximes (**1**) are generated by reaction of alkyl halides with *in situ* generated Co^{I}(dmgH)$_2$py. They are orange crystals that are relatively stable to air and silica gel, but on photolysis they generate R·, which can be trapped in generally high yield or which cyclize in the case of alkenyl radicals.[1]

$$\mathbf{1} + C_6H_5SSC_6H_5 \xrightarrow{h\nu} RSC_6H_5 + R{-}R$$

	RSC_6H_5	R—R
a) $R = CH_3(CH_2)_9$	90%	0
b) $R = CH_3(CH_2)_4$	92%	0

$$CH_2{=}CH(CH_2)_4{-}Co^{III}(dmgH)_2py \xrightarrow[66\%]{(C_6H_5S)_2,\ h\nu,\ C_6H_6} CH_2{=}CH(CH_2)_4SC_6H_5 + \text{cyclopentyl-}CH_2SC_6H_5 \quad (1:12)$$

The R· radicals can also undergo cross coupling with alkenes and with heteroarenes.[2]

1a + quinolinium tosylate (N–H, $^{-}$OTs) $\xrightarrow[87\%]{h\nu,\ C_2H_5OH}$ 4-$C_{10}H_{21}$-quinoline + 2-$C_{10}H_{21}$-quinoline (1:1)

1a + benzimidazolium tosylate ($^{-}$OTs) $\xrightarrow{32-42\%}$ 2-$C_{10}H_{21}$-benzimidazole

[1] B. P. Branchaud, M. S. Meier, and M. N. Malekzadeh, *J. Org.*, **52**, 212 (1987).
[2] B. P. Branchaud and Y. L. Choi, *ibid.*, **53**, 4638 (1988).

Alkynyl(phenyl)iodonium tosylates, $RC{\equiv}CIC_6H_5{\cdot}OTs$, **14**,10–11.

Full details for an improved method for preparation of these polyvalent iodine compounds are available.[1] The method is similar to that used to prepare vinyliodonium tetrafluoroborates (**13**,151), and yields are 60–89%.

Example:

$$BuC{\equiv}CH \xrightarrow[88\%]{1)\ BuLi,\ THF;\ 2)\ ClSi(CH_3)_3} BuC{\equiv}CSi(CH_3)_3 \xrightarrow[60-75\%]{1)\ C_6H_5IO,\ BF_3\cdot O(C_2H_5)_2;\ 2)\ NaOTs} BuC{\equiv}CIC_6H_5\cdot OTs$$

m.p., 77–80°

***(E)*-1,3-*Enynes*.**[2] These salts couple with vinylcopper reagents to give (E)-1,3-enynes in yields of 46–94%.

$$BuMgBr \xrightarrow{1)\ CuBr\cdot S(CH_3)_2;\ 2)\ C_6H_5C{\equiv}CH;\ 3)\ BuC{\equiv}CIC_6H_5\cdot OTs} (C_6H_5)(Bu)C{=}C(H)C{\equiv}CBu\ (58\%) + Bu(Ph)C{=}CH{-}CH{=}C(Ph)Bu\ (\text{trace})$$

[1] P. J. Stang and T. Kitamura, *Org. Syn.*, submitted (1988).
[2] *Idem*, *Am. Soc.*, **109**, 7561 (1987).

1-Alkynyltriisopropoxytitanium, $RC{\equiv}CTi(O\text{-}i\text{-}Pr)_3$. These titanium acetylides are prepared by reaction of 1-alkynes with BuLi and $ClTi(O\text{-}i\text{-}Pr)_3$.

Cleavage of epoxides.[1] These reagents react with unsymmetrical epoxides selectively at the more-substituted carbon atom to provide 2-substituted 3-butyn-1-ols in 33–79% yield. However, the cleavage requires an electron-releasing group at the epoxide, and it takes place with partial or complete retention of configuration.

Example:

$$C_6H_5\text{-epoxide} + C_6H_5C{\equiv}CTi(O\text{-}i\text{-}Pr)_3 \xrightarrow[(70\%)]{} C_6H_5CH(CH_2OH)C{\equiv}CC_6H_5$$

[1] N. Krause and D. Seebach, *Ber.*, **121**, 1315 (1988).

Allene.

Photochemical* [2+2]*cycloaddition to enones.[1] This reaction has been shown to occur mainly from the less-hindered α-side of steroidal enones.[2] Cycloaddition to the cyclopentenone (**1**) is also stereoselective, but the stereoselectivity and the yields are dependent upon the solvent. The highest yields are generally obtained in an apolar solvent (such as hexane).

$$\text{RO-cyclopentenone} + CH_2{=}C{=}CH_2 \xrightarrow{h\nu} \text{RO-bicyclo (CH}_2\text{ adjacent)} + \text{RO-bicyclo (CH}_2\text{ distal)}$$

R	Solvent	Yield	*endo/exo*	
R = *t*-Bu	C_6H_{12}	76%	49:41	95:5
	CH_3CN	58%	38:62	89:11
R = H	C_6H_{12}	27%	63:37	87:13
	CH_3CN	45%	62:38	90:10

[1] W. Stensen, J. S. Svendsen, O. Hofer, and L. K. Sydnes, *Acta Chem. Scand.*, **B42**, 259 (1988).

[2] R. Farwaha, P. de Mayo, J. H. Schauble, and Y. C. Toong, *J. Org.*, **50**, 245 (1985).

Allenylsilanes.

[3+2]*Annelation; azulenes.*[1] A new route to azulenes involves reaction of tropylium cations with allenylsilanes, which generates a vinyl cation (**a**). This cation cyclizes to a cycloheptadienyl cation, which can be converted to a dihydroazulene by elimination of H^+. However, the presence of (excess) tropylium cation favors aromatization to an azulene. Protodesilylation of the allenylsilane can be effected by use of trimethoxymethylsilane as an acid scavenger. Yields are low (~25%)

BF_4^- + H, R'–C(3)=C=C(1)–SiR_3, CH_3 $\xrightarrow[CH_3CN,\ 25°]{CH_3Si(OCH_3)_3}$ [cycloheptatrienyl–CH(R')–C^+=C(CH_3)(SiR_3) BF_4^-] **a**

R_3 = *t*-$BuMe_2$,
R' = $(CH_2)_3C_6H_5$

a → [CH_3, SiR_3, H, R'] **b** $\xrightarrow{57\%}$ CH_3, SiR_3, R'

with allenylsilanes lacking an alkyl group at C_3, and an alkyl group at C_1 is essential for this synthesis.

[1] D. A. Becker and R. L. Danheiser, *Am. Soc.*, **111**, 389 (1989).

Allyltrimethylsilane.

Conjugate addition to cyclopropanes.[1] In the presence of $C_2H_5AlCl_2$, allyltrimethylsilane undergoes conjugate addition to cyclopropanes substituted by *gem*-alkoxycarbonyl groups.

$$(CH_3)_3SiCH_2CH{=}CH_2 + \text{(cyclopropane: } C_2H_5;\ COOC_2H_5,\ COOC_2H_5) \xrightarrow[92\%]{C_2H_5AlCl_2,\ CH_2Cl_2,\ 0°} H_2C{=}CH\text{–}CH_2\text{–}CH(C_2H_5)\text{–}CH_2\text{–}CH(COOC_2H_5)\text{–}COOC_2H_5$$

Review.[2]

[1] R. Bambal and R. D. W. Kemmitt, *J.C.S. Chem. Comm.*, 734 (1988).
[2] A. Hosomi, *Accts. Chem. Res.*, **21**, 200 (1988).

(S)-B-Allyl-2-(trimethylsilyl)borolane (1).

The borolane is prepared by kinetic resolution of (±) B-methoxy-2-trimethylsilylborolane **2** with (1S,2S)-(+)-N-methylpseudoephedrine (0.5 equiv.) to give (+)-**3**. Reaction of (+)-**3** with allylmagnesium bromide provides the borolane (S)-**1**.

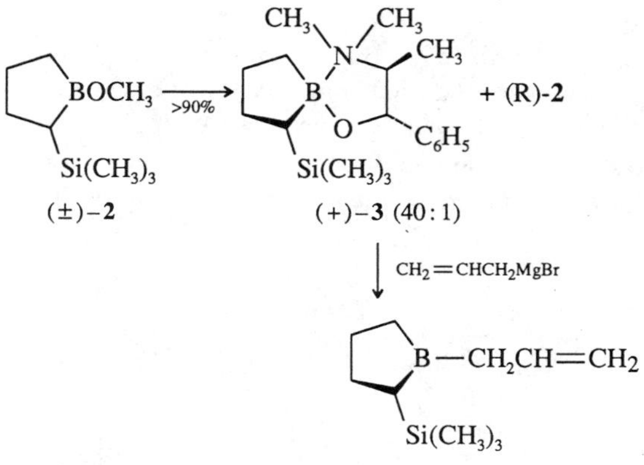

$$(\pm)\text{-}\mathbf{2}\ \xrightarrow{>90\%}\ (+)\text{-}\mathbf{3}\ (40:1) + (R)\text{-}\mathbf{2}$$

BOCH₃ replaced: see scheme labels below.

Asymmetric allylboration of RCHO.[1] (S)-**1** reacts with aliphatic or aryl aldehydes or with α,β-enals to form homoallylic alcohols in 92–97% ee and 80–92% chemical yield. The chemical and optical yields are higher than those obtained with B-allyldiisopinocamphenylborane (**14**,12), with allylboronates modified with tartrates, or with B-allyltrimethylsilylboronates. The high asymmetric induction is believed to result from steric effects rather than electronic effects.

[1] R. P. Short and S. Masamune, *Am. Soc.*, **111**, 1892 (1989).

Alumina

Diels–Alder catalyst.[1] Attempted purification of **1** by chromatography on neutral Al_2O_3 results in an intramolecular Diels–Alder reaction at 20° in 83% yield. The product (**2**) is verrucarol, a typical mycotoxic sesquiterpene.

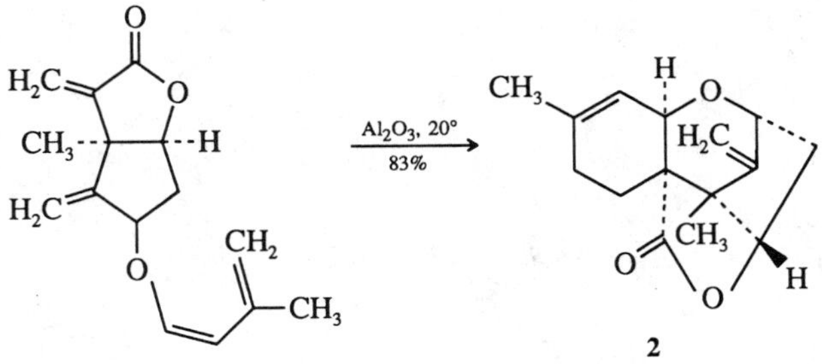

$$\mathbf{1}\ \xrightarrow[83\%]{Al_2O_3,\ 20^\circ}\ \mathbf{2}$$

[1] M. Koreeda, D. J. Ricca, and J. I. Luengo, *J. Org.*, **53**, 5586 (1988).

Aluminum chloride.

De-t-butylation of phenols.[1] *t*-Butyl groups are often used for protection of various positions of arenes, and are commonly removed by a reaction involving a *t*-butyl acceptor plus a Lewis acid. Surprisingly, $AlCl_3$ alone can effect de-*t*-butylation of phenols at room temperature in nitromethane or methylene chloride. Removal of an *ortho-t*-butyl group occurs more readily than that of a *para*-group.

Examples:

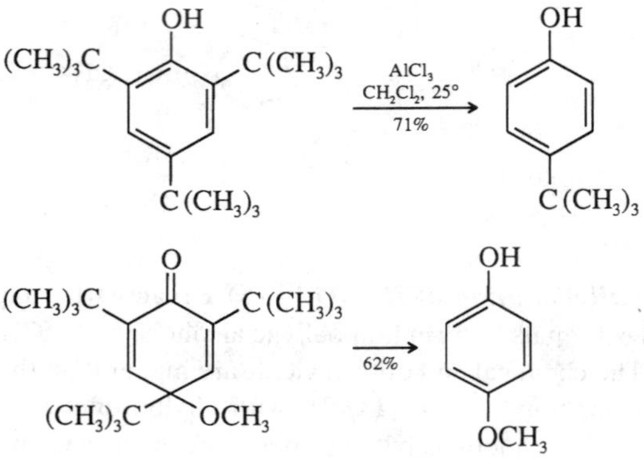

$$\text{2,4,6-tri-}t\text{-butylphenol} \xrightarrow[71\%]{AlCl_3,\ CH_2Cl_2,\ 25°} \text{4-}t\text{-butylphenol}$$

$$\text{2,4,6-tri-}t\text{-butyl-4-methoxycyclohexa-2,5-dienone} \xrightarrow{62\%} \text{4-methoxyphenol}$$

[1] N. Lewis and I. Morgan, *Syn. Comm.*, **18**, 1783 (1988).

Aluminum(III) iodide.

Deoxygenation of epoxides.[1] The reagent effects deoxygenation of epoxides via an intermediate iodohydrin (**11**,30), which can be isolated as the *trans*-isomer in some cases. This reagent can effect selective reactions in the case of some diepoxides.

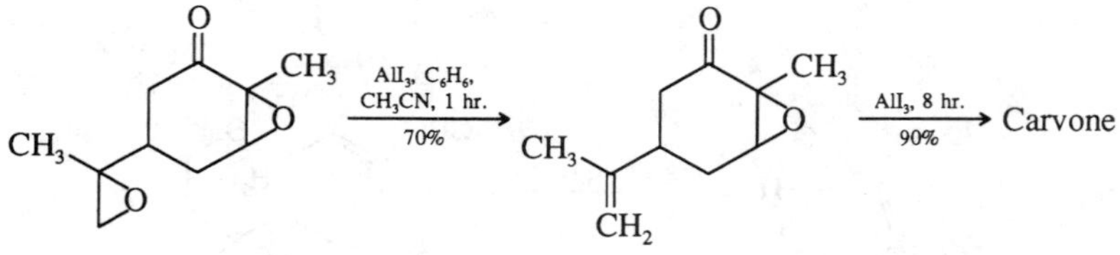

$$\text{carvone diepoxide} \xrightarrow[70\%]{AlI_3,\ C_6H_6,\ CH_3CN,\ 1\ hr.} \text{carvone monoepoxide} \xrightarrow[90\%]{AlI_3,\ 8\ hr.} \text{Carvone}$$

[1] P. Sarmah and N. C. Barua, *Tetrahedron Letters*, **29**, 5815 (1988).

O-(2-Aminobenzoyl)hydroxylamine,

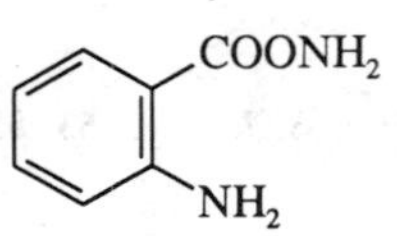

$$2\text{-}H_2NC_6H_4COONH_2 \quad (\mathbf{1}).$$

Preparation from isatoic anhydride.[1]

ArCHO→ArCN.[2] Reaction of aromatic aldehydes with **1** in refluxing ethanol containing $BF_3 \cdot O(C_2H_5)_2$ provides the corresponding nitrile in 78–94% yield.

[1] F. D. Eddy, K. Vaughan and M. F. G. Stevens, *Canad. J. Chem.*, **56**, 1616 (1978).
[2] P. S. N. Reddy and P. P. Reddy, *Syn. Comm.*, **18**, 2179 (1988).

(S)-(−)-1-Amino-2-(1-methoxy-1-methylethyl)pyrrolidine (SADP).

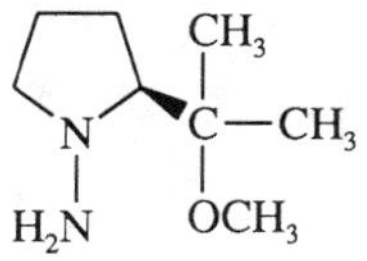

(**1**), b.p. 65°/4.5mm., α_D–15.4°

Hantzsch dihydropyridine synthesis.[1] The original Hantzsch synthesis[2] involves condensation of two equivalents of a keto ester with an aldehyde in the presence of ammonia. In an enantioselective version.[3] the chirality is introduced by use of a chiral hydrazone (**2**) of an alkyl acetoacetate prepared from **1**. The anion of **2** is then treated with Michael acceptors to form adducts (**3**), which cyclize to 4-aryl-1,4-dihydropyridines (**4**), in 64–72% overall yield and in 84–98% ee.

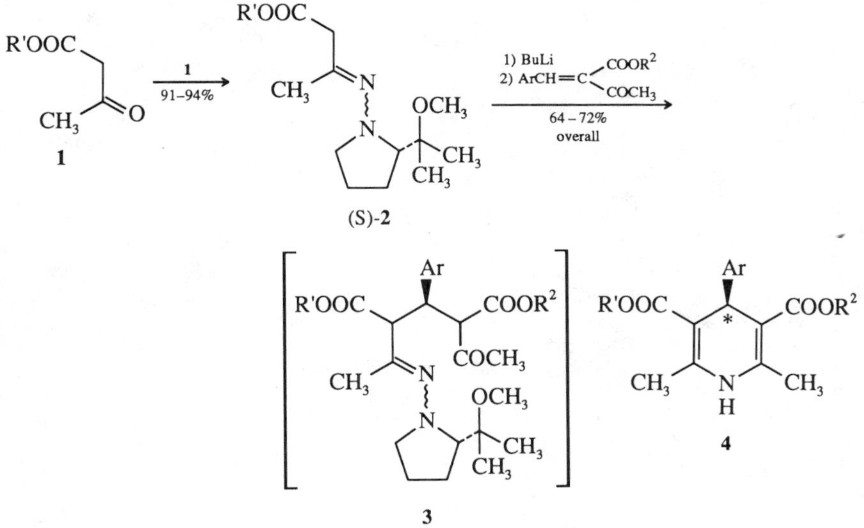

1 → (1, 91–94%) → (S)-2 → (1) BuLi; 2) ArCH=C(COOR²)(COCH₃); 64–72% overall) → [3] → 4

[1] D. Enders, H. Kipphardt, P. Gerdes, L. J. Brēna-Valle, and V. Bhushan, *Bull. Soc. Chim. Belg.*, **97**, 691 (1988).
[2] A. Hantzsch, *Ann.*, **215**, 1,72 (1882); A. Singer and S. M. McElvain, *Org. Syn.*, Coll. Vol. II, 214 (1943).
[3] D. Enders, S. Müller, and A. S. Demir, *Tetrahedron Letters*, **29**, 6437 (1988).

Ammonia–Acetic acid.

Tetracyclization. The Heathcock group[1] has described a remarkably short and efficient synthesis of the skeleton of Daphniphyllum alkaloids (**2**) by reaction of the dialdehyde **1** with gaseous ammonia and then dissolution in acetic acid at 70°. The yield is 77%, based on the diol precursor to **1**. The azadiene **a** and the imine **b** have both been isolated and identified. The conversion of **a** to **b** is an intramolecular Diels–Alder type reaction. The tetracyclization may well be involved in the biosynthesis of alkaloids such as Daphnilactone A (**3**).

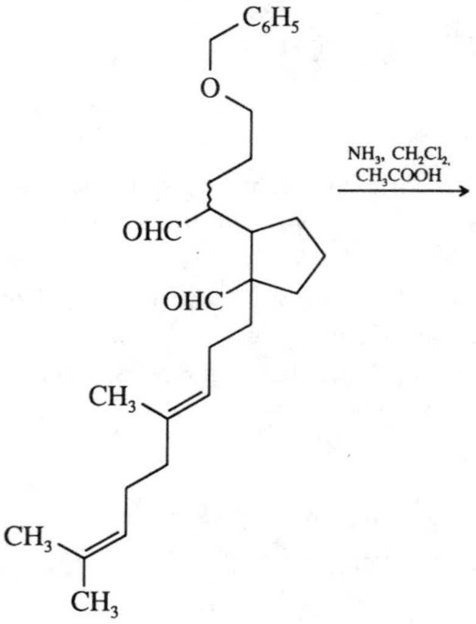

C_6H_5
O
OHC
OHC
CH_3
CH_3
CH_3
1

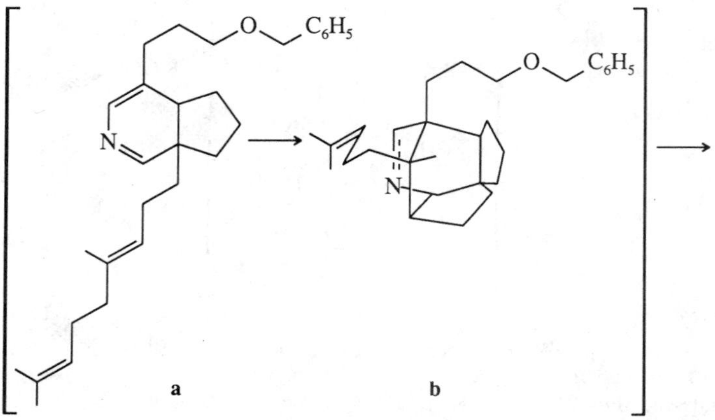

NH_3, CH_2Cl_2, CH_3COOH →

O C_6H_5
N
O C_6H_5
N
a **b**

[1] R. B. Ruggeri, M. M. Hansen, and C. H. Heathcock, *Am. Soc.*, **110**, 8734 (1988).

Ammonium chloride.

Diaryl ketimines, $Ar_2C{=}NH$. The patent literature has described the preparation of $(C_6H_5)_2C{=}NH$ by direct reaction of benzophenone with ammonia, but this route has not been investigated further. Actually, a number of diaryl ketimines can be obtained in >95% yield by reaction of a diaryl ketone with liquid ammonia and a catalytic amount of ammonium chloride in THF in a sealed tube at 120° (60 bar).

[1] G. Verardo, A. G. Giumanini, P. Strazzolini, and M. Poiana, *Syn. Comm.*, **18**, 1501 (1988).

Ammonium formate–Palladium catalyst.

Clemmensen-type reduction.[1] Aromatic ketones can be reduced to the corresponding methylene compounds with ammonium formate on transfer hydrogenation in acetic acid catalyzed by 10% Pd/C. The reduction is usually complete in 10–30 minutes at 110°. Halo and nitro substituents can be reduced under these conditions, and α,β-unsaturated carbonyl groups are reduced to saturated carbonyl groups.

$$C_6H_5COCH_3 \xrightarrow[55\%]{HCOONH_4,\ Pd/C/HOAc} C_6H_5CH_2CH_3$$

$$C_6H_5COC_6H_5 \xrightarrow[92\%]{} (C_6H_5)_2CH_2$$

$$(C_6H_5CH{=}CH)_2CO \xrightarrow[88\%]{} (C_6H_5CH_2CH_2)_2CO$$

Use of ammonium formate as a catalytic hydrogen-transfer reagent has been reviewed.[2]

$RNO_2 \rightarrow RNH_2$.[3] This reduction can be carried out by transfer hydrogenation using ammonium formate catalyzed by Pd/C.[3] The reaction proceeds in 60–90% yield with complete retention of configuration.

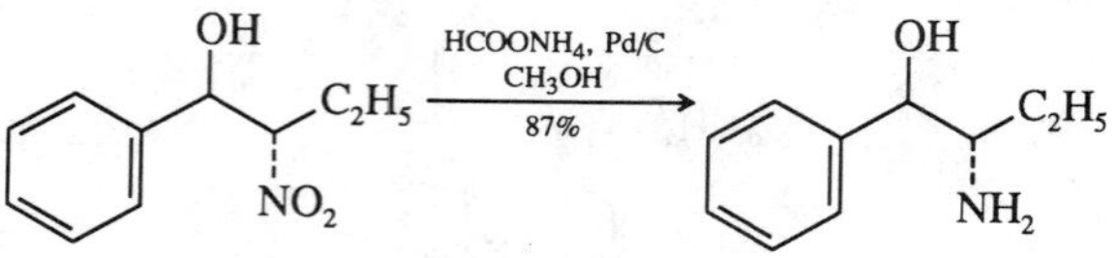

(*syn*/*anti* = 1 : 3.2) (*syn*/*anti* = 1 : 3.2)

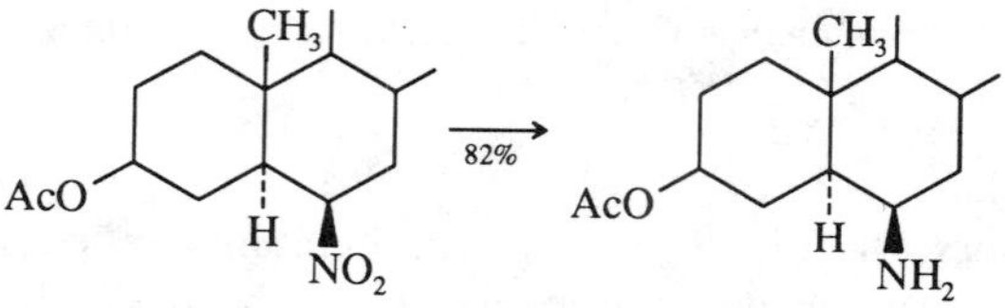

[1] S. Ram and L. D. Spicer, *Tetrahedron Letters*, **29**, 3741 (1988).
[2] S. Ram and R. E. Ehrenkaufer, *Synthesis*, 91 (1988).
[3] S. Ram and R. E. Ehrenkaufer, *ibid.*, 133 (1986); A. G. M. Barrett and C. D. Spilling, *Tetrahedron Letters*, **29**, 5733 (1988).

Ammonium nitrate–Trifluoroacetic anhydride.

N-Nitration.[1] This reaction can be effected with the relatively safe combination of NH_4NO_3 and $(CF_3CO)_2O$ in nitromethane at 0°.

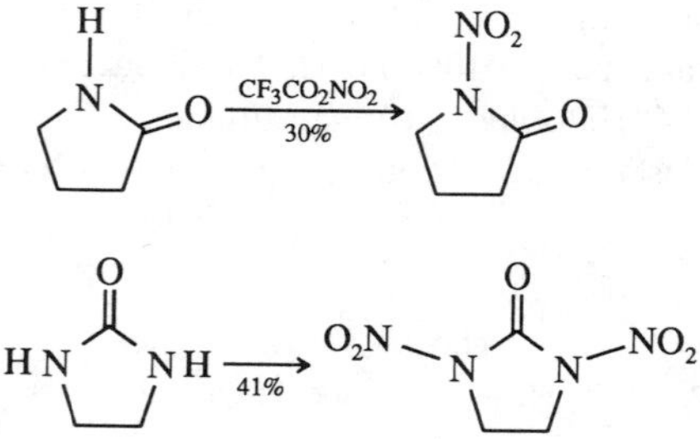

[1] S. C. Suri and R. D. Chapman, *Synthesis*, 743 (1988).

Antimony(V) chloride–Tin(II) iodide–Chlorotrimethylsilane, $SbCl_5$, SnI_2, $ClSi(CH_3)_3$ (**1**).

Cyclization of silyloxy carbonyl compounds.[1] Five- to seven-membered ethers can be obtained from γ-, δ-, and ε-silyloxy aldehydes or ketones by silyl nucleophiles in the presence of these three reagents, the combination of which is more efficient than trityl hexachloroantimonate, $TrSbCl_6$. The silyl nucleophile can be triethylsilane, allyltrimethylsilane, or cyanotrimethylsilane.

$$C_6H_5C(=O)(CH_2)_2CH_2OSi(CH_3)_3 + (C_2H_5)_3SiH \xrightarrow[91\%]{1,\ CH_2Cl_2,\ -78^\circ} \text{2-phenyltetrahydrofuran } (C_6H_5)$$

$$H{-}C(=O)(CH_2)_3CH(OSi(CH_3)_3)C_6H_5 + (CH_3)_3SiCN \longrightarrow \text{2-}C_6H_5\text{-6-CN-tetrahydropyran}$$

cis, 28%
trans, 53%

[1] K. Homma and T. Mukaiyama, *Chem. Letters*, 259 (1989).

Antimony(V) chloride–Tin(II) trifluoromethanesulfonate.

Michael reactions.[1] This system promotes a Michael addition of silyl enol ethers to α,β-unsaturated thiolesters, which are excellent Michael acceptors. Trityl salts are less effective, as are Lewis acids in combination with $SnCl_2$.

$$C_2H_5SC(=O)CH{=}CHCH_3 + C_2H_5SC(OSi(CH_3)_3){=}CHCH_3 \xrightarrow[75\%]{SbCl_5,\ Sn(OTf)_2,\ CH_2Cl_2,\ -78^\circ} C_2H_5SC(=O)CH(CH_3)CH(CH_3)CH_2C(=O)SC_2H_5$$

(*anti*/*syn* = 87 : 13)

[1] S. Kobayashi, M. Tamura, and T. Mukaiyama, *Chem. Letters*, 91 (1988).

Arenesulfenyl isocyanates, ArSNCO.

These isocyanates (**1**), unlike the stable pentafluorobenzenesulfenyl isocyanate, cannot be isolated, but can be trapped as arylsulfenylcarbamates or -ureas.

Example:

$$C_2H_5SCl + AgNCO \longrightarrow [C_6H_5SNCO]\ (\mathbf{1}) \xrightarrow[47\%]{CH_3OH} C_6H_5SNHCOOCH_3$$

$$\mathbf{1} \xrightarrow[36\%]{HN(CH_3)_2} C_6H_5SNHC(=O)N(CH_3)_2$$

[1] T. Nagase, K. Akama, and S. Ozaki, *Chem. Letters*, 1385 (1988).

Arene(tricarbonyl)chromium complexes.

Nucleophilic addition.[1] The reaction of (anisole)$Cr(CO)_3$ (**1**) with $LiC(CH_3)_2CN$, followed by oxidative decomplexation, gives almost exclusively the *meta*-adduct, regardless of the reaction time, the temperature, or the presence of HMPT. This regiospecificity is not usual, however. The behavior of (naphtha-

$Cr(CO_3)$ OCH₃ **1** — 1) $LiC(CH_3)_2CN$, −30°; 2) I_2 → $(CH_3)_2CCN$, OCH_3 (m/o = 97 : 3)

lene)$Cr(CO)_3$ or of (5,8-dimethoxynaphthalene)$Cr(CO)_3$ (**2**) is more typical. In the latter case, addition of the same nucleophile results in both 1,4,5- and 1,4,6-trisubstituted naphthalene, (**3**) and (**4**). In this case the ratio of **3**:**4** depends upon the temperature, the presence of HMPT or TMEDA, and the cation (Li or K).

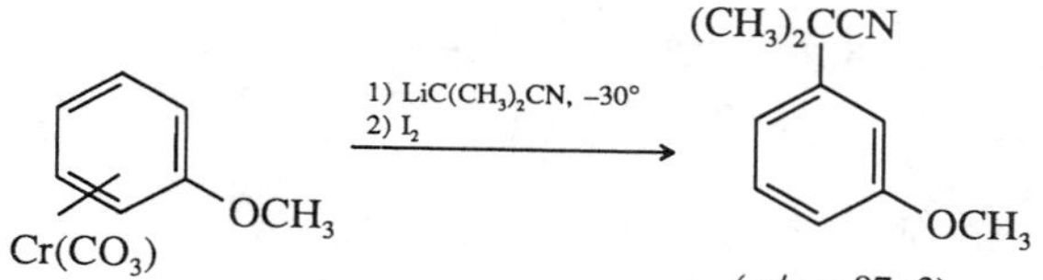

OCH_3, $Cr(CO)_3$, OCH_3 **2**

$MC(CH_3)_2CN$, THF ⇌ **3**: OCH_3, $C(CH_3)_2CN$, OCH_3, $Cr(CO_3)_3$

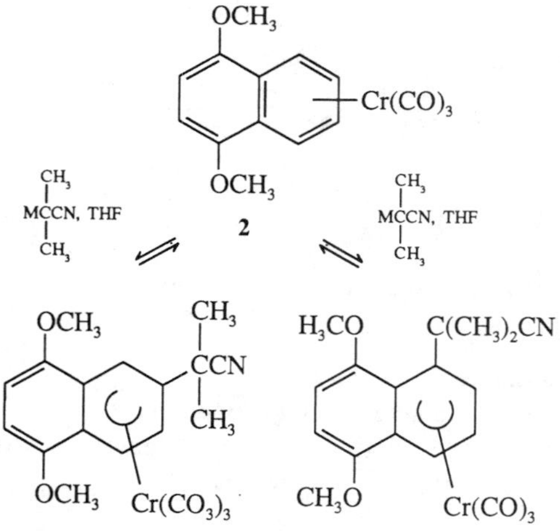

$MC(CH_3)_2CN$, THF ⇌ **4**: H_3CO, $C(CH_3)_2CN$, CH_3O, $Cr(CO)_3$

[1] E. P. Kündig, V. Desobry, D. P. Simmons, and E. Wenger, *Am. Soc.*, **111**, 1804 (1989).

Azidotrimethylsilane–Aluminum isopropoxide.

2-Trimethylsiloxy azides.[1] This system reacts rapidly with epoxides to give trimethylsiloxy azides by attack at the less-substituted carbon. The products are precursors to azidohydrins and β-amino alcohols.

$$HOH_2C\text{-}\underset{O}{\triangle} + (CH_3)_3SiN_3 \xrightarrow[59\%]{Al(O\text{-}i\text{-}Pr)_3} HOH_2C\text{-}\underset{(CH_3)_3SiO}{CH}\text{-}CH_2N_3$$

[1] M. Emziane, P. Lhoste, and D. Sinou, *Synthesis*, 541 (1988).

Azidotrimethylsilane–Trifluoromethanesulfonic acid.

Aromatic amination.[1] Electrophilic amination of arenes can be effected with aminodiazonium tetrachloroaluminate, a reagent prepared *in situ* from $N_3Si(CH_3)_3$, $AlCl_3$, and hydrogen chloride. Higher yields are obtained with aminodiazonium triflate, $NH_2N_2^+OSO_2CF_3^-$, prepared *in situ* from $N_3Si(CH_3)_3$ and F_3CSO_3H in the presence of the arene. Yields of aminoarenes are $>90\%$ from arenes that are not deactivated by electron-withdrawing groups. The isomer distribution is typical for electrophilic aromatic substitutions.

[1] G. A. Olah and T. D. Ernst, *J. Org.*, **54**, 1203 (1989).

B

Barium ferrate hydrate, $BaFeO_4 \cdot H_2O$ (**1**).

Preparation: $Na_2FeO_4 + Ba(NO_3)_2 \rightarrow$ **1** (dark purple).

Oxidation.[1] This reagent is similar to barium manganate. It also couples aromatic amines to azo compounds in moderate yield and converts thiols to disulfides in 85–100% yield.

[1] H. Firouzabadi, D. Mohajer, and M. Entezari-Moghadam, *Bull. Chem. Soc. Japan*, **61**, 2185 (1988).

Barium manganate, 8, 21; **9**, 23; **10**, 16; **12**, 38.

ArCHO → ArCOOH.[1] Aromatic aldehydes can be oxidized to the corresponding acid by this reagent (excess) at 25–40° in CH_2Cl_2; yields 30–92%. Aliphatic aldehydes polymerize under these conditions.

[1] R. G. Srivastava and P. S. Venkataramani, *Syn. Comm.*, **18**, 2193 (1988).

Benzeneseleninic anhydride, $(C_6H_5SeO)_2O$ (**1**).

Oxidation of phenols (**10**,23–24).[1] The earlier report of oxidation of phenols with **1** mainly to *o*-quinones has been confirmed; recent reports of oxidation to *p*-quinones has been shown to involve contamination of **1** with benzeneseleninic acid, $C_6H_5SeO_2H$ (**2**), which is *para*-selective for this oxidation. Solvents also can play a role. THF and benzene are the solvents of choice for **1**, and CH_2Cl_2 is the best solvent for *para*-oxidation with **2**. In acetic anhydride, both **1** and **2** are *ortho*-selective. In addition, phenylselenoquinones can be formed in oxidations with both **1** and **2**, but this reaction can be inhibited by addition of indole.

[1] D. H. R. Barton, J.-P. Finet, and M. Thomas, *Tetrahedron*, **44**, 6397 (1988).

Benzeneseleninyl chloride, $C_6H_5\overset{\overset{\displaystyle O}{\|}}{Se}Cl$ (**1**).

Chlorination.[1] In combination with $AlCl_3$, this reagent chlorinates alkenes to provide vinyl chlorides and chlorinates arenes, but not benzene, to provide aryl chlorides. Anisole, diphenyl ether, and N,N-dimethylaniline are chlorinated exclusively in the *para*-position. Bromination can be effected by the combination of **1** with aluminum bromide.

[1] N. Kamigata, T. Satoh, M. Yoshida, H. Matsuyama, and M. Kameyama, *Bull. Chem. Soc. Japan*, **61**, 449, 2226 (1988).

Benzenesulfenyl chloride.

Reaction with silyl ketene acetals; epoxy alcohols.[1] C_6H_5SCl reacts with silyl ketene acetals to give α-sulfenyl-β-hydroxy esters (**2**) with *anti*-selectivity. The

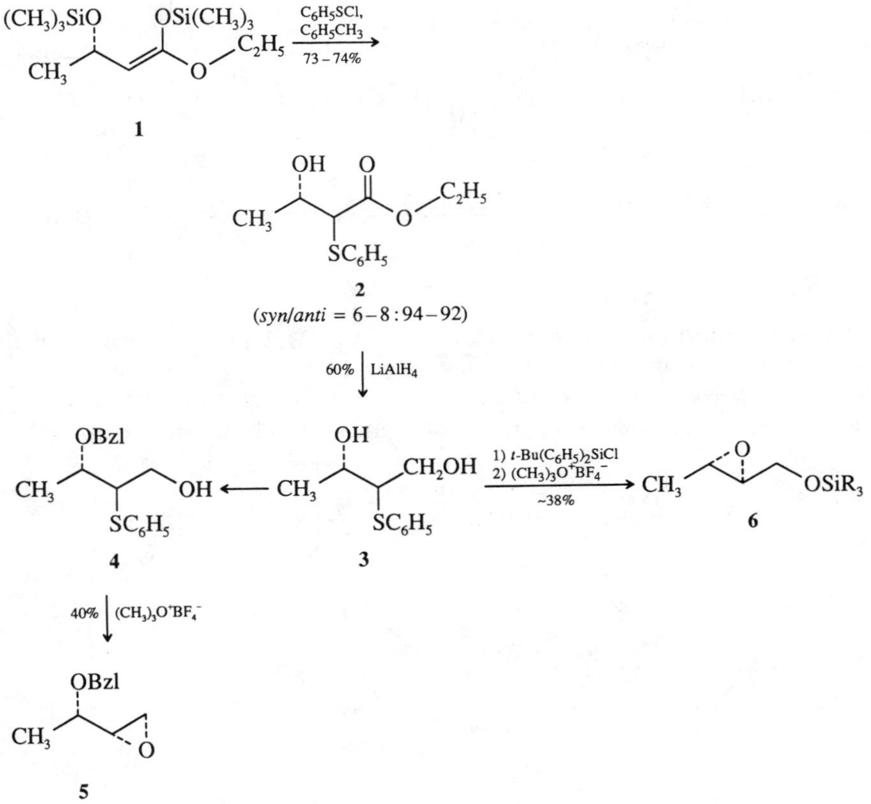

$(CH_3)_3SiO$ $OSi(CH_3)_3$ CH_3 O C_2H_5

1

C_6H_5SCl, $C_6H_5CH_3$ → 73–74%

OH O CH_3 O C_2H_5 SC_6H_5

2

(*syn/anti* = 6–8 : 94–92)

60% ↓ $LiAlH_4$

OBzl CH_3 OH SC_6H_5 **4** ← OH CH_3 CH_2OH SC_6H_5 **3**

1) t-$Bu(C_6H_5)_2SiCl$ 2) $(CH_3)_3O^+BF_4^-$ → ~38%

CH_3 O $OSiR_3$ **6**

40% ↓ $(CH_3)_3O^+BF_4^-$

OBzl CH_3 O

5

Scheme I

product (**2**) from 1-ethoxy-1,3-bis(trimethylsilyloxy)butene-1 (**1**) can be converted into either *syn*-3,4-epoxybutane-2-ol (**5**) or *trans*-2,3-epoxybutane-1-ol (**6**), Scheme I.

[1] G. Guanti, L. Banfi, E. Narisano, and S. Thea, *Chem. Letters*, 1683 (1988).

Benzenetellurinic anhydride, $(C_6H_5\overset{O}{\overset{\|}{Te}})_2O$ (**1**).

Oxytellurinylation.[1] Arenetellurinic anhydrides oxidize 1-alkenes to *vic*-diacetates in refluxing acetic acid. The actual reagent is assumed to be ArTe(O)OAc.

The reaction with γ- or δ-hydroxylated 1-alkenes results in a cyclic ether. *p*-Methoxybenzenetellurinic anhydride (**2**) is the preferred reagent for this cyclofunctionalization.

2, HOAc → 45% + 22%

[1] N. X. Hu, Y. Aso, T. Otsubo, and F. Ogura, *Tetrahedron Letters*, **28**, 1281 (1987).

Benzenetellurinyl trifluoroacetate, $C_6H_5Te(O)OCOCF_3$ (**1**), is prepared by reaction of benzenetellurinic anhydride with trifluoroacetic anhydride.[1]

Allylsilanes → allylamines.[1] The conversion is effected with **1** catalyzed by BF_3 etherate and is believed to involve an intermediate allyl telluroxide (**a**), which

$CH_3(CH_2)_5CH{=}CHCH_2Si(CH_3)_3$ (E/Z = 84 : 16) $\xrightarrow[ClCH_2CH_2Cl]{1,\ BF_3 \cdot O(C_2H_5)_2}$ [$RCH{=}CHCH_2Te(O)C_6H_5$] **a**

a + piperidine → **2** (71%) + **3** (20%)

reacts with arylamines generally without rearrangement to give **3**, but with alkylamines with rearrangement to give **2**. The reaction with a cyclic amine such as piperidine gives a mixture of both possible products as shown in the formulation.

[1] N. X. Hu, Y. Aso, T. Otsubo, and F. Ogura, *Tetrahedron Letters*, **29**, 4949 (1988).

Benzothiazole, (**1**).

α-Hydroxy carbonyl compounds.[1] This heterocycle on metallation (BuLi) at −78° provides the 2-lithio derivative (**2**,100% yield), which is stable at temperatures below −50°. Reaction of the anion with aldehydes or ketones results in 2-(α-hydroxyalkyl)benzothiazoles (70–90% yield), which can be converted into α-hydroxycarbonyl compounds, as shown for a particular case in equation (I).

[1] H. Chikashita, M. Ishibaba, K. Ori, and K. Itoh, *Bull. Chem. Soc. Japan*, **61**, 3637 (1988).

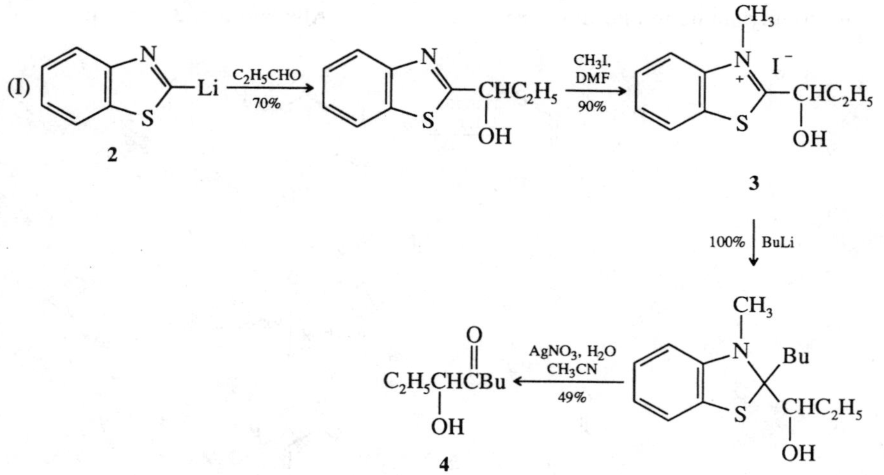

2-Benzoylthio-3-nitropyridine (1). The reagent is generated *in situ* by reaction of 3,3′-dinitro-2,2′-dipyridyl disulfide with benzoic acid and triphenylphosphine in CH_2Cl_2.

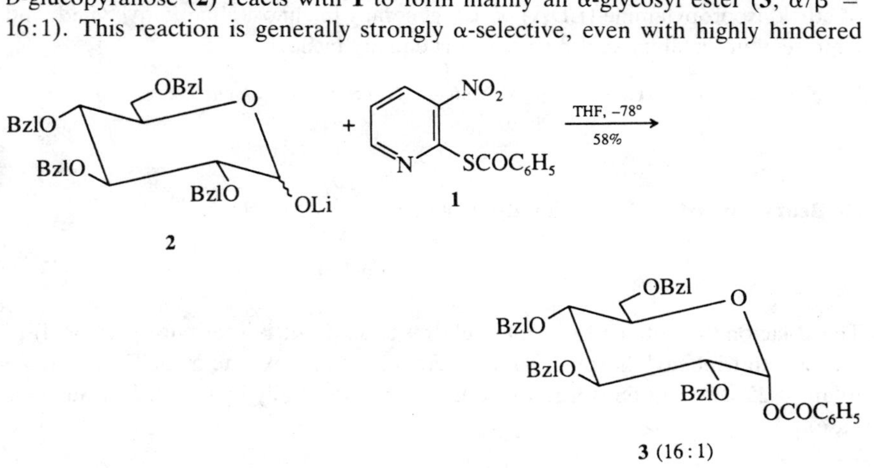

α-Glycosyl esters.[1] The anomeric lithium alkoxide of 2,3,4,5-tetra-O-benzyl-D-glucopyranose (**2**) reacts with **1** to form mainly an α-glycosyl ester (**3**, α/β = 16:1). This reaction is generally strongly α-selective, even with highly hindered systems and is free from epimerization of α-substituted carbohydrate carboxylic acids. It is applicable to other 2-acylthio-3-nitropyridines.

[1] A. G. M. Barrett and B. C. B. Bezuidenhoudt, *Heterocycles*, **28**, special issue, 209 (1989).

Benzyl chloroformate ($ClCO_2CH_2C_6H_5$), **1**, 109; **2**, 59.

Protection of uracils with a benzyloxycarbonylmethyl group.[1] Reaction of the uracil (**2**) with formalin gives a viscous oil, which reacts with **1** in acetonitrile and triethylamine to afford **3** in 85% overall yield. Alkylation of **3** in the presence of ethyldiisopropylamine (EDA) occurs generally in almost quantitative yield, and deprotection (catalytic hydrogenation) is equally facile.

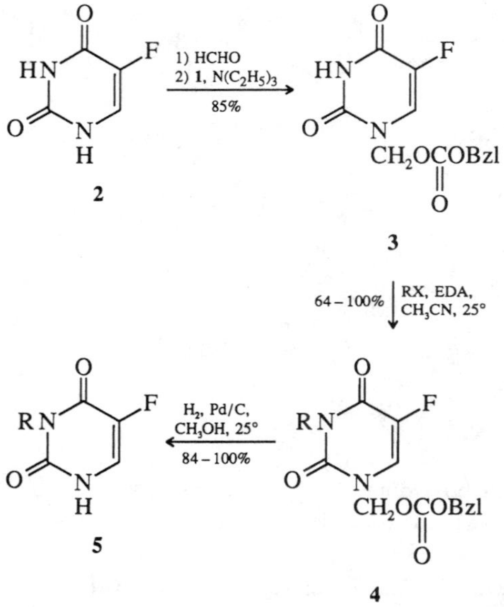

[1] T. Nagase, K. Seike, K. Shiraishi, Y. Yamada, and S. Ozaki, *Chem. Letters*, 1381 (1988).

N-(Benzyloxycarbonyl)-L-serine β-lactone, 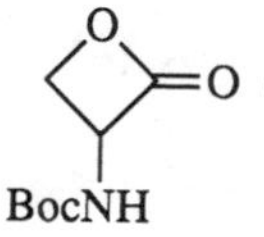 (**1**).

The β-lactone is obtained in 47% yield by reaction of N-(benzyloxycarbonyl)-L-serine with triphenylphosphine and dimethyl azodicarboxylate in THF. It is stable to air at 25° and can be stored for some time; it is rapidly hydrolyzed by aqueous base.

HO–CH₂–CH(HNBoc)–COOH $\xrightarrow[47\%]{P(C_6H_5)_3,\ DMAD}$ **1**

N-Protected α-amino acids.[1] A wide variety of nucleophiles attack this lactone at the β-carbon to provide optically pure N-protected α-amino acids. The preparation of N-Boc-β-(pyrazol-1-yl)-L-alanine (**2**) is typical. This unusual α-amino acid occurs in watermelon seeds.

1 + pyrazole $\xrightarrow[40-70\%]{CH_3CN,\ 52-54°}$ **2**

[1] S. V. Pansare, G. Huyer, L. D. Arnold, and J. C. Vederas, *Org. Syn.*, submitted (1989).

Benzyloxymethylenetriphenylphosphorane, $(C_6H_5)_3P{=}CHOBzl$ (**1**).

Pseudo-α-D-glycopyranose (4).[1] Carbohydrates in which the ring oxygen is replaced by a methylene group are known as pseudo sugars and can have useful physiological properties. A new method for conversion of α-D-glucose to **4** involves first rearrangement to the protected 5-hydroxycyclohexanone **2**. This ketone reacts with the phosphorane **1** to give the benzyloxyvinyl ether **3**, which on Pd-catalyzed

2 $\xrightarrow[55\%]{1}$ **3** $\xrightarrow[75\%]{H_2,\ Pd/C,\ CH_3OH}$ **4** + **5** (65:35)

6 $\xrightarrow{H_2,\ Pd/C,\ CH_3OH}$ **7**

hydrogenation furnishes **4** as the major product. Surprisingly, the hydrogenation of **6** gives a pseudo-β-L-pyranose (**7**) as the major product.

[1] D. H. R. Barton, S. D. Géro, J. Cléophax, A. S. Machado, and B. Quiclet-Sire, *J.C.S. Chem. Comm.*, 1184 (1988).

(S)-*o*-[N-(Benzylproloylamino)benzophenone, (N-Bzl-pyrrolidin-2-yl)$C(=O)NHC_6H_4C(=O)C_6H_5$ (**1**).

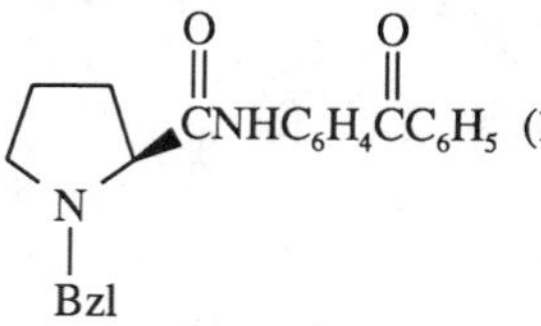

Schiff complex with glycine and Ni(NO₃)₂. This reagent and $Ni(NO_3)_2$ react with glycine in methanol to form the chiral complex **2**.[1]

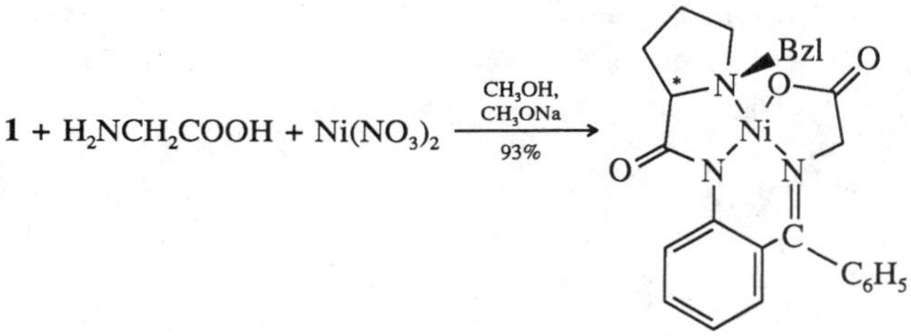

$$\mathbf{1} + H_2NCH_2COOH + Ni(NO_3)_2 \xrightarrow[93\%]{CH_3OH,\ CH_3ONa} \mathbf{2}$$

2

β-Hydroxy-α-amino acids.[1] The complex **2** can serve as a substitute for glycine in a reaction of aldehydes or ketones to provide *threo*-β-hydroxy-α-amino acids. Thus the reaction of paraformaldehyde with **2** provides (S)-serine in 80–98% ee.

$$\mathbf{2} + HCHO \longrightarrow \text{complex} \xrightarrow[67-77\%]{CH_3COOH} HOCH_2CH(NH_2)COOH$$

(S)- Serine (89% ee)

Prolines.[2] The complex **2** undergoes base-catalyzed (CH_3ONa) Michael addition to α,β-enals and -enones. Reaction of **2** with acrolein furnishes a dihydropyrrole-2-carboxylic acid which is reduced to (S)-proline.

[1] Y. N. Belokon, A. G. Bulychev. S. V. Vitt, Y. T. Struchkov, A. S. Batsanov, T. V. Timofeeva, V. A. Tsyryapkin, M. G. Ryzhov, L. A. Lysova, V. I. Bakhmutov, and V. M. Belikov, *Am. Soc.*, **107**, 4252 (1985).

[2] Y. N. Belokon, A. G. Bulychev, V. A. Pavlov, E. B. Fedorova, V. A. Tsyryapkin, V. A. Bakhmutov, and V. M. Belikov, *J.C.S. Perkin I*, 2075 (1988).

$$2 + CH_2{=}CHCHO \xrightarrow{CH_3ONa}$$ (N-benzyl-2,3-dihydropyrrole-2-carboxylic acid, N–Bzl, COOH)

73% $NaBH_3CN$ ↓

(proline, N–H, COOH)

(>95% ee)

Benzyltrimethylammonium dichloroiodate, $Bzl(CH_3)_3\overset{+}{N}Cl_2I^-$ (**1**), yellow needles, m.p. 125–126°. The salt is prepared by reaction of $Bzl(CH_3)_3N^+\ Cl^-$ in water with ICl in CH_2Cl_2 at 25°.

Chlorination of ArCOCH₃.[1] The reagent effects α-chlorination of aryl methyl ketones.

$$\text{(I) } ArCOCH_3 \xrightarrow[95-99\%]{1,\ ClCH_2CH_2Cl,\ CH_3OH} ArCOCH_2Cl + Bzl(CH_3)_3N^+Cl^- + HCl + I_2$$

[1] S. Kajigaeshi, T. Kakinami, M. Moriwaki, S. Fujisaki, K. Maeno, and T. Okamoto, *Chem. Letters*, 2109 (1987); *idem*, *Synthesis*, 545 (1988).

Benzyltrimethylammonium tetrachloroiodate, $(CH_3)_3\overset{+}{N}BzlICl_4^-$ (**1**), yellow needles, m.p. 106–125°.

The salt is prepared by reaction by $(CH_3)_3\overset{+}{N}BzlI^-$ with CCl_3I in CH_2Cl_2 (96% yield).

Side-chain chlorination of alkylarenes.[1] The reagent effects benzylic chlorination of alkylarenes in CCl_4 in the presence of AIBN. The co-product is $(CH_3)_3\overset{+}{N}BzlICl_2^-$.

$$C_6H_5CH_2CH_3 \xrightarrow[53\%]{1,\ AIBN} C_6H_5CHClCH_3 + \text{di- and trichlorinated products}$$

$$(C_6H_5)_3CH \xrightarrow[90\%]{} (C_6H_5)_3CCl$$

[1] S. Kajigaeshi, T. Kakinami, M. Moriwaki, T. Tanaka, and S. Fujisaki, *Tetrahedron Letters*, **29**, 5783 (1988).

1,1′-Bi-2,2′-naphthol (BINOL, **1**).

Cleavage and resolution of epoxides.[1] The aluminum reagent obtained by reaction of (R)-**1** with diethyl- or dimethylaluminum chloride shows slight reactivity in reactions with epoxides, but the ate complex (**2**), prepared from **1**, $(C_2H_5)_2AlCl$, and lithium butoxide, converts cyclohexene oxide into the chlorohydrin (**3**) in 40% ee.

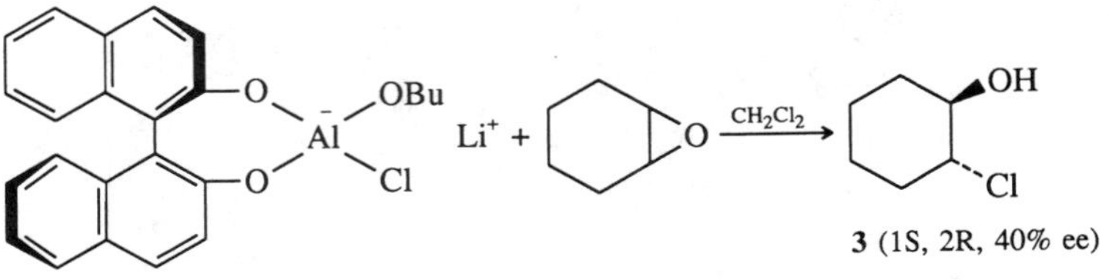

3 (1S, 2R, 40% ee)

2

When the keto epoxide **4** is treated with **2** for 5 hours at −30°, most of the epoxide is cyclized to the acetal **5**, but the recovered keto epoxide (**4**) is completely resolved, and can be used for the synthesis of (−)-C_{16}-juvenile hormone (**6**).

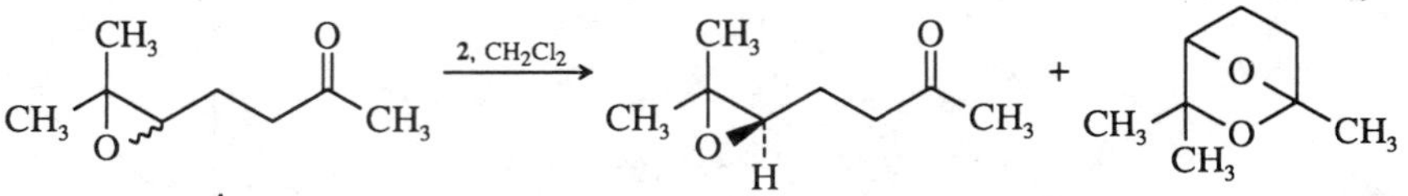

4

(R)-**4** (>95% ee)

5

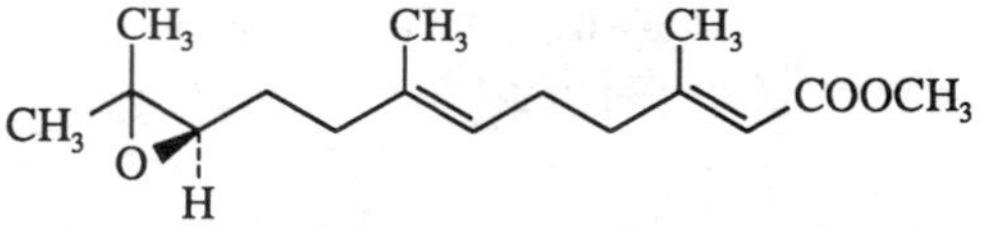

(−)-**6**

[1] Y. Naruse, T. Esaki, and H. Yamamoto, *Tetrahedron*, **44**, 4747 (1988).

(R)-(+)-1,1′-Bi-2,2′-naphthol (BINOL)–**Dichloro- or dibromodiisopropoxytitanium (1)**. These reagents form a complex **1** with the probable formula (R)-**1a** or (R)-**1b**.

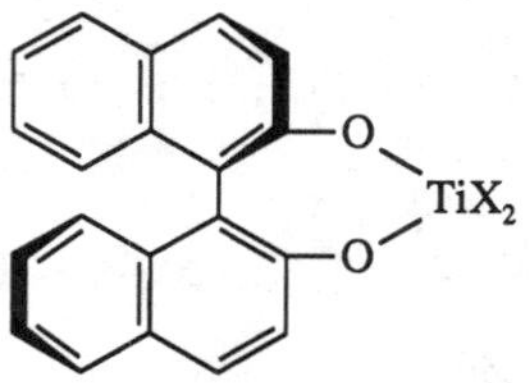

(R)−**1a**, X = Cl
(R)−**1b**, X = Br

Asymmetric glyoxalate ene reactions.[1] In the presence of 4 Å molecular sieves (MS), methyl glyoxalate undergoes enantioselective ene reactions with 1,1-disubstituted alkenes when catalyzed by **1a** or **1b**. The former is superior in reactions involving a methyl hydrogen, whereas the latter is superior in reactions involving a methylene hydrogen. The reaction can provide α-hydroxy esters in 70–95% yield and 85–98% ee.

$$(CH_3)_2C{=}CH_2 + HC(O)COOCH_3 \xrightarrow{4\text{Å MS, } CH_2Cl_2} CH_3C({=}CH_2)CH_2CH(OH)COOCH_3$$

1a	72%	R, 95% ee
1b	87%	R, 83% ee

$$\text{methylenecyclohexane } (=CH_2) + HC(O)COOCH_3 \longrightarrow \text{1-cyclohexenyl}{-}CH_2CH(OH)COOCH_3$$

1a	82%	R, 83% ee
1b	89%	R, 98% ee

[1] K. Mikami, M. Terada, and T. Nakai, *Am. Soc.*, **111**, 1940 (1989).

Birch reduction.

Reductive alkylation. N-*t*-Butylnaphthalenesulfonamides can undergo Birch reductive alkylation in acceptable yields. The products undergo pyrolysis to alkylnaphthalenes.[1]

CH_3O-naphthalene-SO_2NH-*t*-Bu $\xrightarrow[58\%]{1)\ BuLi,\ NH_3,\ Li;\ 2)\ CH_3I}$ CH_3O-dihydronaphthalene(CH_3)(SO_2NH-*t*-Bu) $\xrightarrow[77.5\%]{\Delta}$ CH_3O-naphthalene-CH_3

Birch reduction of phenanthrene or 9-alkylphenanthrenes results in tetrahydro derivatives, but 9,10-dihydro derivatives are obtained in high yield by Li/NH_3 reduction of phenanthrenes substituted by a 9-trimethylsilyl group (equation I). The trimethylsilyl group of **1** can be removed in quantitative yield with Bu_4NF in

(I) 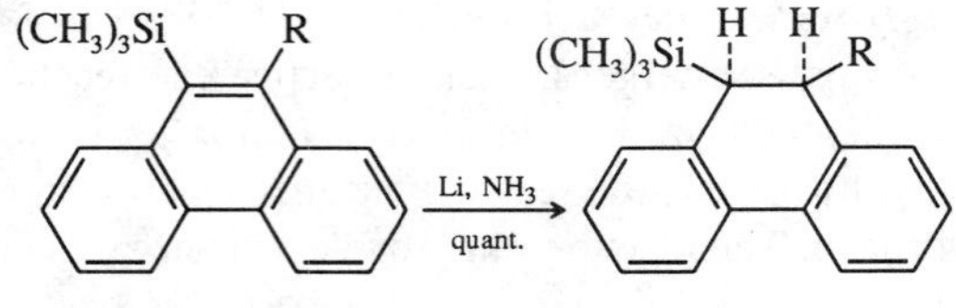 $(CH_3)_3Si$, R → Li, NH_3, quant. → $(CH_3)_3Si$, H, H, R

R = H, CH_3, C_2H_5 **1**

THF, but if an alkyl halide is present, a *trans*-9,10-dialkylphenanthrene is obtained (equation II).[2]

(II) 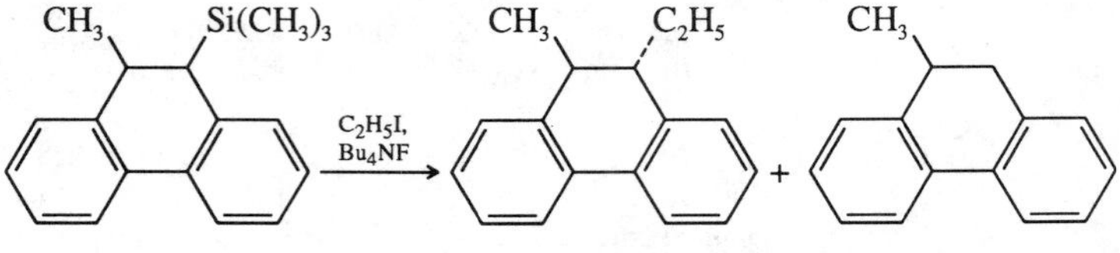 CH_3, $Si(CH_3)_3$ → C_2H_5I, Bu_4NF → CH_3, C_2H_5 (78%) + CH_3 (22%)

Reduction of benzocyclobutenes.[3] Birch reduction of **1**, available from *m*-hydroxybenzaldehyde, provides the enone **2** after hydrolysis of the intermediate. This enone is a useful precursor to the cyclobutane monoterpenes grandisol (**4**) and lineaton (**5**).

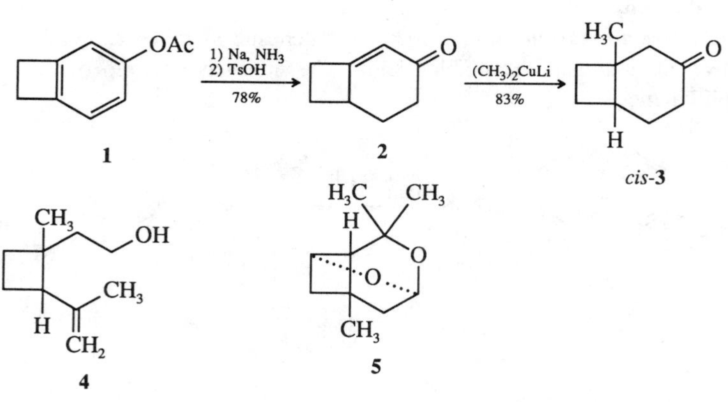

1 (OAc) → 1) Na, NH_3 2) TsOH, 78% → **2** (O) → $(CH_3)_2CuLi$, 83% → *cis*-**3** (H_3C, O, H)

4 (CH_3, OH, CH_3, H, CH_2) **5** (H_3C, CH_3, H, O, O, CH_3)

[1] L. Gottlieb and H. J. E. Loewenthal, *Tetrahedron Letters*, **29**, 4473 (1988).
[2] P. W. Rabideau and Z. Marcinow, *ibid.*, **29**, 3761 (1988).
[3] T. Kametani, T. Toya, K. Ueda, M. Tsubuki, and T. Honda, *J.C.S. Perkin I*, 2433 (1988).

Bis(acetonitrile)dichloropalladium(II).

Rearrangement of allylic acetates.[1] This Pd(II) catalyst effects rearrangement of (1E,4Z)-3-acetoxy-1,4-dienes to (2E,4Z)-1-acetoxy-2,4-dienes at 25° in 0.5–2 hours in 90–96% yield. This catalyzed rearrangement favors formation of the (2E,4Z)-1-acetoxy-2,4-diene. It is the last step in a synthesis of the acetate (**1**) of methyl α-dimorphecolate.

$$\text{R'-CH=CH-CH(OAc)-CH=CH-R}^2 \xrightarrow[90-96\%]{\text{Pd(II), THF}} \text{R'-CH=CH-CH=CH-CH(OAc)-R}^2$$

(R = H, CH_3, Bu)

$$CH_3O_2C(CH_2)_7\text{-CH=CH-CH(OAc)-CH=CH-}(CH_2)_4CH_3 \xrightarrow[97\%]{\text{Pd(II),}} CH_3OOC(CH_2)_7\text{-CH(OAc)-CH=CH-CH=CH-}C_5H_{11}$$

1

[1] M. M. L. Crilley, B. T. Golding, and C. Pierpoint, *J.C.S. Perkin I*, 2061 (1988).

Bis(benzonitrile)dichloropalladium(II).

Desulfonylative coupling of $ArSO_2Cl$.[1] Desulfonylative coupling of $ArSO_2Cl$ with acrylic esters catalyzed by palladium can proceed in high yield when carried out with solid–liquid phase transfer catalysts, the best of which is benzyltrioctylammonium chloride, $BzlOct_3NCl$.

$$ArSO_2Cl + CH_2{=}CHCOOBu \xrightarrow[90\%]{\text{Pd(II), }K_2CO_3\text{, }BzlOct_3NCl} ArCH{=}CHCOOBu$$

(Ar = 1-Naphthyl)

[1] M. Miura, H. Hashimoto, K. Itoh, and M. Nomura, *Tetrahedron Letters*, **30**, 975 (1989).

(R)-Bis-(1,1′-binaphthalenyl) 2,2′-diselenide,

$Se)_2$ (**1**)

Methoxyselenenylation of alkenes.[1] This reaction can be effected by reaction of **1** with Br_2 in CH_2Cl_2 to form a red solid that is considered to be a selenenyl bromide. This intermediate reacts with an alkene in CH_3OH to give a methoxyselenenylated alkene. The reaction with styrene gives only the Markovnikov product in a yield of 49% and 49% de.

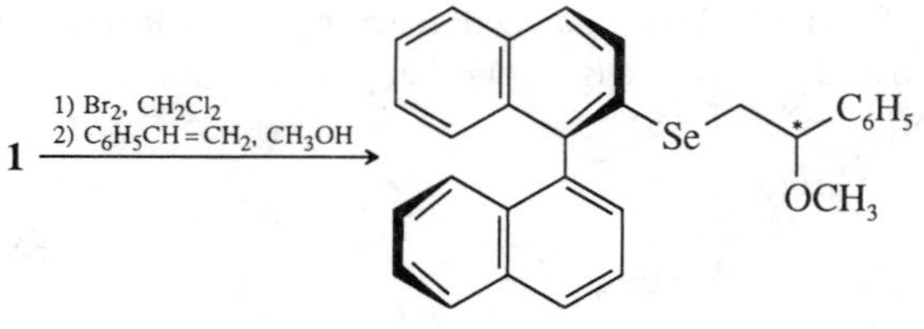

$$\mathbf{1} \xrightarrow[\text{2) } C_6H_5CH{=}CH_2,\ CH_3OH]{\text{1) } Br_2,\ CH_2Cl_2}$$

Se C₆H₅ OCH₃

(49% de)

[1] S. Tomoda and M. Iwaoka, *Chem. Letters*, 1895 (1988).

Bis(*sym*-collidine)iodine(I) perchlorate (Iodonium di-*sym*-collidine) perchlorate, $^{+}I(C_8H_{11}N)_2\ ClO_4^-$ (**1**), **10**, 212–213.

Iodocyclization.[1] This reagent converts various unsaturated alcohols into ethers and unsaturated carboxylic acids into lactones. It is particularly useful for synthesis of 2-(1-iodomethyl)oxiranes from allylic alcohols and of 2-(1-iodomethyl)oxetanes from homoallylic alcohols.

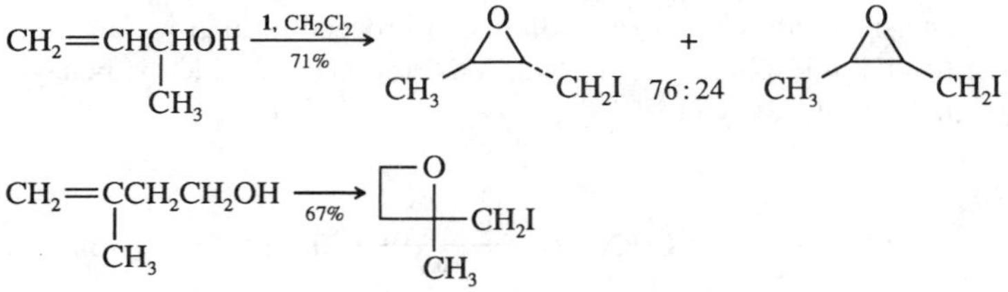

$$CH_2{=}CHCH(CH_3)OH \xrightarrow[71\%]{\mathbf{1},\ CH_2Cl_2} \text{CH}_3\text{–oxirane–CH}_2\text{I (cis)} + \text{CH}_3\text{–oxirane–CH}_2\text{I (trans)} \quad 76:24$$

$$CH_2{=}C(CH_3)CH_2CH_2OH \xrightarrow[67\%]{} \text{2-methyl-2-(iodomethyl)oxetane}$$

[1] R. D. Evans, J. W. Magee, and J. H. Schauble, *Synthesis*, 862 (1988).

Bis(cyclooctadiene)nickel(0), Ni(COD)$_2$.

Bicyclic α-pyrans and -pyrones. The Ni(0)-catalyzed reaction of an aldehyde with a terminally dialkyl-substituted 1,7-octadiyne results in bicyclic α-pyrans.[1] This reaction is highly dependent on the length of the chain connecting the triple bonds and on the presence of the terminal substituents.

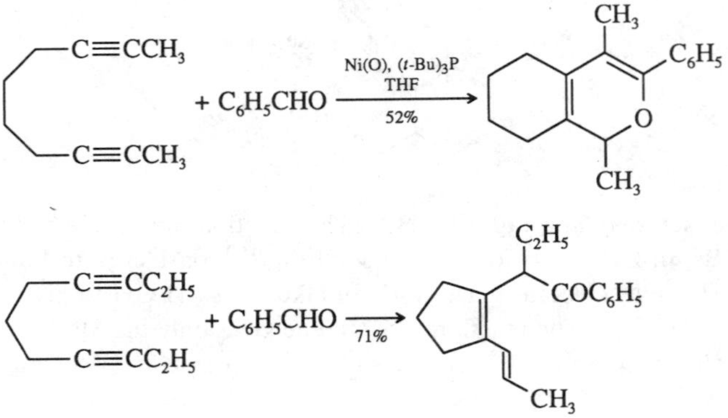

$$\text{(CH}_2)_4(C{\equiv}CCH_3)_2 + C_6H_5CHO \xrightarrow[52\%]{\text{Ni(O), }(t\text{-Bu})_3P,\ \text{THF}} \text{bicyclic α-pyran (CH}_3\text{, C}_6H_5\text{, CH}_3)$$

$$\text{(CH}_2)_3(C{\equiv}CC_2H_5)_2 + C_6H_5CHO \xrightarrow[71\%]{} \text{cyclopentene with CH(C}_2H_5)COC_6H_5 \text{ and CH=CHCH}_3$$

This nickel(0) catalyst in combination with a trialkylphosphine catalyzes a cycloaddition of 1,6- and 1,7-diynes (**1**) and CO_2 to provide bicyclo α-pyrones (**2**) at the expense of dimerization of **1** to **3**.[2] Cycloaddition to **2** is highly dependent on the structure of the phosphine. In general, tricyclohexylphosphine is the most suitable phosphine for cyclization to the α-pyrone.

$C{\equiv}C{-}R$ / $C{\equiv}C{-}R$ (**1**) + CO_2 $\xrightarrow{Ni(COD)_2,\ R_3P}$ **2** (50–85%) + **3** (0–45%) [$(CH_2)_4C{\equiv}CR$]

Allylnickel homoenolates.[3] 2-Vinyl-1,3-dioxolan-4-ones (**1**), available by condensation of acrolein with (R)- or (S)-2-hydroxyalkanoic acids, react with $Ni(COD)_2$ to form an allylnickel complex which on reaction with $ClSi(CH_3)_3$ forms the dark red complex **2**. In the presence of light and a polar solvent this complex undergoes

R(OH)COOH + $CH_2{=}CHCHO$ ⟶
(R = cyclohexyl)

1 (R, O, $CH{=}CH_2$) $\xrightarrow[78\%]{1)\ Ni(COD)_2;\ 2)\ ClSi(CH_3)_3}$ **2** (Cy, O, C=O, TMSO, ClNi)

2 + RX $\xrightarrow[72-82\%]{h\nu,\ DMF}$ **3** (E/Z = 9:1) (Cy, O, R, C=O, TMSO) $\xrightarrow[95\%]{H_3O^+}$ $OHCCH_2CH_2R$

3 $\xrightarrow[95\%]{C_6H_5CH(OCH_3)_2,\ TMSOTf}$ **4** (94% de) (C_6H_5, OCH_3, CH, Cy, O, R)

alkylation to provide (E)-enol ethers (**3**), and these serve as a homoenolate equivalent of a new type. The complex also undergoes enantioselective aldol reaction with acetals.

[3+2]*Cycloaddition of methylenecyclopropanes with 2-cyclopentenone* (cf. **13**, 35).[4] Methylenecyclopropanes react with 2-cyclopentenone in the presence of this Ni(0) catalyst, triphenylphosphine and triethylborane (Lewis acid) to form 6-methylenebicyclo[3.3.0]octane-2-ones.

R = *n*-C_5H_{11} + 2-cyclopentenone —Ni(O)/P(C_6H_5)$_3$, B(C_2H_5)$_3$, 68%→ *trans*-**1** + *cis*-**1** (85 : 15)

methylenecyclopropane (CH_2, CH_3) + 2-cyclopentenone —73%→ (*endo*) (CH_3, CH_2) + (*exo*) (CH_3, CH_2) (63 : 37)

[1] T. Tsuda, T. Kiyoi, T. Miyane, and T. Saegusa, *Am. Soc.*, **110**, 8570 (1988).
[2] T. Tsuda, S. Morikawa, R. Sumiya, and T. Saegusa, *J. Org.*, **53**, 3140 (1988).
[3] D. J. Krysan and P. B. Mackenzie, *Am. Soc.*, **110**, 6273 (1988).
[4] P. Binger and B. Schäfer, *Tetrahedron Letters*, **29**, 4539 (1988).

Bis(cyclopentadienyl)dihydridozirconium, Cp_2ZrH_2, **14**, 37.

Meerwein–Ponndorf–Verley reductions. Zirconocene or hafnocene can catalyze reduction of carbonyl compounds with isopropanol. This method is useful for preferential reduction of keto aldehydes to hydroxy ketones and of α,β-enones or -enals to allylic alcohols.[1]

$CH_3COCH_2CHO + (CH_3)_2CHOH \xrightarrow[95\%]{Cp_2ZrH_2} CH_3COCH_2CH_2OH$

$CH_3CH{=}CHCHO + (CH_3)_2CHOH \xrightarrow[91\%]{} CH_3CH{=}CHCH_2OH$

[1] T. Nakano, S. Umano, Y. Kino, Y. Ishii, and M. Ogawa, *J. Org.*, **53**, 3752 (1988).

Bis(cyclopentadienyl)diphenyltitanium, $Cp_2Ti(C_6H_5)_2$. Preparation.[1]

Hydrosilylation of ketones.[2] Diphenylsilane or methylphenylsilane in combination with diphenyltitanocene can convert ketones into alkoxysilanes. The active species is probably a titanium hydride complex, possibly $(Cp_2TiH)_2H$.

$$CH_3(CH_2)_4COCH_3 + (C_6H_5)_2SiH_2 + Cp_2Ti(C_6H_5)_2 \xrightarrow[91\%]{120°} CH_3(CH_2)_4CH(OSiH(C_6H_5)_2)CH_3$$

[1] L. Summers, R. H. Uloth, and A. Holmes, *Am. Soc.*, **77**, 3604 (1955).
[2] T. Nakano and Y. Nagai, *Chem. Letters*, 481 (1988).

Bis(dibenzylideneacetone)palladium(0)–Tri(2-furyl)phosphine, $Pd(dba)_2$–(2-furyl)$_3$P (1).

Cross-coupling of allyl halides with unsaturated tin reagents.[1] 3-Chloromethylcephems (**2**) are inert to vinyltin reagents under Stille's original conditions (**12**, 56). A study of various solvents and ligands finds that tri(2-furyl)phosphine enhances the rate and leads to generally good yields for this coupling.

$$\mathbf{2} + CH_3CH{=}CHSnBu_3 \xrightarrow[78\%]{\mathbf{1},\ THF,\ 65°} \mathbf{3}$$

(**2**: R^1NH-cephem with 3-CH_2Cl and $COOR^2$; **3**: R^1NH-cephem with 3-$CH_2CH{=}CHCH_3$ and $COOR^2$)

[1] V. Farina, S. R. Baker, D. A. Benigni, and C. Sapino, Jr., *Tetrahedron Letters*, **29**, 5739 (1988).

p-Bis(diphenylhydrosilyl)benzene (1).

Preparation:

$$2(C_6H_5)_2SiHCl + Mg + Br{-}C_6H_4{-}Br \xrightarrow[68\%]{ether} (C_6H_5)_2SiH{-}C_6H_4{-}SiH(C_6H_5)_2$$

1, m.p. 108.5°

Deoxygenation of ROH.[1] Acetates, primary, secondary, or tertiary, are deoxygenated when heated with **1** in the presence of a radical initiator, di-*t*-butyl peroxide (DTBP). Yields using $(C_6H_5)_3SiH$ in place of **1** are definitely lower. The presence of a double bond in the substrate can lower yields owing possibly to hydrosilylation.

$$C_6H_5CH_2C(OAc)(CH_3)_2 \xrightarrow[71\%]{\text{1, DTBP, } 140^\circ} C_6H_5CH_2CH(CH_3)_2$$

[1] H. Sano, T. Takeda, and T. Migita, *Chem. Letters*, 119 (1988).

(R)- and (S)-2,2′-Bis(diphenylphosphine)-1,1′-binaphthyl (**1**, BINAP)

Asymmetric hydrogenation of allylic alcohols (**14**, 39–40).[1] Mammalian dolichols (**2**) are terminal dihydropolyisoprenols which are involved in glycoprotein synthesis. They contain one terminal chiral primary allylic alcohol group. The polyprenols **3** present in plants correspond to dolichols except that they lack the terminal double bond considered to be (Z). They can be obtained by hydrogenation of **2** catalyzed by (bistrifluoroacetate)ruthenium(II) and (S)-**1**, which affects only the terminal double bond to provide (S)-**3** in >95% ee.

$$(CH_3)_2C{=}CH_2(CH_2C(CH_3){=}CH_2)_nCH_2C(CH_3){=}CHCH_2OH \ (Z) \xrightarrow[100\%]{H_2,\ [(S)-1]\,Ru(CF_3CO_2)_2}$$

2, n = 13–15 (E+Z)

$$(CH_3)_2C{=}CH_2(CH_2C(CH_3){=}CH_2)_nCH_2CH(CH_3)CH_2CH_2OH$$

(S)-**3** (>95% ee)

Asymmetric hydroboration.[2] Hydroboration of alkenes with catecholborane catalyzed by Rh(COD)Cl·2BINAP or $[Rh(COD)Cl]_2$·2DIOP and followed by oxidation provides optically active alcohols in 20–75% ee; Rh-BINAP systems are somewhat more enantioselective than Rh-DIOP systems.

Norbornene → HO-norbornane (exo-norborneol)

[Rh(COD)Cl]•2DIOP, –25° — 1R, 2R, 57% ee
[Rh(COD)Cl]•2BINAP, –25° — 1R, 2R, 64% ee

$$CH_3C(C_6H_5){=}CH_2 \longrightarrow CH_3CH(C_6H_5)CH_2OH$$

R, 27% ee

[1] B. Imperiali and J. W. Zimmerman, *Tetrahedron Letters*, **29**, 5343 (1988).
[2] K. Burgess and M. J. Ohlmeyer, *J. Org.*, **53**, 5178 (1988).

(2S,3S)-(−)-Bis(diphenylphosphino)butane [(S,S)-Chiraphos],

$(C_6H_5)_2P\overset{*}{C}H(CH_3)-\overset{*}{C}H(CH_3)P(C_6H_5)_2$ (**1**). Supplier: Strem.

Asymmetric cyclization to tricyclic ergolines.[1] The ergoline system of ergot alkaloids can be obtained in fairly high enantioselectivity by cyclization of **2** catalyzed by Pd(0) complexed with a chiral ligand and a base. The ligands BINAP,

AcOCH$_2$ NO$_2$ → (Pd(0), L, base) → CH$_2$ NO$_2$

2		**3**
+ (−)-**1**, DME, KF	28%	70% ee
+ (−)-**1**, THF, K_2CO_3	62%	66% ee

DIPAMP, and Norphos provide **3** with some enantioselectivity (26–47% ee). The most effective ligand is **1** in combination with KF, but the yield is low. The best practical route to optically active **3** is the use of **1** as ligand and K_2CO_3 as base.

[1] J. P. Genet and S. Grisoni, *Tetrahedron Letters*, **29**, 4543 (1988).

[1,2-Bis(diphenylphosphino)ethane]cyclopentadienyl(iodomethane)ruthenium trifluoromethanesulfonate, Cp(dppe)Ru(ICH_3) triflate (**1**).

Preparation:

CpRu(dppe)I (**2**) $\xrightarrow[85\%]{CF_3SO_3CH_3,\ CH_2Cl_2}$ $[CpRu(dppe)(ICH_3)]^+ CF_3SO_3^-$ (**1**)

(C$_6$H$_5$)$_2$P, Ru, P(C$_6$H$_5$)$_2$, I, ICH$_3$

C-Methylation.[1] This triflate methylates enamines and enolates as well as other nucleophiles.

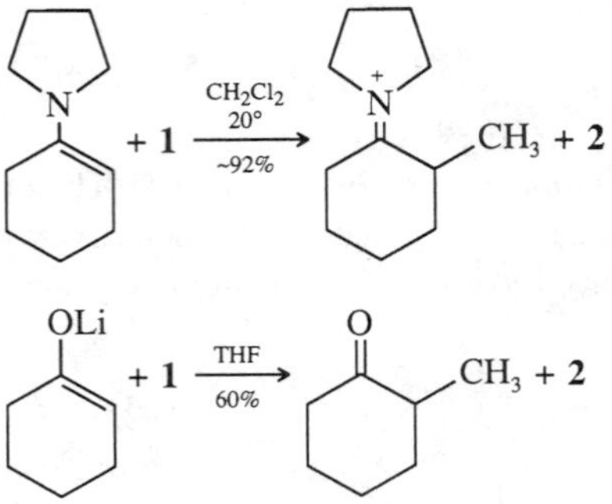

N + 1 $\xrightarrow[\sim 92\%]{CH_2Cl_2,\ 20°}$ $\overset{+}{N}$ CH_3 + 2

OLi + 1 $\xrightarrow[60\%]{THF}$ O CH_3 + 2

[1] R. J. Kulawiec and R. H. Crabtree, *Organometallics*, **7**, 1891 (1988).

Bis(1,1,1,3,3,3-hexafluoro-2-propyl) hydrogen phosphite (1).
Preparation:

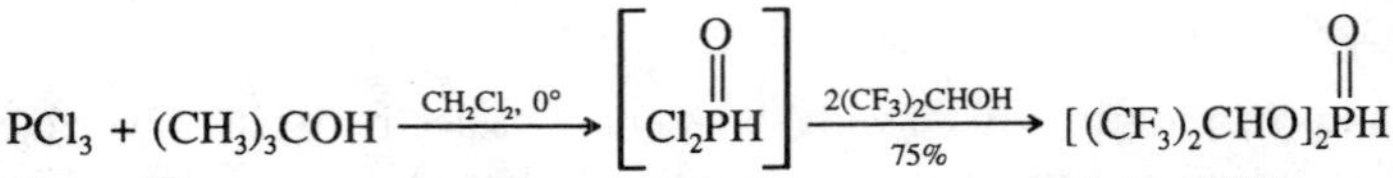

$$PCl_3 + (CH_3)_3COH \xrightarrow{CH_2Cl_2,\ 0°} \left[Cl_2\overset{O}{\overset{\|}{P}}H \right] \xrightarrow[75\%]{2(CF_3)_2CHOH} [(CF_3)_2CHO]_2\overset{O}{\overset{\|}{P}}H$$

1, b.p. 35°/1 mm

Phosphonylation.[1] This reagent is useful for preparation of protected deoxyribonucleoside-3-hydrogen phosphonates (**2**), which can be used for synthesis of deoxyribonucleotides.

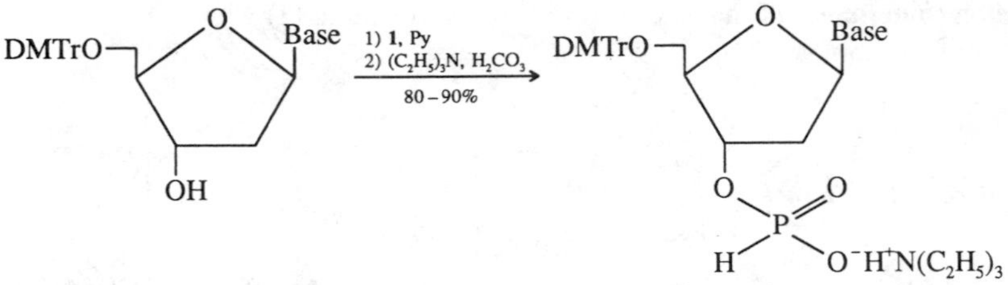

DMTrO, O, Base, OH $\xrightarrow[80-90\%]{1)\ \mathbf{1},\ Py;\ 2)\ (C_2H_5)_3N,\ H_2CO_3}$ DMTrO, O, Base, O, P, O, H, $O^-H^+N(C_2H_5)_3$

2

[1] H. Takaku, S. Yamakage, O. Sakatsume, and M. Ohtsuki, *Chem. Letters*, 1675 (1988).

1,6-Bis(dimethylamino)pyrene, (1).

Decarboxylation.[1] This reaction can be carried out via N-acyloxyphthalimides, which undergo loss of CO_2 when irradiated (350–450 nm) in aqueous isopropanol containing *t*-BuSH (proton donor) and **1**, which serves as a sensitizer. Other sensitizers are less useful.

$$RCO_2N(\text{phthalimide}) + \mathbf{1} + t\text{-BuSH} \xrightarrow[i\text{-PrOH, } H_2O]{h\nu} \underset{(72-98\%)}{RH} + \underset{(\sim 10\%)}{(t\text{-BuS})_2} + HN(\text{phthalimide})$$

[1] K. Okada, K. Okamoto, and M. Oda, *Am. Soc.*, **110**, 8736 (1988).

2,4-Bis(*p*-methoxyphenyl)-1,3-dithiaphosphetane-2,4-disulfide (Lawesson's reagent).

ROH → RSH.[1] This transformation can be effected with this reagent (**1**) in moderate to high yield. The yield is quantitative in the case of triphenylmethanol $[C_6H_5)_3COH]$, but some dehydration is observed with tertiary alcohols possessing an α-methyl or α-benzyl group.

[1] T. Nishio, *J.C.S. Chem. Comm.*, 205 (1989).

Bismuth(III) chloride, $BiCl_3$.

Aldol reaction.[1] Silyl enol ethers react with aldehydes at 25° to give aldols in the presence of $BiCl_3$ (5 mole %). The classical version (Mukaiyama, **6**, 590–591) of this reaction usually requires a full equivalent of $TiCl_4$ as the promotor. The $BiCl_3$ version permits use of ketones as well as aldehydes in reactions carried out at 25°, but a longer time is required and yields are only moderate (20–65%). Both versions show only slight diastereoselectivity.

[1] H. Ohki, M. Wada, and K. Akiba, *Tetrahedron Letters*, **29**, 4719 (1988).

Bis(1,3-propanediamino)copper(II) chloride,

$$\left[Cu(H_2N(CH_2)_3NH_2)_2\right]^{2+}\ 2\,Cl^-\ (\mathbf{1}).$$

Oxidative coupling of hydroquinones.[1] Oxygen can effect coupling and oxidation of hydroquinones to biquinones in 50–75% yield in the presence of this complex.

$$2\ \text{(catechol)} \xrightarrow[75\%]{O_2,\ \mathbf{1},\ C_2H_5OH,\ HOAc} \text{biquinone}$$

$$2\ \text{(1,4-naphthalenediol)} \xrightarrow{52\%} \text{2,2'-bi-1,4-naphthoquinone}$$

[1] S. M. Paraskevas, D. Konstantinidis, and G. Vassilara, *Synthesis*, 897 (1988).

Bis(pyridine)iodine(I) tetrafluoroborate, $I(py)_2BF_4$.

The compound is obtained by reaction of iodine and pyridine with $HgO \cdot HBF_4$ supported on silica.

1,1-*Diodo-1-alkenes.*[1] This reagent (**1**) in combination with a wide variety of nucleophiles reacts with 1-iodoalkynes to afford 2-substituted 1,1-diiodo-1-alkenes (**2**) in 50–85% yield. The products can be converted to a variety of alkenes by reaction of both the iodine atoms.[2]

$$C_6H_5C{\equiv}CI + \mathbf{1} + (CH_3)_2CHOH \xrightarrow[80\%]{CH_2Cl_2} (C_6H_5)((CH_3)_2CHO)C{=}CI_2\ (\mathbf{2}) \xrightarrow[57\%]{1)\ sec\text{-}BuLi\ \ 2)\ CH_3CHO} (C_6H_5)((CH_3)_2CHO)C{=}C(CHOHCH_3)I$$

$$\mathbf{2} \xrightarrow[\text{2) } CH_3Li]{\text{1) } sec\text{-BuLi}} \left[(C_6H_5)((CH_3)_2CHO)C{=}CLi_2 \right] \xrightarrow[84\%]{CH_3I} (C_6H_5)((CH_3)_2CHO)C{=}C(CH_3)_2$$

Intramolecular cyclization of homoallylbenzenes.[3] This reagent effects intramolecular cyclization of homoallylbenzenes to iodine-substituted bicyclic compounds. It also effects cyclization of 1,5-dienes to iodocyclohexenes.

$$C_6H_5CH_2CH_2CH{=}CH_2 \xrightarrow[80\%]{\mathbf{1},\ HBF_4,\ CH_2Cl_2,\ -90^\circ} \text{2-iodotetralin}$$

$$C_6H_5CH_2CH_2CH{=}CHC_6H_5 \xrightarrow{85\%} \text{2-iodo-1-phenyltetralin}$$

$$CH_2{=}CH(CH_2)_2CH{=}CH_2 \xrightarrow[70\%]{\mathbf{1},\ CF_3SO_3H,\ CH_2Cl_2,\ -60^\circ} \text{4-iodocyclohexene}$$

[1] J. Barluenga, J. M. Gonzalez, P. J. Campos, and G. Asensio, *Angew. Chem. Int. Ed.*, **24**, 319 (1985).
[2] J. Barluenga, M. A. Rodriguez, P. J. Campos, and G. Asensio *Am. Soc.*, **110**, 5567 (1988).
[3] J. Barluenga, J. M. Gonzalez, P. J. Campos, and G. Asensio, *Angew. Chem. Int. Ed.*, **27**, 1546 (1988).

Bis(tributyltin) oxide, $(Bu_3Sn)_2O$.

Cleavage of esters.[1] Pivaloyloxymethyl esters have been used for protection of N^7- or N^9-adenine and of β-lactams, including penicillanic acids. Deprotection is a problem in the case of β-lactam derivatives because of the ready cleavage of the ring. The pivaloyloxymethyl esters of penicillanic acids can be cleaved in 55–60% yield by bis(tributyltin) oxide in ether at 25°, probably by way of an organic tin intermediate. This method is also suitable for cleavage of simple esters, even those containing peptide bonds.

[1] E. G. Mata and O. A. Mascaretti, *Tetrahedron Letters*, **29**, 6893 (1988).

Bis(tributyltin) oxide–Silver nitrate, $(Bu_3Sn)_2O$–$AgNO_3$.

$RCH_2X \rightarrow RCH_2OH$.[1] Primary alkyl bromides or iodides react with these two reagents to give the corresponding primary alcohols in moderate to high yield. Optimum ratios for this reaction are $(Bu_3Sn)_2O/Ag/\text{halide} = 2.2:1.1:1.0$. Alkyl chlorides can be precursors if first treated with NaI (3 equiv.). The reaction is not useful in the case of secondary halides because of elimination.

$$RCH_2X \xrightarrow[DMF,\ 20-90^\circ]{(Bu_3Sn)_2O,\ AgNO_3} [RCH_2OSnBu_3] \xrightarrow[28-84\%]{SiO_2\ or\ H_2O} RCH_2OH$$

[1]M. Gingras and T. H. Chan, *Tetrahedron Letters*, **30**, 279 (1989).

[Bis(trifluoroacetoxy)iodo]pentafluorobenzene, $C_6F_5I[OCOCF_3]_2$ (**1**).

This hypervalent iodine reagent is prepared by reaction of C_6F_5I with trifluoroacetic anhydride.

Oxidative cleavage of alkynes.[1] This reagent (3 equiv.) cleaves 1-alkynes to the norcarboxylic acids in 55–80% yield.

[1] R. M. Moriarty, R. Penmasta, A. K. Awasthi, and I. Prakash, *J. Org.*, **53**, 6124 (1988).

1,1-Bis(trimethylsilyl)ethylene (1).

Peterson reactions.[1] Reaction of Li in THF with **1** leads to the 1,4-dianion (**2**) of 1,1,4,4-tetrakis(trimethylsilyl)butane in 96% yield. This dianion reacts with paraformaldehyde to form 1,5-diene **3**, which can be converted to **4** by bromination–bromodesilylation.

$$\underset{\mathbf{1}}{[(CH_3)_3Si]_2C{=}CH_2} \xrightarrow[96\%]{Li,\ THF} \underset{\mathbf{2}}{[(CH_3)_3Si]_2\bar{C}CH_2CH_2\bar{C}[Si(CH_3)_3]_2}$$

$$\mathbf{2} \xrightarrow[75\%]{HCHO} \underset{\mathbf{3}}{CH_2{=}C(Si(CH_3)_3)CH_2CH_2C(Si(CH_3)_3){=}CH_2}$$

$$\mathbf{3} \xrightarrow[44\%]{1)\ Br_2;\ 2)\ NaOC_2H_5} \underset{\mathbf{4}}{CH_2{=}C(Br)CH_2CH_2C(Br){=}CH_2}$$

The dianion **2** can be alkylated as expected. Reaction with dihalosilanes results in silacyclopentanes (equation I).

$$(I)\quad \mathbf{2} + H_2SiCl_2 \xrightarrow[41\%]{THF} \text{2,2,5,5-}[Si(CH_3)_3]_2\text{-substituted silacyclopentane } (SiH_2 \text{ ring})$$

[1] M. Kira, T. Hino, Y. Kubota, N. Matsuyama, and H. Sakurai, *Tetrahedron Letters*, **29**, 6939 (1988).

Bis(trimethylsilyl) peroxide, $[(CH_3)_3SiO]_2$, **11**, 67; **12**, 63.

Electrophilic hydroxylation.[1] Reaction of this peroxide (**1**) with Grignard reagents affords the corresponding silyloxy derivatives in 70–90% yield. The reaction with vinyl Grignard reagents results in silyl enol ethers or α-silyloxy ketones.

$$RC(MgBr){=}CH_2 \xrightarrow{1} RC(OSi(CH_3)_3){=}CH_2 \xrightarrow[57\%]{C_6H_5CHO,\ TiCl_4} RC(=O)CH_2CH(OH)C_6H_5$$

$$\text{1-cyclohexenyl-MgBr} \xrightarrow[61\%]{1} \text{2-(trimethylsilyloxy)cyclohexanone } (OSi(CH_3)_3)$$

Oxidation of vinyllithiums.[2] Vinyllithiums are oxidized to silyl enol ethers by bis(trimethylsilyl) peroxide (**1**) with retention of configuration. Yields can be increased by use of bis(*t*-butyldimethylsilyl) peroxide (**2**).

$$C_6H_5CH{=}CHBr \xrightarrow{1)\ t\text{-BuLi};\ 2)\ 1 \text{ or } 2} C_6H_5CH{=}CHOSiMe_2R$$

(E)	**1**	11%	(E, 100%)
(Z)	**1**	18%	(Z, 100%)
(Z)	**2**	35%	(Z, 100%)

This transformation can be carried out by oxidation with (+)-camphorylsulfonyloxaziridine (**13**, 64–65) followed by trapping with $ClSi(CH_3)_3$, but this route is accompanied by partial inversion of geometry.

α-Hydroxy carboxylic acids or amides.[3] Reaction of the dianion (LDA) of these acids or anions of amides with **1** provides the corresponding α-hydroxy compound in 30–60% yield. The reagent fails to hydroxylate the lithium anion of cyclohexanone or of ethyl phenylacetate.

[1] L. Camici, P. Dembech, A. Ricci, G. Seconi, and M. Taddei, *Tetrahedron*, **44**, 4197 (1988).
[2] F. A. Davis, G. Sankar Lal, and J. Wei, *Tetrahedron Letters*, **29**, 4269 (1988).
[3] M. Pohmakotr and C. Winotai, *Syn. Comm.*, **18**, 2141 (1988).

Bis(trimethylsilyl) peroxide/Trifluoromethanesulfonic acid, $[(CH_3)_3SiO]_2/CF_3SO_3H$.

Aromatic hydroxylation.[1] This reaction can be effected in moderate yield with 30% H_2O_2 in combination with HF/BF_3 etherate (**11**, 256). A more effective and convenient reagent is bis(trimethylsilyl) peroxide activated by triflic acid.

[1] G. A. Olah and T. D. Ernst, *J. Org.*, **54**, 1204 (1989).

Bis(trimethylsilyl) selenide, $[(CH_3)_3Si]_2Se$ (**1**).[1]

Intramolecular* [4 + 2]*cycloaddition of selenaldehydes.[2] The reaction of diunsaturated aldehydes such as **2** with **1** in the presence of a catalytic amount of BuLi gives selenaldehydes (**a**) which undergo intramolecular Diels–Alder reactions to give selenabicyclics **3** in modest yield.

2, n = 3 —(**1**, BuLi, THF, 55°)→ [**a**] —(63%)→ **3**

(*cis*/*trans* = 65 : 35)

[1] Preparation: M. R. Detty, and M. D. Seidler, *J. Org.*, **47**, 1354 (1982).
[2] S. Murai and N. Sonoda, *Tetrahedron Letters*, **29**, 6965 (1988).

1,2-Bis(trimethylstannyl)-1-alkenes, $R((CH_3)_3Sn)C{=}CHSn(CH_3)_3$ (**1**).

The reagent is obtained, solely or mainly, as the (Z)-isomer by addition of hexamethylditin to 1-alkynes catalyzed by Pd(0). It couples with allylic, benzylic, or aryl halides in the presence of $BzlPdCl[P(C_6H_5)_3]_2$ selectively by replacement of the C_1-$Sn(CH_3)_3$ group. The stannyl group at C_1 is also replaced by halogen on reaction of (Z)-**1** with NIS or NBS.[1]

[1] T. N. Mitchell, K. Kwetkat, D. Rutschow, and U. Schneider, *Tetrahedron*, **45**, 969 (1989).

3,3′-Bis(triphenylsilyl)binaphthol–Trimethylaluminum (**14**, 46–47).

Ene reaction. The ene reaction of aldehydes with alkenes catalyzed by (R)-**1** and 4-Å molecular sieves can provide homoallylic alcohols with marked enantioselectivity (equations I and II).

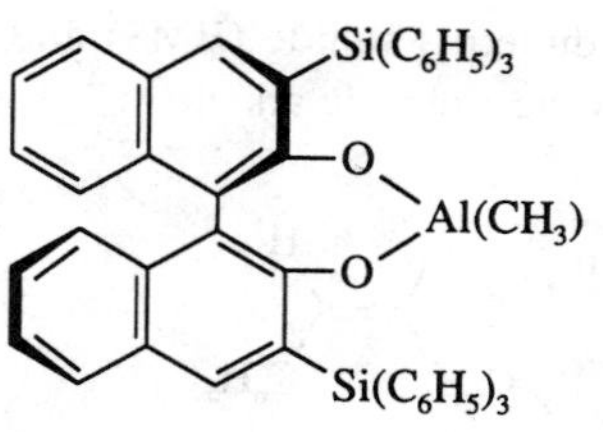

$Si(C_6H_5)_3$, O, $Al(CH_3)$, O, $Si(C_6H_5)_3$

(R)−**1**

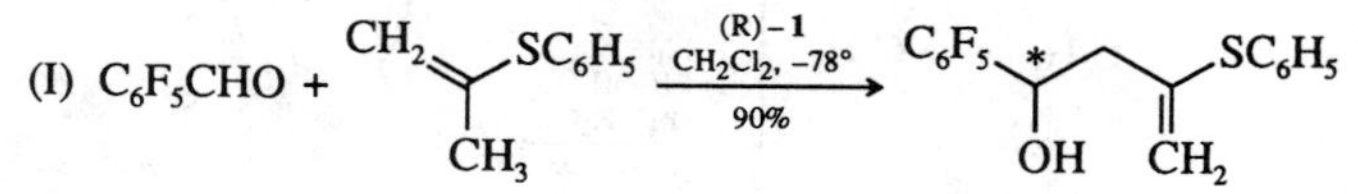

(I) C_6F_5CHO + $CH_2=C(CH_3)SC_6H_5$ $\xrightarrow[90\%]{(R)-1,\ CH_2Cl_2,\ -78°}$ $C_6F_5\overset{*}{C}H(OH)CH_2C(=CH_2)SC_6H_5$

88% ee

(II) $Cl_3CCHO + CH_2{=}C(CH_3)_2 \xrightarrow[70\%]{} Cl_3C\overset{*}{C}H(OH)CH_2C(=CH_2)CH_3$

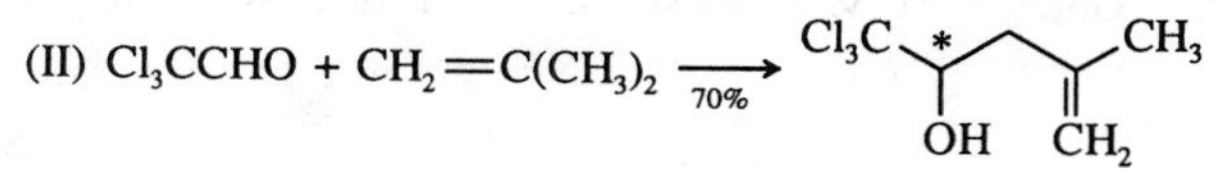

(S), 74% ee

[1] K. Maruoka, Y. Hoshino, T. Shirasaka, and H. Yamamoto, *Tetrahedron Letters*, **29**, 3967 (1988).

Blue tetrazolium, 1, 61.

Oxidation of steroid enones.[1] 3-Keto-Δ^4-steroids are oxidized by blue tetrazolium (**1**) in an alkaline medium to 2,6-diketo-Δ^4-steroids in 70–80% yield.

Other examples:

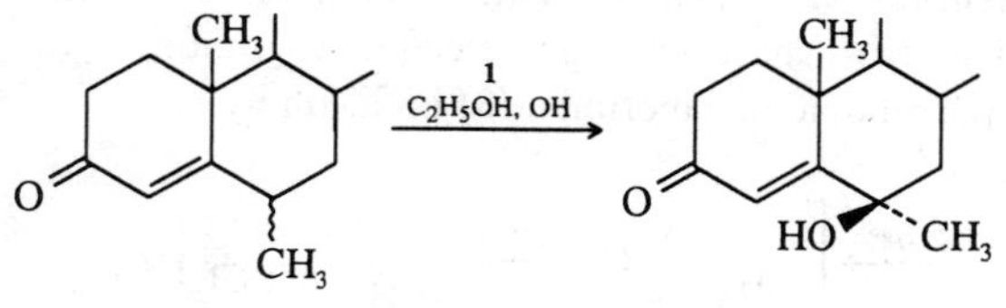

CH_3, O, CH_3 $\xrightarrow{1,\ C_2H_5OH,\ OH}$ CH_3, O, HO, CH_3

α−CH_3 65%

β−CH_3 27%

CH_3, O, OH $\xrightarrow{55-75\%}$ CH_3, O, O, O $\rightleftharpoons$ CH_3, O, O, OH

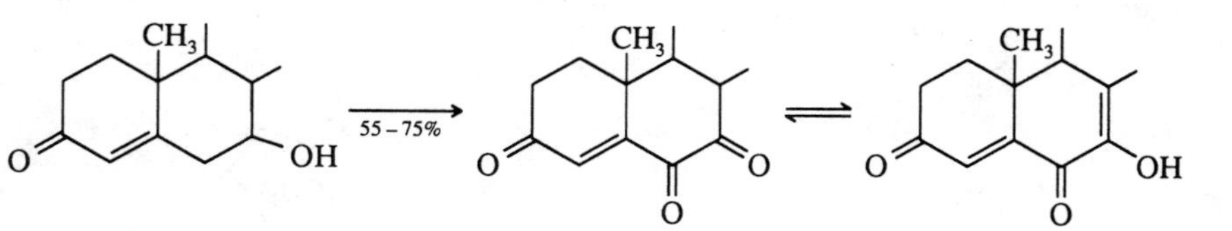

[1] J. Jasiczak, *J.C.S. Chem. Comm.*, 2687 (1988).

9-Borabicyclo[3.3.1]nonane (9-BBN).

Enamines → alkenes.[1] Hydroboration of (E)-enamines of aldehydes or ketones with 9-BBN followed by methanolysis (retention) affords alkenes in 60–80%

yield. The same reaction, but with borane·dimethyl sulfide (BMS) and including a final oxidation with alkaline H_2O_2, affords the isomeric alkene.

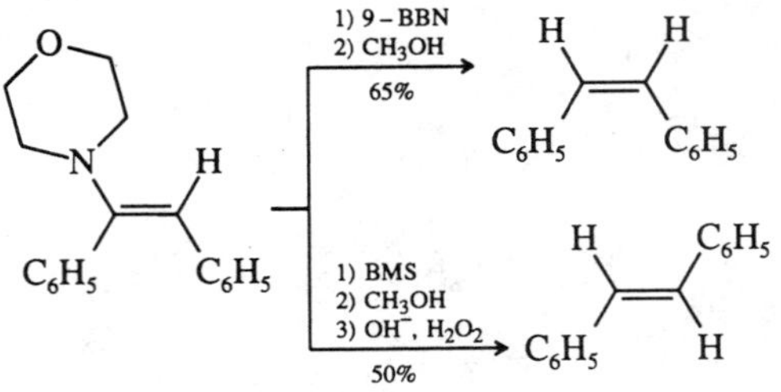

1) 9-BBN
2) CH_3OH
65%

1) BMS
2) CH_3OH
3) OH^-, H_2O_2
50%

[1] B. Singaram, C. T. Goralski, M. V. Rangaishenvi, and H. C. Brown, *Am. Soc.*, **111**, 384 (1989).

Borane–Dimethyl sulfide.

Reduction of ozonides to alcohols.[1] This reagent (3 equiv.) reduces ozonides of alkenes to alcohols in CH_2Cl_2 at 22° in generally high yield without reduction of ester groups.

[1] L. A. Flippin, D. W. Gallagher, and K. Jalali-Araghi, *J. Org.*, **54**, 1430 (1989).

Boron(III) chloride.

Cyclization of ω-azidoalkeneboronic esters.[1] Treatment of these esters **1** with BCl_3 provides chloro derivatives (**a**). These undergo selective migration of the ω-aminoalkyl group to give a pyrrolidine or piperidine (**3**) on alcoholysis.

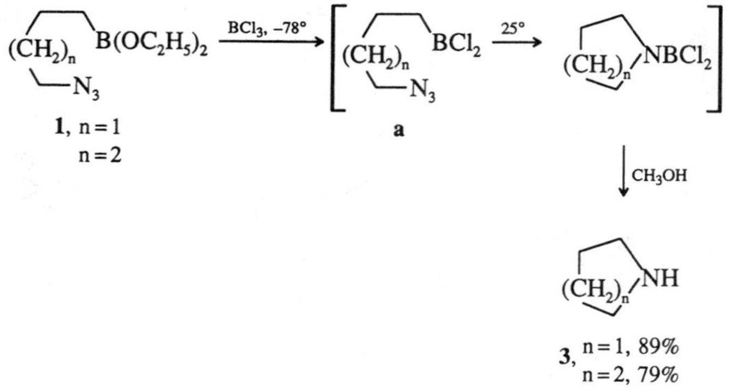

$(CH_2)_n$ $B(OC_2H_5)_2$ / N_3 —(BCl_3, −78°)→ [$(CH_2)_n$ BCl_2 / N_3 (**a**) —(25°)→ $(CH_2)_n$ $NBCl_2$] —(CH_3OH)→ $(CH_2)_n$ NH

1, n = 1
n = 2

3, n = 1, 89%
n = 2, 79%

[1] J. M. Jego, B. Carboni, M. Vaultier, and R. Carrie, *J.C.S. Chem. Comm.*, 142 (1989).

Boron trifluoride etherate.

3-Acyltetrahydrofurans. These products can be obtained by BF_3-catalyzed condensation of (Z)-4-hydroxy-1-alkenylcarbamates with aldehydes or ketones in high diastereo- and enantioselectivity.[1] Use of a (Z)-*anti*-precursor (**1**) results in *cis, trans, trans*-tetrasubstituted tetrahydrofurans from aldehydes with high selectivity.

OH
CH_3 CH_3 + $(CH_3)_2CHCHO$ $\xrightarrow[80\%]{BF_3 \cdot O(C_2H_5)_2}$
BzlO CH_3 $OCON(i\text{-}Pr)_2$
anti-**1**

CH_3 $COCH_3$
CH_3
$C(CH_3)_3$
BzlO O
2

1 + $(CH_3)_2C{=}O$ $\xrightarrow[80\%]{}$ CH_3 ... $COCH_3$, CH_3, CH_3, BzlO, O

Allylic alcohols from sulfones.[2] Polish chemists have extended the Julia synthesis of alkenes (**11**, 474) to a synthesis of allylic alcohols. In the presence of 1 equiv. of BF_3 etherate, α-alkoxy aldehydes react with lithiated sulfones to form adducts that are converted to allylic alcohols on reduction with sodium amalgam. This reaction was developed specifically for a synthesis of prostaglandins from Corey's lactone–aldehyde, but should have wider application.

R$\diagdown$$SO_2C_6H_5$ + $HCOCHC_5H_{11}$ (OSiR_3) $\xrightarrow[30-90\%]{1)\ LDA,\ THF;\ 2)\ BF_3 \cdot O(C_2H_5)_2}$ R–CH($C_6H_5SO_2$)–CH(OH)–CHC_5H_{11} (OSiR_3)

75–85% ↓ 1) CH_3SO_2Cl; 2) Na/Hg, Na_2HPO_4

R–CH=CH–CH(C_5H_{11})(OSiR_3)

Claisen condensation of acetoacetate dianion.[3] In the presence of BF_3 etherate the dianion of acetoacetate undergoes Claisen condensation with the tetramethyldiamides of dicarboxylic acids **1** to form polyketides (**2**, equation I). The

$$\text{(I)}\ CH_3\overset{O}{\overset{\|}{C}}CH_2COOCH_3 + (CH_2)_n(CON(CH_3)_2)_2 \xrightarrow[71\%]{1)\ NaH,\ BuLi,\ THF;\ 2)\ BF_3\cdot O(C_2H_5)_2} (CH_2)_n(CO CH_2\overset{O}{\overset{\|}{C}}CH_2COOCH_3)_2$$

1, n = 4 → **2** (n = 4)

2 (n = 4) $\xrightarrow[78\%]{Ca(OAe)_2,\ 80°}$ **3**

2, n = 2 $\xrightarrow[67\%\ (overall)]{HCl}$ **4**

(**3** and **4**: OH, $COOCH_3$, CH_2COOCH_3 substituted benzo-fused cyclic ketones)

product (**2**) obtained from tetramethylsuccinamide undergoes acid-catalyzed cyclization to the aromatic system **4**; but aromatization of **2**, from an adipamide, is effected with treatment with calcium acetate at 80°.

This BF_3-promoted Claisen condensation can be used for a general synthesis of 3,5-dioxoalkanoates from esters (equation II).

$$\text{(II)}\ RCOOCH_2(C_2H_5) + CH_3\overset{O}{\overset{\|}{C}}CH_2COOCH_3 \xrightarrow{1)\ NaH,\ BuLi,\ 2)\ BF_3,\ -78°} R\overset{O}{\overset{\|}{C}}CH_2\overset{O}{\overset{\|}{C}}CH_2COOCH_3$$

R	Yield
$R = n\text{-}C_9H_{19}$	60%
$R = ClCH_2$	43%
$R = C_6H_5$	72%

Photocyclization of arylimines.[4] The photocyclization of stilbenes to phenanthrenes (**13**, 231) can be extended to arylimines such as **1** when BF_3 etherate is used as activator for the imines.

$h\nu$, $BF_3 \cdot O(C_2H_5)_2$, CH_2Cl_2, 38–45%

1, R = H, CH_3;
R' = H, OCH_3;
X = CH, N

[1] D. Hoppe, T. Krämer, C. F. Erdbrügger, and E. Egert, *Tetrahedron Letters*, **30**, 1233 (1989).
[2] B. Achmatowicz, E. Baranowska, A. R. Daniewski, J. Pankowski, and J. Wicha, *Tetrahedron*, **44**, 4989 (1988).
[3] M. Yamaguchi, K. Shibato, H. Nakashima, and T. Minami, *ibid.*, **44**, 4767 (1988).
[4] C. M. Thompson and S. Docter, *Tetrahedron Letters*, **29**, 5213 (1988).

Bromine.

Bromolactonization.[1] Reaction of unsaturated acids with Br_2 and thallium(I) carbonate results in bromine-substituted lactones in moderate yield. Yields generally are higher than those obtained with NBS in CH_2Cl_2, THF, or DMF.

CH_2COOH — Br_2, Tl_2CO_3, CH_2Cl_2, 62% →

30:70

[1] R. C. Cambie, P. S. Rutledge, R. F. Somerville, P. D. Woodgate, *Synthesis*, 1009 (1988).

Bromine–Methanol.

RCHO → RCOOCH₃. This conversion can be effected by oxidation of aldehydes with Br_2 in CH_3OH via a hemiacetal (equation I). The method is applicable to optically pure aldehydes obtained from sugars. Yields are moderate in the reaction with aromatic aldehydes.[1]

$$\text{(I)}\quad RCHO \underset{}{\overset{CH_3OH}{\rightleftharpoons}} \left[\begin{array}{c} OH \\ | \\ RCHOCH_3 \end{array} \right] \xrightarrow[70-95\%]{Br_2,\ H_2O,\ NaHCO_3} RCOOCH_3$$

This reaction can be extended to primary alcohols and to *vic*-diols, both of which can be oxidized to aldehydes.[2]

$\xrightarrow[N(C_2H_5)_3]{(COCl)_2/DMSO}$ [RCHO $\xrightarrow{CH_3OH}$ RCH(OH)OCH$_3$] $\xrightarrow[84\%]{Br_2}$ ... $COOCH_3$ (α_D + 17°)

NaIO$_4$ → [OHC–C$_5$H$_8$–CHO] $\xrightarrow[87\%]{1)\ CH_3OH;\ 2)\ Br_2}$ CH_3OOC–C$_5$H$_8$–$COOCH_3$

[1] D. R. Williams, F. D. Klingler, E. E. Allen, F. W. Lichtenthaler, *Tetrahedron Letters*, **29**, 5087 (1988).
[2] F. W. Lichtenthaler, P. Jarglis, and K. Lorenz, *Synthesis*, 790 (1988).

Bromine fluoride, BrF. BrF is conveniently formed from Br_2 in Cl_3CF and used directly.

Aromatic bromination.[1] BrF readily undergoes both ionic and radical reactions with arenes, but addition of C_2H_5OH suppresses radical reaction and permits electrophilic bromination. Simple arenes give mono- and dibromo derivatives. BrF is more useful for bromination of deactivated arenes as shown.

$$C_6H_5CN \xrightarrow{BrF} m\text{-}BrC_6H_4CN\ (81\%) + o\text{-}BrC_6H_4CN\ (10\%)$$

$$C_6H_5CHO \xrightarrow[94\%]{} m\text{-}BrC_6H_4CHO$$

[1] S. Rozen, M. Brand, and R. Lidor, *J. Org.*, **53**, 5545 (1988).

(E)-(2-Bromoethenyl)diisopropoxyborane Br–CH=CH–B(O-*i*-Pr)$_2$ (**1**).

Preparation:

$$\text{Br}\diagdown\!\!=\!\!\diagup\text{BBr}_2 + (\text{CH}_3)_2\text{CHOH} \xrightarrow[86\%]{C_5H_{12}} \textbf{1}, \text{ b.p. } 73°/12 \text{ mm.}$$

(E)-1,3-Dienes. In the presence of $Cl_2Pd[P(C_6H_5)_3]_2$ this borane reacts with a vinylzinc chloride to form an intermediate that couples with an alkyl chloride in the presence of a base (LiOH) to give an (E)-1,3-diene.

$$CH_2{=}C(Si(CH_3)_3)ZnCl \xrightarrow{\textbf{1},\ Pd(II)} [CH_2{=}C(Si(CH_3)_3){-}B(O\text{-}i\text{-}Pr)CH{=}CHBr] \xrightarrow[89\%]{C_6H_5CH_2Cl} CH_2{=}C(Si(CH_3)_3)CH{=}CHCH_2C_6H_5$$

E, 99%

$$\text{Bu}CH{=}CH\text{ZnCl} + C_6H_5CH_2Cl \xrightarrow[76\%]{\textbf{1},\ Pd(II)} \text{Bu}CH{=}CH{-}CH{=}CHCH_2C_6H_5$$

E, 96%

(E)-(2-Bromoethenyl)dibromoborane, $BrCH{=}CHBBr_2$, can be used in place of **1** to couple alkylzinc chlorides with an alkyl chloride to form a monoene, but this reagent is unsatisfactory for coupling of alkenylzinc chlorides.[1]

[1] S. Hyuga, N. Yamashina, S. Hara, and A. Suzuki, *Chem. Letters*, 809 (1988).

9-Bromo-9-phenylfluorene, (9-Br, 9-C_6H_5 fluorene) **(1).**

N-Protected α-amino aldehydes.[1] Boc- and Cbo-N-protected α-amino aldehydes can undergo racemization under usual synthetic applications, but protection with a phenylfluorenyl (PhFl) group provides considerable configurational stability, even to silica gel. A typical preparation is shown for N-(PhFl)-L-alaninal (**2**).

$$CH_3CH(NH_2)COOSi(CH_3)_3 \xrightarrow[84\%]{1,\ N(C_2H_5)_3,\ Pb(NO_3)_2,\ CHCl_3/CH_3CN} CH_3CH(NHPhFl)COOH \xrightarrow{87\%}$$

$$CH_3CH(NHPhFl)-C(=O)-N(\text{isoxazolidine}) \xrightarrow[90\%]{1)\ LiAlH_4;\ 2)\ KHSO_4} CH_3CH(NHPhFl)-CH(=O)$$

(66% overall)

2

H
N
CHO
C_6H_5
CH_3

N-Protected α-amino ketones.[2] N-(-Phenylfluorenyl)alaninal (**2**) is readily converted to a variety of N-PhFl-α-amino ketones (**3**) by reaction of a Grignard reagent (excess) followed by oxidation with NCS and $S(CH_3)_2$ (Corey–Kim reagent). Deprotonation and alkylation of these protected α-amino ketones occurs at the α′-carbon atom to provide a mixture of diastereomeric α′-alkyl branched α-amine ketones. These ketones on hydrogenolysis in methanol followed by periodate oxidation give α-alkyl substituted carboxylic acids in high enantiomeric purity.

$$\mathbf{2} \xrightarrow[90\%]{BuMgBr} CH_3CH(NHPhFl)CH(OH)Bu \xrightarrow[80\%]{NCS,\ S(CH_3)_3} CH_3CH(NHPhFl)C(=O)Bu\ (\mathbf{3})$$

$$\mathbf{3} \xrightarrow{1)\ KHMDS;\ 2)\ CH_3I} CH_3CH(NHPhFl)C(=O)CH(CH_3)Pr\ (\mathbf{4}) \xrightarrow[25\%\ overall]{1)\ H_2,\ HCl;\ 2)\ H_5IO_6} HOC(=O)CH(CH_3)Pr$$

(R)-**5** (99% ee)

[1] W. D. Lubell and H. Rapoport, *Am. Soc.*, **109**, 236 (1987).
[2] *Idem, ibid*, **110**, 7447 (1988).

N-Bromosuccinimide.

Deglycosidation.[1] Treatment of *n*-pentenyl glycosides (either α or β) with NBS in 1% aqueous acetonitrile results in liberation of the anomeric hydroxyl group. Brominolysis in the presence of an alcohol results in a glycoside or a disaccharide (in the case of sugar-OH.) Protected benzylic or ester groups are not affected by this brominolysis; a protecting benzylic ether group on C_2 slows this reaction, whereas a protecting ether group has little effect on the rate. This effect can be used for regioselective coupling of glycosides to a trisaccharide. Thus coupling a 2-benzylated pentenyl glycoside with a 2-acetylated pentenyl glycoside with iodonium perchlorate (**10**, 212–213) results in a disaccharide formed from the 2-benzylated pentenyl glycoside as the glycosyl donor. The acetyl groups of the dissacharide are then replaced by benzyl groups, which now activate the reducing end with the *n*-pentenyl anomeric center in glycosidation with another sugar alcohol to give a trisaccharide.

[1] D. R. Mootoo, V. Date, and B. Fraser-Reid, *Am. Soc.*, **110**, 2662 (1988); D. R. Mootoo, P. Konradsson, U. Udodong, and B. Fraser-Reid, *ibid.*, **110**, 5583 (1988).

Bromotrimethylsilane. This silane can be generated *in situ* from $ClSi(CH_3)_3$ and anhydrous NaBr in dry DMF at 25°.

Silyl enol ethers.[1] Reaction of carbonyl compounds with *in situ* generated $BrSi(CH_3)_3$ and triethylamine results mainly in the thermodynamic silyl ether, usually the (Z)-isomer.

Examples:

$$CH_3\overset{O}{\overset{\|}{C}}CH_2CH_3 \xrightarrow[95\%]{BrSi(CH_3)_3,\ N(C_2H_5)_3,\ DMF} CH_3C(OSi(CH_3)_3)=CHCH_3$$

(Z/E = 9 : 1)

$$CH_3\overset{O}{\overset{\|}{C}}CH_2COOC_2H_5 \xrightarrow[100\%]{} CH_3C(OSi(CH_3)_3)=CHCOOC_2H_5$$

(E/Z = 100 : 0)

$$H\overset{O}{\overset{\|}{C}}CH_2Bu \xrightarrow[73\%]{} HC(OSi(CH_3)_3)=CHBu$$

(Z/E = 94 : 6)

[1] S. Ahmad, M. A. Khan, and J. Iqbal, *Syn. Comm.*, **18**, 1679 (1988).

(3R)-Butane-1,3-diol, $CH_3CH(OH)CH_2CH_2OH$ (**1**). The diol is prepared[1] by lithium aluminum hydride reduction of poly(3-hydroxybutyric acid), available from Aldrich, Fluka.

5-Alkyl-3,5-dihydroxypentanoic acids; chiral hydroxylactones.[2] The acetal (**2**) of an aldehyde formed from (R)-**1** couples with 1,3-bis(trimethylsilyloxy)-1-methoxy-1,3-butadiene (**3**), to give a single isomer (**4**). The chiral auxiliary is

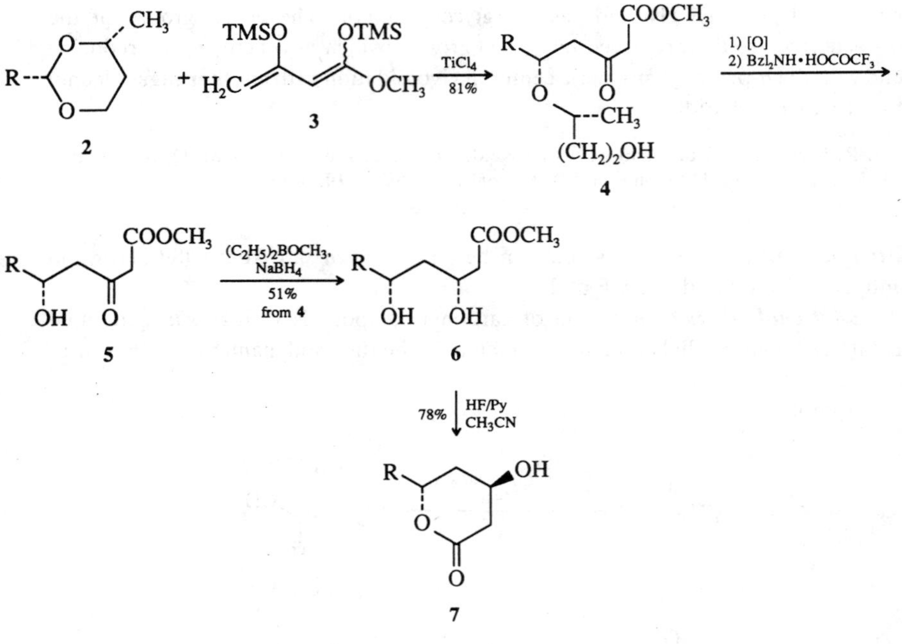

CH_3 TMSO OTMS $COOCH_3$
R O O + H_2C OCH_3 $\xrightarrow[81\%]{TiCl_4}$ R O O CH_3 $(CH_2)_2OH$ $\xrightarrow{\text{1) [O] 2) } Bzl_2NH \cdot HOCOCF_3}$
2 **3** **4**

R $COOCH_3$ OH O $\xrightarrow[\text{51\% from 4}]{(C_2H_5)_2BOCH_3,\ NaBH_4}$ R $COOCH_3$ OH OH
5 **6**

78% HF/Py CH_3CN

R OH O O
7

removed by oxidation with periodinane (**12**, 378–379) followed by β-elimination promoted by dibenzylammonium trifluoroacetate. The resulting aldol (**5**) is then reduced diastereoselectively to the *syn*-1,3-diol **6**. Saponification and lactonization gives the lactone **7**. Lactones of this type are present in compounds inhibiting biosynthesis of cholesterol.

[1] D. Seebach and M. Züger, *Helv.*, **65**, 495 (1982).
[2] W. S. Johnson, A. B. Kelson, and J. D. Elliott, *Tetrahedron Letters*, **29**, 3757 (1988).

(2S,4S)-N-(*t*-Butoxycarbonyl)-4-(dicyclobutylphosphine)-2-[(diphenylphosphino)methyl]pyrrolidine (BCPM, **1**). Preparation.[1]

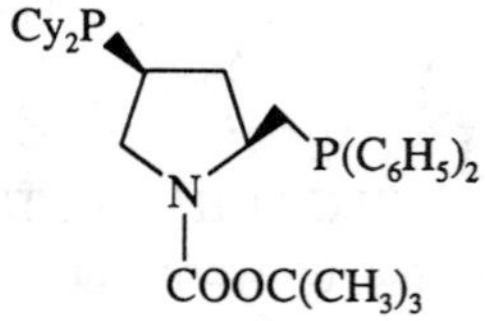

Cy_2P $P(C_6H_5)_2$ N $COOC(CH_3)_3$

1, m.p. 174°, α_D −41°

Asymmetric hydrogenation of > C=O. This bisphosphine is the most effective ligand for asymmetric hydrogenation of ketopantolactone (**2**) to provide R-(−)-pantolactone (**3**).[1]

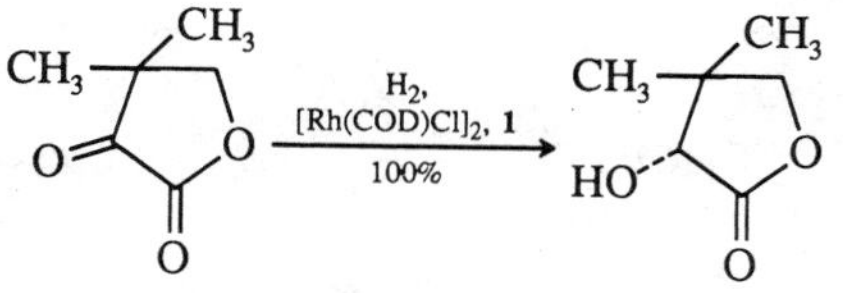

H_2, $[Rh(COD)Cl]_2$, **1**, 100%

2 → (R)−**3**, 92% ee

This chiral bisphosphine in combination with rhodium complexes also effects efficient asymmetric reduction of suitably substituted carbonyl compounds such as α-aminoacetophenones.[2,3]

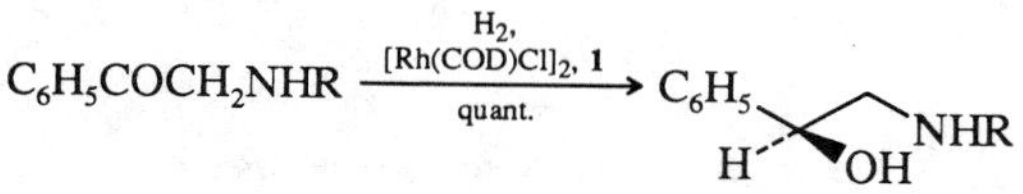

$C_6H_5COCH_2NHR$ —(H_2, $[Rh(COD)Cl]_2$, **1**, quant.)→ $C_6H_5CH(OH)CH_2NHR$

(S, 85–97% ee)

[1] H. Takahashi, M. Hattori, M. Chiba, T. Morimoto, and K. Achiwa, *Tetrahedron Letters*, **27**, 4477 (1986).
[2] H. Takeda, T. Tachinami, M. Aburatani, H. Takahashi, T. Morimoto, and K. Achiwa, *ibid.*, **30**, 363 (1989).
[3] T. Takeda, T. Tachinami, M. Aburatani, H. Takahashi, T. Morimoto, and K. Achiwa, *ibid.*, **30**, 367 (1989).

(2S,4S)-N-(*t*-Butoxycarbonyl)-4-[bis(4′-methoxy-3′,5′-dimethylphenyl)phosphino]-2-{[bis(4′-methoxy-3′,5′-dimethyl)-phosphinyl]methyl}pyrrolidine [1, (2S,4S)-MOD-BPPM)].

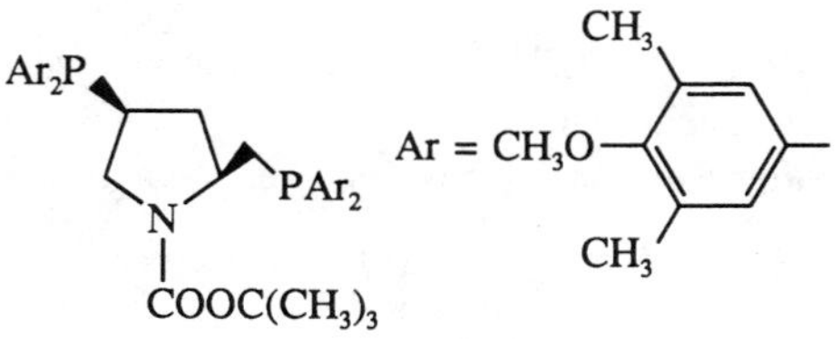

Ar_2P, PAr_2, N, $COOC(CH_3)_3$

1

Ar = 4-CH_3O-3,5-$(CH_3)_2C_6H_2$

Asymmetric hydrogenation. Simpler ligands of this type, but still based on a 4-diphenylphosphinyl-2-(diphenylphosphino)methylpyrrolidine (PPPM), were first reported in 1976 (**8**, 57–59) to be useful for asymmetric hydrogenation of 2-acetamidoacrylic acids. Since then a *para*-methoxy group on the phenyl groups has been found to accelerate the rate, whereas methyl groups increase enantioselectivity. The present ligand (**1**) has been shown to be particularly useful for a rhodium-catalyzed hydrogenation of (Z)-2-acetamidoacrylic acids to the corresponding (R)-amino acid derivatives under mild conditions and with high enantioselectivity.

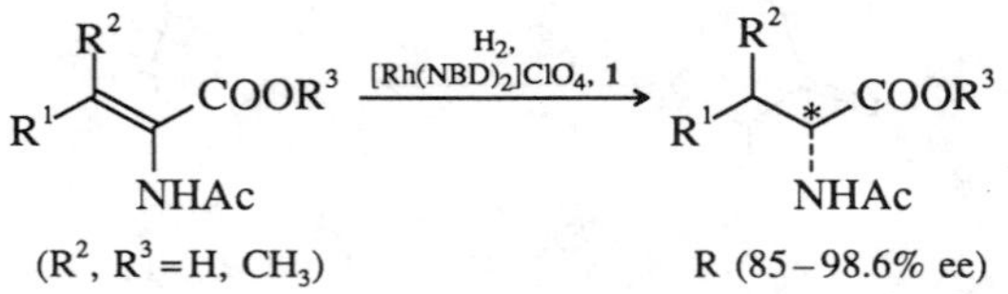

H_2, $[Rh(NBD)_2]ClO_4$, **1**

(R^2, R^3 = H, CH_3) R (85–98.6% ee)

[1] H. Takahashi and K. Achiwa, *Chem. Letters*, 305 (1989).

4-*t*-Butylcatechol, $(CH_3)_3C$–C_6H_3–OH (**1**), OH

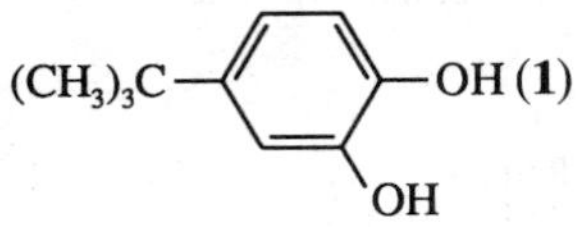

Cyclization of enynones to methylenecyclopentenones. This cyclization can be effected under thermal conditions, particularly in the case of activated (Z)-enynones, but in only moderate yield. Some acceleration is observed with reagents known to promote enolization (LiCl and collidine *p*-toluenesulfonate), but dramatic effects can be obtained with phenols such as **1**. Vitamin E is marginally more

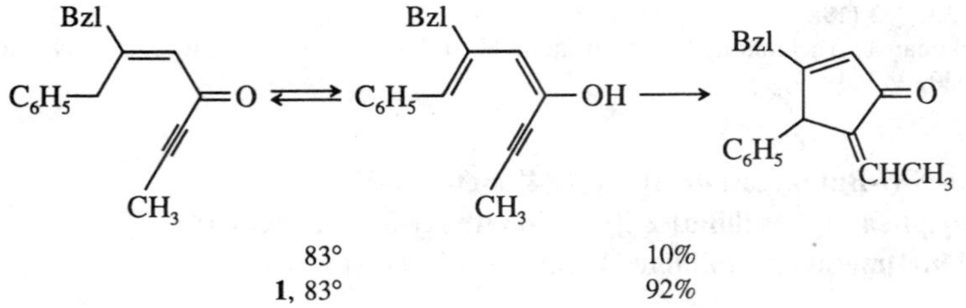

83°		10%
1, 83°		92%

effective than the simpler phenol **1**. The report attributes the effectiveness of phenols to a single electron-transfer reaction.

See Collidine *p*-toluenesulfonate for conversion of enynones to phenols.

[1] P. A. Jacobi, L. M. Armacost, J. I. Kravitz, and M. J. Martinelli, *Tetrahedron Letters*, **29**, 6869 (1988).

***t*-Butyldimethylsilyl trifluoromethanesulfonate, 10**, 63; **11**, 90–91; **13**, 86.

Diels–Alder catalyst.[1] In the presence of this triflate (**1**) and 2,6-lutidine, chromones (**2**) undergo [4 + 2]cycloaddition to α,β-enones to give *cis*-fused adducts in generally high yield (75–95%).

2, R^1 = H, CH_3, C_2H_5 + $CH_2{=}CHCCH_2R^2$ (R^2 = H, CH_3) —(1, base, CH_2Cl_2, 80°; 92–94%)→ [product with R^1, H, R^2, $OSiR_3$]

[1] Y. Lee, H. Iwasaki, Y. Yamamoto, K. Ohkata, and K. Akiyla, *Heterocycles*, **29**, 35 (1989).

t-Butyldiphenylsilyllithium.

This hindered reagent can be prepared by reaction of *t*-Bu$(C_6H_5)_2$SiCl with Li in THF at 0°. It adds to ketones to form α-hydroxysilanes. Evidently a Brook rearrangement to a silyl ether is unfavorable because of steric hindrance. The corresponding silylcuprate, $[(C_6H_5)_2$-*t*-BuSi$]_2$CuLi, undergoes conjugate addition to enones and unsaturated esters in good yield.[1]

[1] P. Cuadrado, A. M. González, B. González, and F. J. Pulido, *Syn. Comm.*, **19**, 275 (1989).

t-Butyl hydroperoxide–(+)-DET–Ti-(O-*i*-Pr)$_4$

Sharpless epoxidation of alkenylsilanols.[1] Allylic silanols also undergo highly enantioselective Sharpless epoxidation. This reaction furnishes simple epoxides such as styrene oxide in high optical purity. Thus reaction of *trans*-β-lithiostyrene (**1**) with $ClSi(CH_3)_2H$ gives **2**, which can be oxidized to the alkenylsilanol **3**. Sharpless epoxidation of **3** gives the epoxide **4**, which is converted to styrene epoxide **5** by cleavage with fluoride ion. The stereochemistry of epoxidation of **3** is similar to that of the corresponding allylic alcohol.

$C_6H_5CH{=}CHLi$ (**1**) —($ClSiH(CH_3)_2$, 92%)→ $C_6H_5CH{=}CHSiH(CH_3)_2$ (**2**) —(Ag_2O, $AgNO_3$, THF, 71%)→ $C_6H_5CH{=}CHSi(CH_3)_2OH$ (**3**)

3 —((*t*-BuOOH), (+) – DET, Ti(O-*i*-Pr)$_4$, 50%)→ **4** [C_6H_5-epoxide-$Si(CH_3)_2OH$]

4 —($(C_2H_5)_4NF$, CH_3CN, 90%)→ C_6H_5-epoxide **5** (85%–95% ee)

[1] T. H. Chan, L. M. Chen, and D. Wang, *J.C.S. Chem. Comm.*, 1280 (1988).

_t_-Butyl hydroperoxide–Dichlorotris(triphenylphosphine)ruthenium(II).

Oxidative dealkylation of* tert*-amines.[1] Oxidation of *tert*-N-methylamines with *t*-BuOOH (2 equiv.) catalyzed by this ruthenium complex results in α-(*t*-butyldioxy)alkylamines. N-Methyl groups are oxidized selectively in the presence of

$$C_6H_5N(CH_3)-CH_3 \xrightarrow[C_6H_6]{t\text{-BuOOH, Ru(II)}} C_6H_5N(CH_2OOC(CH_3)_3)CH_3$$

$$\xrightarrow[93\%]{H_3\overset{+}{O}} C_6H_5N(H)CH_3 + t\text{-BuOH} + H_2O$$

other alkyl or alkenyl groups. Amides are also oxidized at the α-position of the nitrogen.

$$\text{N-}COCH_3\text{-tetrahydroisoquinoline} \xrightarrow[87\%]{t\text{-BuOOH, Ru(II)}} \text{1-}OOC(CH_3)_3\text{-N-}COCH_3\text{-tetrahydroisoquinoline}$$

Oxidation of homoallylic amines results in *cis*-fused bicyclic amines via an iminium ion (**a**).

$$H_3C-N(C_6H_5)\text{-CH}_2CH_2\text{-cyclohexenyl} \longrightarrow \left[H_2C{=}\overset{+}{N}(C_6H_5)\text{-CH}_2CH_2\text{-cyclohexenyl}\right] \xrightarrow[85\%]{CF_3COOH} \text{OH, H-decahydroisoquinoline-N-}C_6H_5$$

a

[1] S.-I. Murahashi, T. Naota, and K. Yonemura, *Am. Soc.*, **110**, 8256 (1988).

_t_-Butyl hydroperoxide–Dioxobis(2,4-pentanedionate)molybdenum, 11, 91; **12**, 89.

Cleavage of* vic*-diols.[1] C—C bonds substituted by hydroxy groups are cleaved by this system to ketones and carboxylic acids in high yield. Di-*tert*-diols are cleaved exclusively to ketones. The carboxylic acids are derived from aldehydes. Chlorobenzene and chloroform are the solvents of choice.

$$(C_6H_5)_2C(OH)-C(OH)(C_6H_5)_2 \xrightarrow[96\%]{(CH_3)_3COOH,\ MoO_2(acac)_2} (C_6H_5)_2C{=}O$$

$$C_6H_5CH(OH)-CH_2OH \xrightarrow[83\%]{} C_6H_5COOH$$

[1] K. Naneda, K. Morimoto, and T. Imanaka, *Chem. Letters*, 1295 (1988).

t-Butyl hydroperoxide–Pyridinum chlorochromate.

Oxidation of alcohols or methylene groups.[1] In the presence of PCC or pyridinium fluorochromate (PFC)[2] as catalyst, *t*-BuOOH can oxidize alcohols in good yield. This combination also effects oxidation of a benzylic or a propargylic methylene group to a keto group in moderate yield; for example: indane → 1-indanone, 46–63% yield; fluorene → fluorenone, 53% yield.

[1] J. Muzart, *New J. Chem.*, **13**, 9 (1989).
[2] M. N. Bhattacharjee, M. K. Chaudhuri, and S. Purkayastha, *Tetrahedron*, **43**, 5389 (1987).

t-Butyl hydroperoxide–Pyridinium dichromate.

2,5-*Cyclohexadien*-1-*ones*. This system (1:1) is effective for oxidation of 3,3-disubstituted-1,4-cyclohexadienes to 2,5-cyclohexadien-1-ones.[1]

$$\text{3-}CH_3\text{-3-}CH_3OOC\text{-}1\text{-}CH_3\text{-cyclohexa-1,4-diene} \xrightarrow[74\%]{t\text{-BuOOH},\ PDC,\ 25^\circ} \text{4-}CH_3OOC\text{-4-}CH_3\text{-3-}CH_3\text{-cyclohexa-2,5-dien-1-one}$$

[1] A. G. Schultz, A. G. Taveras, and R. E. Harrington, *Tetrahedron Letters*, **29**, 3907 (1988).

t-Butyl hydroperoxide–Vanadyl acetylacetonate.

Cyclopentenones.[1] Vinyl allenes with a directing hydroxyl substituent are oxidized by *t*-butyl hydroperoxide and $VO(acac)_2$ directly to cyclopentenones in 40–70% yield.

$$PrCH{=}CHCH{=}C{=}CHCH_2OH \xrightarrow{(CH_3)_3COOH,\ VO(acac)_2} [\text{epoxide: } CH_2OH,\ (CH_2)_2CH_3] \rightarrow \text{cyclopentenone with } CH_2OH \text{ and Pr}$$

[1] S. J. Kim and J. K. Cha, *Tetrahedron Letters*, **29**, 5613 (1988).

***t*-Butyl 3-keto-4-hexenoate,** $CH_3CH{=}CHC(O)CH_2COOC(CH_3)_3$ (**1**). Preparation.[1]

cis-*Decalins*.[2] The enolate of this Nazarov reagent (**1**), conveniently generated with Cs_2CO_3, reacts with 2-methoxycarbonyl-2-cyclohexenone (**2**) to form a product that is cyclized with TsOH in C_6H_6 to the diketones *exo*- and *endo*-**3** in a ratio that is dependent on the solvent. In a nonpolar solvent *exo*-**3** is favored,

2, E = $COOCH_3$ + **1** (CH_3, O, $CO_2C(CH_3)_3$) $\xrightarrow{1)\ Cs_2CO_3;\ 2)\ TsOH,\ C_6H_6}$ *exo*-**3** + *endo*-**3**

$CHCl_3$	99.5 ≥ 0.5
DMF	55:46

whereas in DMF or CH_3CN the ratio is about 1:1. The report presents evidence that a [4 + 2]cycloaddition is involved at least in the nonpolar solvent and that the reaction may involve a sequential double Michael addition in polar solvents.

[1] D. Liotta, C. Barnum, R. Puleo, G. Zima, C. Bayer, and H. S. Kezar, *J. Org.*, **46**, 2920 (1981).

[2] J.-F. Lavalleé and P. Deslongchamps, *Tetrahedron Letters*, **29**, 5117 (1988).

Butyllithium.

Cyclopropanation.[1] The chiral bis(β-phenylthio)carboxamide (**1**) on treatment with BuLi (4 equiv.) cyclizes to a mixture of the two cyclopropanes **2** and **3** in the ratio 1:3, which is readily separated by chromatography thanks to the free hydroxyl group. Higher diastereoselectivity obtains by similar cyclopropanation of

CH_3 CH_3 NHCO SC_6H_5 SC_6H_5 OH CH_3 **1** —4 BuLi, 0°, 94%→

H SC_6H_5 RNHC(=O) H + 1:3 H H RNHC(=O) SC_6H_5

(1S, 2R) – **2** (1R, 2S) – **2**

the triisopropylsilyl ether of **1** (1:11). Removal of the chiral auxiliary (*d*-camphor) is effected by N-*t*-butoxycarbonylation (**12**, 282) and hydrolysis.

α-Oxy-o-xylylene **(a).**[2] Treatment of a benzocyclobutenol or the acetate (**1**) with BuLi at 0° generates an α-oxy-*o*-xylylene (**a**), which can be trapped by dienophiles. Thus, the reaction of **a** with methyl maleate gives *endo*- and *exo*-**2** as the

OH(Ac) **1** —BuLi, THF, –78 to –25°→ [OLi, CH_2] a —(2) $CH_3O_2CCH{=}CHCO_2CH_3$, 0°, 44%→

OH $COOCH_3$ $COOCH_3$ + 60:17 OH $COOCH_3$ $COOCH_3$

endo –**2** *exo* –**2**

major adducts in the ratio 60:17. This result contrasts with the thermal reaction of **1** at 110° with the same dienophile to give **2** as the only adduct (31% yield) but in an *endo*/*exo* ratio of 1:4. Another example of the value of this method is shown in equation (I).

(I) OH(OAc), C_6H_5 —BuLi→ [OLi, C_6H_5] + maleic anhydride —0°, 76%→

OH, H, O, O, H, C_6H_5, O + (72:20) OH, H, O, O, H, C_6H_5, O

Dehydrobromination.[3] One approach to trialkylsiloxyalkynes involves dehydrobromination of **1**, prepared from tribromoethanol, with BuLi (2 equiv.) to form (Z)-2-bromovinyl silyl ethers **2**, which can be converted by LDA followed by alkylation into siloxyalkynes (**4**). These products undergo cycloaddition with vinyl

$Br_3CCH_2OSiR_3$ —2 BuLi→ Br–CH=CH–$OSiR_3$ (**2**) —1) 2 LDA, 2) C_2H_5OH, 96%→ $HC{\equiv}COSiR_3$ (**3**)

1, R_3 = t-Bu_2CH_3 56%
= i-Pr_3 61%

2 —30–70%, 1) 2 LDA, 2) CH_3I→ $CH_3C{\equiv}COSiR_3$ (**4**)

ketenes (**a**), generated from cyclobutenones (**5**) to produce resorcinol derivatives (**6**).

CH_3, O, R (**5**, R = CH_2OMOM) —140°→ [CH_3, C=O, R, CH_2] (**a**) —**4**, R_3 = i-Pr, 59%→ OH, CH_3, R_3SiO, R, CH_3 (**6**)

Cyclization of α-alkoxylithiums.[4] A useful route to tetrahydrofurans involves lithiation of (tributylstannyl)methyl ethers. Another useful route to α-alkoxyli-

$$C_6H_{13}\text{-CH(CH}_2\text{CH=CH}_2\text{)-O-CH}_2SnBu_3 \xrightarrow[54\%]{BuLi} \text{2-}C_6H_{13}\text{-4-}CH_3\text{-tetrahydrofuran}$$

(*cis*/*trans* = 11:1)

thiums is the reaction of lithium naphthalenide (LN) with α-(phenylthio)ethers.

$$CH_2\text{=CH-CH}_2CH_2\text{-O-CH(R)}SC_6H_5 \xrightarrow[52\%]{LN} \text{2-R-3-}CH_3\text{-tetrahydrofuran}$$

(*trans*/*cis* = 7:1)

anti-*Selective* [2,3]*Wittig rearrangement*[5] (cf., **12**, 60). Rearrangement of α-(silyloxy)allyl ethers (**1**) provides the silyloxy 1,2-diol with high *anti*-selectivity regardless of the geometry of the double bond of **1**.

$$C_6H_5CH_2\text{-O-CH}_2\text{-CH=CH-OSi(CH}_3)_2C(CH_3)_3 \xrightarrow[THF,\ -78°]{BuLi} H_2C\text{=CH-CH(OSiR}_3\text{)-CH(OH)}C_6H_5$$

1		**2**
(Z/E) = 71:29	73%	*anti*/*syn* = 82:18
= 93:7	81%	77:23

[1] K. Tanaka, I. Funaki, A. Kaji, K. Minami, M. Sawada, and T. Tanaka, *Am. Soc.*, **110**, 7185 (1988).
[2] W. Choy and H. Yang, *J. Org.*, **53**, 5796 (1988).
[3] R. L. Danheiser, A. Nishida, S. Savariar, and M. P. Trova, *Tetrahedron Letters*, **29**, 4917 (1988).
[4] C. A. Broka and T. Shen, *Am. Soc.*, **111**, 2981 (1989).
[5] E. Nakai and T. Nakai, *Tetrahedron Letters*, **29**, 5409 (1988).

Butyllithium–Hexamethylphosphoric triamide.

Nitroalkane dianions. Seebach *et al.*[1] have shown that reaction of nitroethane with BuLi (2 equiv.) and excess HMPT generates the α,α-dianion as shown by reaction with benzaldehyde to form **3** as the major product. If the sequence is reversed, addition of nitroethane to a solution of BuLi and HMPT, then the α,β-dianion is formed as shown by reaction with benzaldehyde to form **4**.[2] Thus the ratio of 3:4 is controlled by HMPT. In its absence, **3** is exclusive product (57%). Similar results obtain with 1-nitropropane.

$$C_2H_5NO_2 \xrightarrow[76\%]{1)\ BuLi,\ HMPT\ (2:1)\quad 2)\ C_6H_5CHO} C_6H_5CH(OH)CH(CH_3)NO_2 + C_6H_5CH(OH)CH_2CH_2NO_2$$

3 5:95 4

Spirocyclization of an epoxy sulfone.[3] A short synthesis of nitramide (**3**) involves as the key step the reaction of the epoxy sulfone **1** with 2 equiv. of BuLi in THF/HMPA (9:1) to form the spirocycle **2** (73% yield). Both the sulfone and sulfonamide groups are cleaved by freshly prepared sodium amalgam (**7**, 326–327).

CH_2NHTs (cyclohexenyl) $\xrightarrow{\sim 60\%}$ **1** (epoxide, $CH_2N(Ts)(CH_2)_3SO_2C_7H_7$) $\xrightarrow[73\%]{2BuLi,\ THF/HMPT,\ -20 \rightarrow 25^\circ}$ **2** ($C_7H_7SO_2$, NTs, OH) $\xrightarrow[76\%]{Na(Hg),\ Na_2HPO_4,\ CH_3OH}$ **3** (NH, OH)

[1] D. Seebach, A. K. Beck, T. Mukhopadhyay, and E. Thomas, *Helv.*, **65**, 1101 (1982).
[2] K. Yamada, S. Tanaka, S. Kohmoto, and M. Yamamoto, *J.C.S. Chem. Comm.*, 110 (1989).
[3] D. Tanner, H. H. Ming, and M. Bergdahl, *Tetrahedron Letters*, **29**, 6493 (1988).

Butyllithium–Potassium *t*-butoxide.

α-Alkylation of allylsilanes.[1] Deprotonation of (E)- or (Z)-allylsilane **1** with this base followed immediately by alkylation (CH_3I) provides the α-methyl substituted allylsilane (**2**) with the same configuration as the starting silane. However,

(I) Pr–CH=CH–CH$_2$–$Si(CH_3)_3$ (E)-**1** $\xrightarrow[60\%]{1)\ BuLi,\ KOC(CH_3)_3,\ -50^\circ\quad 2)\ CH_3I,\ -50^\circ}$ Pr–CH=CH–CH(CH_3)–$Si(CH_3)_3$ **2**(Z/E = 5:95)

(Z)-**1** $\xrightarrow{63\%}$ **2**(Z/E = 98:2)

alkylation with BuI of the anion of either (E)- or (Z)-**1** provides 20–25% of the (E)-α-substituted vinylsilane, (E)-Pr(Bu)CHCH=CHSi$(CH_3)_3$, which is easily separated from the major product.

[1] A. Mordini, G. Palio, A. Ricci, and M. Taddei, *Tetrahedron Letters*, **29**, 4991 (1988).

Butyllithium–Tetramethylethylenediamine (TMEDA).

o-*Lithiation of aryl thiols.* Three laboratories[1,2,3] report *ortho*-directed lithiation of aryl thiols. Thus lithium 2-lithiobenzenethiolate is formed in quantitative yield by reaction of thiophenol with BuLi (2 equiv.) and TMEDA (2 equiv.) in cyclohexane at 0 → 25°.[1] However, a thiol group is less powerful as an *ortho*-directing group than a methoxy group. Thus *ortho*-metalation of ArSH provides a general route to substituted arenethiols.[2] This *ortho*-metalation can be extended

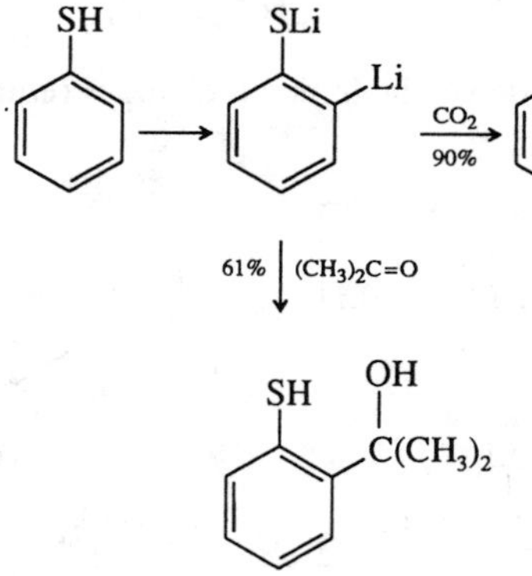

to related systems such as 2-pyridinethiol and 2-(phenylthio)tetrahydropyran (**1**).[3]

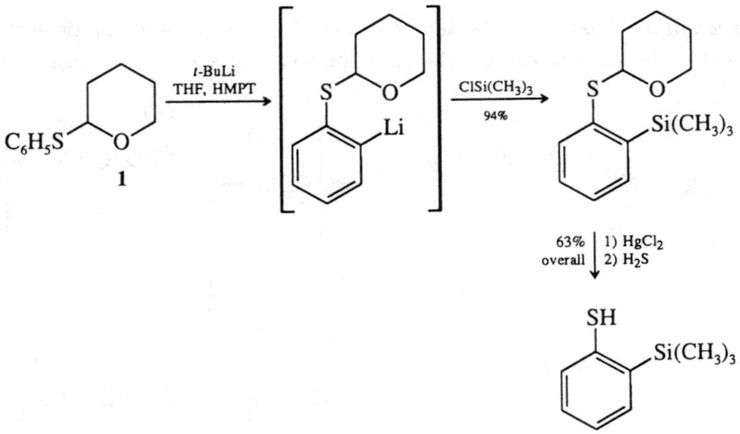

[1] G. D. Figuly, C. K. Loop, and J. C. Martin, *Am. Soc.*, **111**, 654 (1989).
[2] K. Smith, C. M. Lindsay, and G. J. Pritchard, *ibid.*, **111**, 665 (1989).
[3] E. Block, V. Eswarakrishnan, M. Gernon, G. Ofori-Okai, C. Saha, K. Tang, and J. Zubieta, *ibid.*, **111**, 658 (1989).

t-Butyllithium.

Lithiation of vinylamines. The vinylic amine **1** is lithiated by *t*-BuLi at the tertiary vinyl position, presumably because of chelation with the dimethylamino group (equation I).[1] Various allylic and methylallylic amines also can undergo

(I) CH_3, N, N, H_3C, CH_3 $\xrightarrow{t\text{-BuLi}}$ CH_3, N, Li, N, CH_3, CH_3

lithiation with *n*- and *t*-BuLi to provide vinylic lithiated dianions (equations II and III).[2]

(II) C_6H_5NH, CH_2 $\xrightarrow{\text{BuLi},\ t\text{-BuLi}}$ [Li, C_6H_5N·······Li] $\xrightarrow[88\%]{(CH_3)_2S_2}$ C_6H_5NH SCH_3

(III) C_6H_5NH, CH_3, CH_2 $\longrightarrow$ Li, C_6H_5N, Li, CH_2 $\xrightarrow[71\%]{(CH_3)_2C=O}$ OH, C_6H_5NH $C(CH_3)_2$, CH_2

Protection of amides.[3] N-Benzyl groups have been used for protection of amides, but in contrast to benzylamines, benzylamides resist hydrogenolysis. A

$C_6H_5C(=O)-N(CH_3)-CH_2C_6H_4OCH_3\text{-}p$ $\xrightarrow{1)\ t\text{-BuLi};\ 2)\ O_2}$ $[C_6H_5C(=O)-N(CH_3)-CH(OH)C_6H_4OCH_3\text{-}p]$

$\downarrow$ 79%

$C_6H_5C(=O)-NH(CH_3)$ + $OHCC_6H_4OCH_3\text{-}p$

new method for deprotection is generation of the benzylic carbanion (*t*-BuLi), which is oxidized (O_2 or MoOPH) to the benzaldehyde with release of the debenzylated amide. Yields are moderate to reasonably good.

1,3-SiR$_3$ to Br rearrangement.[4] A new route to hindered chiral binaphthols involves a 1,3-rearrangement of SiR_3 with *t*-BuLi.

(R)-1 →[ClSi-*t*-BuPh$_2$, 95%] (R)-2 →[*t*-BuLi, 80%] (R)-3

[1] G. Stork, C. S. Shiner, C.-W. Cheng, and R. L. Polt, *Am. Soc.*, **108**, 304 (1986).
[2] J. Barluenga, F. J. Fañanás, F. Foubelo, and M. Yus, *J.C.S. Chem. Comm.*, 1135 (1988).
[3] R. M. Williams and E. Kwarst, *Tetrahedron Letters*, **30**, 451 (1989).
[4] K. Maruoka, T. Itoh, Y. Araki, T. Shirasaka, and H. Yamamoto, *Bull Chem. Soc. Japan*, **61**, 2975 (1988).

Butyltrichlorotin, $BuSnCl_3$ (**1**).

Etherification.[1] In the presence of $BuSnCl_3$, 1,4- and 1,5-diols are converted into acyclic and cyclic ethers. The actual catalyst is an organoalkoxytin dihalide, $BuSn(OR)Cl_2$. This dehydration of 1,5-heptadiene-4-ol (**1**) results in three acyclic ethers with **2** being the major product.

$$HO(CH_2)_nOH \xrightarrow[94-98\%]{BuSnCl_3,\ 150-200^\circ} \text{cyclic ether } (CH_2)_n\text{, O}$$

n = 4,5 → n = 1,2

$$CH_3CH{=}CHCH(OH)CH{=}CH_2 \xrightarrow[75\%]{BuSnCl_3,\ 80^\circ} (CH_3CH{=}CHCH(CH_2CH{=}CH_2))_2O + 2\ \text{isomers}$$

1 → **2** (40–50%)

[1] G. Tagliavini, D. Marton, and D. Furlani, *Tetrahedron*, **45**, 1187 (1989).

2-Butynal, $CH_3C{\equiv}CCHO$.

Phenol annelation.[1] A key step in a synthesis of the aglycone (**4**) of pseudopterosin A is a reaction of the silyl enol ether **2** with 2-butynal (**1**) and $(CH_3)_3SiOTf$ at −78° followed by PCC oxidation of the resulting propargylic alcohol to give **3** in 61% overall yield.

[1] E. J. Corey and P. Carpino, *Am. Soc.*, **111**, 5472 (1989).

C

Camphanoyl chloride (1). Supplier: Aldrich.

Preparation from camphoric acid:

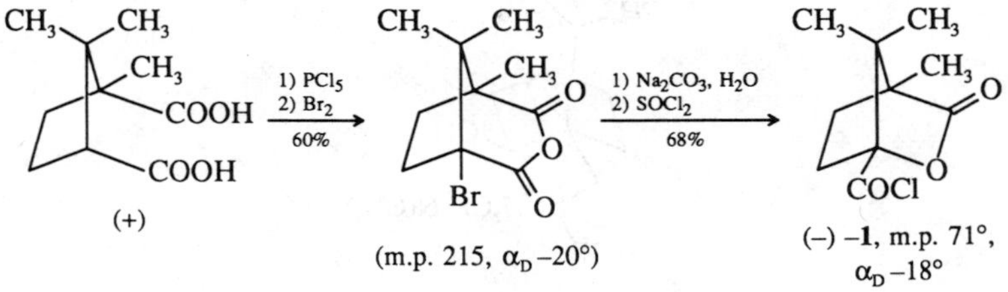

CH3 CH3 CH3 COOH COOH
(+)
1) PCl5
2) Br2
60%
Br
(m.p. 215, αD −20°)
1) Na2CO3, H2O
2) SOCl2
68%
COCl
(–) –1, m.p. 71°,
αD −18°

Resolution of alcohols.[1] The reagent is widely used for resolution of chiral alcohols via the diastereometric esters of (−)-camphoric acid formed from **1**. Camphorates generally have high melting points and can be resolved by fractional crystallization or chromatography.

[1] D. Kappes and H. Gerlach, *Org. Syn.*, submitted (1989).

Camphorquinone, 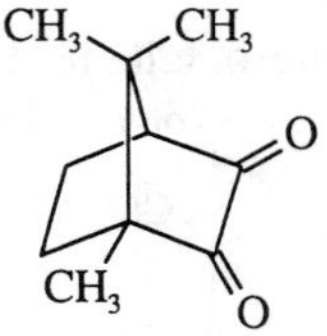 (–)–**1**. Both (+)- and (−)-**1** are available from Aldrich.

Optically active homoallylic alcohols.[1] The allylborane reagent **2**, prepared from (−)-**1** in five steps, adds to aldehydes to give optically active homoallylic alcohols in 88–96% ee.

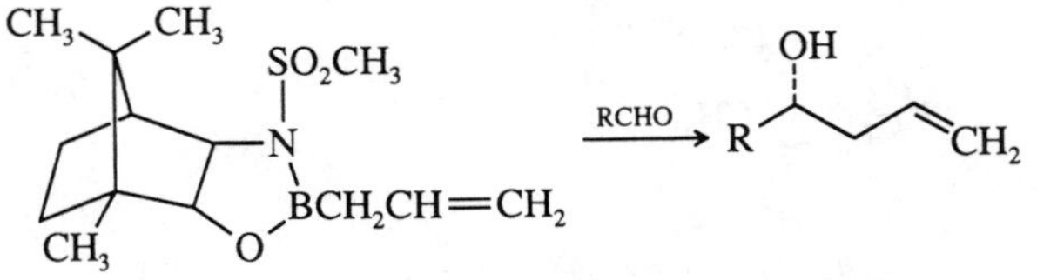

SO2CH3
N
BCH2CH=CH2
CH3
RCHO
OH
R
CH2

2

$R = C_2H_5$	92%	92% ee
$= C(CH_3)_3$	80%	88% ee
$= C_6H_5$	91%	88% ee

[1] M. T. Reetz and T. Zierke, *Chem. Ind.*, 663 (1988).

Camphor-10-sulfonic acid.

Asymmetric addition of $(CH_2{=}CH)_2Zn$ to aldehydes. The chiral tridendate ligand **1**, derived from (+)-camphor-10-sulfonic acid, effects highly enantioselective addition of R_2Zn to aldehydes. Significantly, even divinylzinc adds to aryl and aliphatic aldehydes to furnish (R)-allylic alcohols in 82–96% ee.

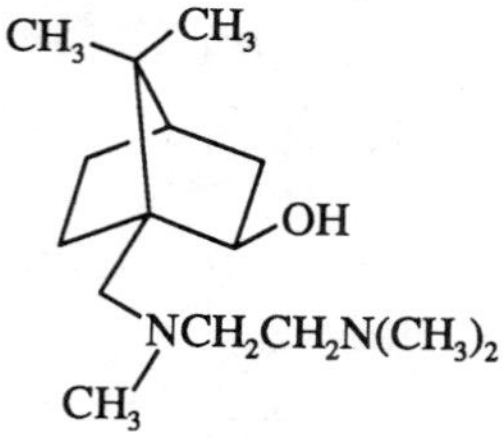

CH_3 CH_3
OH
$NCH_2CH_2N(CH_3)_2$
CH_3

1

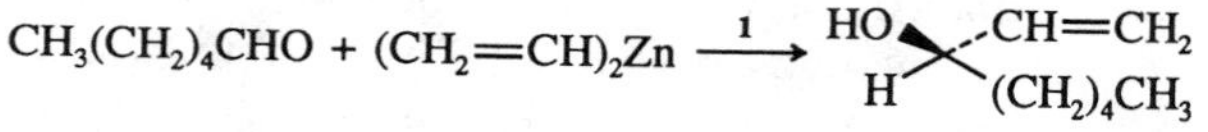

$$CH_3(CH_2)_4CHO + (CH_2{=}CH)_2Zn \xrightarrow{1} HO(H)C(CH{=}CH_2)(CH_2)_4CH_3$$

(R), >96% ee

[1] W. Oppolzer and R. N. Radinov, *Tetrahedron Letters*, **29**, 5645 (1988).

Camphor-10-sulfonyl chloride.

Enantioselective fluorination. The N-fluoro sultams **1** and **2** are prepared from the imine available from (+)-camphor-10-sulfonyl chloride.

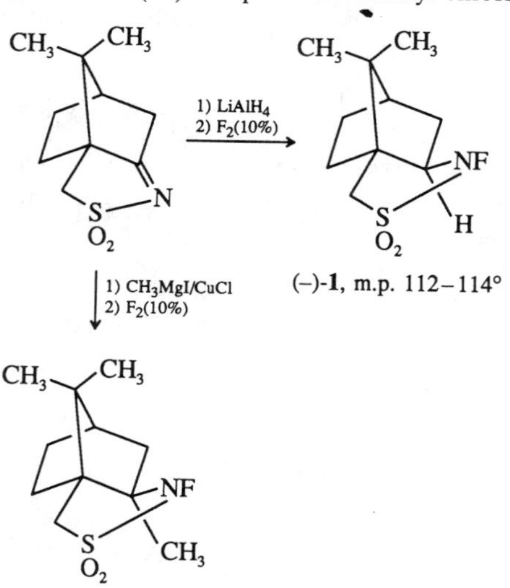

CH_3 CH_3
N
S
O_2

1) $LiAlH_4$
2) F_2(10%)

CH_3 CH_3
NF
H
S
O_2

(−)-**1**, m.p. 112–114°

1) $CH_3MgI/CuCl$
2) F_2(10%)

CH_3 CH_3
NF
CH_3
S
O_2

(+)-**2**, m.p. 151–154°

These are the first reagents known to effect enantioselective fluorination of enolates of carbonyl compounds. The highest enantioselectivity is observed in the reaction of (−)-**1** with a β-keto ester (equation I).

(I) 2-(ethoxycarbonyl)cyclopentanone $\xrightarrow[63\%]{\text{1) NaH, ether; 2) (–)-1}}$ 2-fluoro-2-($COOC_2H_5$)cyclopentanone (70% ee)

[1] E. Differding and R. W. Lang, *Tetrahedron Letters*, **29**, 6087 (1988).

(Carbonyl)chlorobis(triphenylphosphine)rhodium, $ClRh(CO)[P(C_6H_5)_3]_2$ (**1**).

Hydrostannation of 1-alkynes.[1] Tributyltin hydride adds spontaneously to terminal alkynes to form a 1:1 mixture of *cis*- and *trans*-1-alkenylstannanes. However, use of this rhodium catalyst and a radical inhibitor (**2**, galvinoxyl) provides 2-alkenylstannanes as the major product.

$$RC{\equiv}CH + Bu_3SnH \xrightarrow{1,\,2} R(Bu_3Sn)C{=}CH_2 + RCH{=}CHSnBu_3$$

R	Yield	Ratio	
R = C_6H_5	99%	78:22	(E/Z = 91:9)
R = $AcOCH_2$	71%	73:27	(E > 99)
R = $Si(CH_3)_3$	91%	76:24	(E, 100%)

[1] K. Kikukawa, H. Umekawa, F. Wada, and T. Matsuda, *Chem. Letters*, 881 (1988).

Carbonylhydridotris(triphenylphosphine)rhodium, $RhH(CO)[P(C_6H_5)_3]_3$ (**1**).

Regioselective hydroformylation.[1] The rhodium-catalyzed hydroformylation of (Z)-*t*-butyldiphenylsilylalkenes is an excellent route to β-silyl aldehydes (equation I). A bulky silyl group is essential for this regiocontrol.

(I) (Z)-*t*-$BuPh_2Si$CH=CHBu $\xrightarrow[80\%]{CO,\ H_2,\ 1,\ 80^\circ}$ $R_3SiCH_2CH(CHO)Bu + R_3SiCH(CHO)CH_2Bu$ (96:4)

This reaction can be used to obtain aldols (equation II).

(II) (Z)-R_3SiCH=CHCH(OH)CH_3 $\xrightarrow{83\%}$ $R_3SiCH_2CH(CHO)CH(OH)CH_3$ (98:2)

Hydroformylation of the 1,2-bisilylalkene **2** is accompanied by Brooke rearrangement to provide a silyl enol ether (**3**).

$$R_3Si\text{-}CH{=}CH\text{-}Si(CH_3)_3 \ (\mathbf{2}) \xrightarrow{60\%} R_3SiCH_2\text{-}CH{=}CH\text{-}OSi(CH_3)_3 \ (\mathbf{3})$$

[1] M. M. Doyle, W. R. Jackson, and P. Perlmutter, *Tetrahedron Letters*, **30**, 233 (1989).

Cerium(IV) amonium nitrate (CAN).

***Addition of* 1,3-*diones to vinyl acetate.*[1]** Use of CAN for oxidative addition of carbonyl compounds to alkenes has been reported in several patents. The method can be extended to synthesis of acyl or alkoxycarbonylfurans.

$$CH_3COCH_2COCH_3 + CH_2{=}CHOAc \xrightarrow[56\%]{1)\ CAN,\ CH_3CN;\ 2)\ PyH^+TsO^-,\ C_6H_5CH_3} \text{2-AcO-4-COCH}_3\text{-5-CH}_3\text{-furan}$$

$$HCOCH_2COOC_2H_5 + CH_2{=}CHOAc \xrightarrow[9\%]{} \text{3-COOC}_2H_5\text{-furan}$$

$$\text{5-methylcyclohexane-1,3-dione} + CH_3CH{=}CHOAc \xrightarrow[30\%]{} \text{evodone (3,6-dimethyl-6,7-dihydro-4(5H)-benzofuranone)}$$

Yields are lower with propenyl acetate than with vinyl acetate, but even so, the reaction is a viable route to monoterpene evodone (last example).

***α-Iodination of ketones.*[2]** Iodine in combination with CAN converts ketones into α-iodo ketones at 25–50° in HOAc or CH_3OH, with substitution at the more substituted α-position. This method is superior to an earlier one using I_2–$Cu(OAc)_2$ in HOAc (**10**, 211).

$$\text{decalone} \xrightarrow[80\%]{I_2,\ CAN,\ CH_3OH,\ 75°} \alpha\text{-iodo decalone}$$

$$BuCOCH_3 \xrightarrow[70\%]{I_2,\ CAN,\ 25°} PrCH(I)COCH_3 + BuCOCH_2I \quad (74:26)$$

Oxidative demethylation of tetramethoxynaphthalenes.[3] Demethylation of 1,4,5,8-tetramethoxynaphthalenes with CAN (in H_2O) can lead to two types of products (equation I). If R is an electron-releasing group, quinone **2** is formed

(I) **1** $\xrightarrow[\text{CH}_3\text{CN, 25°}]{\text{CAN}}$ **2** + **3**

	2	3
1, R = H	70%	—
R = CH_2OH	75%	13%
R = CHO	—	85%

mainly; but if R is an electron-withdrawing group, quinone **3** is formed. Both types of quinones are stable to CAN, but can be demethylated with AgO–40% HNO_3 or with $AlCl_3$.

2 $\xrightarrow{\text{AgO–40\% HNO}_3}$ **4**

R = H	53%
R = CH_2OH	52%

3 $\xrightarrow[\text{CH}_2\text{Cl}_2]{\text{AlCl}_3}$ **5**

R = CHO 43%

4-Oxo aldehyde dimethyl acetals.[4] The reaction of a ketone with vinyl acetate and CAN results in a radical species that is converted to an α-(2-acetoxy-2-nitrosylethyl) ketone. Solvolysis in methanol results in the dimethyl acetal of a 4-oxo aldehyde. Similar reactions of ketones with isopropenyl acetate and Mn(III) acetate provide 1,4-diketones (**6**, 356).

Sulfoxides.[5] Oxidation of sulfides to sulfoxides can be effected with CAN (2 equiv.) in CH_2Cl_2/H_2O catalyzed by Bu_4NBr at 25° (90–100% yields). A Pummerer-type rearrangement does not occur under these conditions.

$$\text{cyclopentanone} + CH_2{=}CHOAc \xrightarrow{\text{CAN, } NaHCO_3} \left[\text{2-}(CH_2CH(ONO_2)OAc)\text{cyclopentanone} \right] \xrightarrow[68\%]{CH_3OH,\ \Delta} \text{2-}(CH_2CH(OCH_3)_2)\text{cyclopentanone}$$

[1] E. Baciocchi and R. Ruzziconi, *Syn. Comm.*, **18**, 1841 (1988).
[2] C. A. Horiuchi and S. Kiji, *Chem. Letters*, 31 (1988).
[3] Y. Tanoue and A. Terada, *Bull. Chem. Soc. Japan*, **61**, 2039 (1988).
[4] E. Baciocchi and R. Ruzziconi, *J. Org.*, **51**, 1645 (1986); *Org. Syn.*, submitted (1989).
[5] E. Baciocchi, A. Piermattei, and R. Ruzziconi, *Syn. Comm.*, **18**, 2167 (1988).

Cerium(III) chloride.

Functionalized allylsilanes.[1] A Grignard reagent when activated by $CeCl_3$ reacts with esters,[1] ester acetals or lactones[2] to give functionalized allylsilanes.

$$C_6H_5CH{=}CHCOOC_2H_5 \xrightarrow[CeCl_3,\ -60\rightarrow 25°]{(CH_3)_3SiCH_2MgCl} \left[C_6H_5CH{=}CHC(OH)[CH_2Si(CH_3)_3]_2 \right] \xrightarrow[91\%]{SiO_2,\ -HOSi(CH_3)_3} C_6H_5CH{=}CHC({=}CH_2)CH_2Si(CH_3)_3$$

$$(CH_3)_3SiCH_2MgCl + O{=}\text{(tetrahydrofuran-2-one)} \xrightarrow[18\%]{} HOCH_2(CH_2)_2C({=}CH_2)CH_2Si(CH_3)_3$$

$$+ CeCl_3 \quad 74\%$$

$ArCH_2OH \rightarrow ArCHO$.[3] Platinum-catalyzed oxygenation of *m*-hydroxybenzyl alcohol (**1**) gives a mixture of **2** and **3**. If $CeCl_3$ is also present in traces, the further oxidation of **2** to **3** is markedly depressed, possibly by acetalization of **2**.

Carbonylation of benzyl chloride.[4] This reaction can be effected with cobalt carbonyl or a palladium(0) catalyst under phase-transfer conditions. $CeCl_3$ promotes this phase-transfer process and also permits the use of nickel(II) cyanide as the

$$\text{HO-C}_6\text{H}_4\text{-CH}_2\text{OH (1)} + O_2 \xrightarrow[\text{CH}_3\text{OH, H}_2\text{O}]{\text{Pt/C, Bi}_2(\text{SO}_4)_3} \text{HO-C}_6\text{H}_4\text{-CHO (2)} + \text{HO-C}_6\text{H}_4\text{-COOH (3)}$$

	2	3
	(71%)	(18%)
+ $CeCl_3$	88%	1%

metal catalyst. In fact, $Ni(CN)_2$ is inactive in the absence of $CeCl_3$, which also affects the rate of reaction. Tetrabutylammonium hydrogen sulfate and cetyltrimethylammonium bromide are the most efficient phase-transfer catalysts.

$$C_6H_5CH_2Cl + CO \xrightarrow[89\%]{Ni(CN)_2,\ CeCl_3,\ C_{16}H_{33}\overset{+}{N}(CH_3)_3\overset{-}{Br}} C_6H_5CH_2COOH$$

[1] B. A. Narayanan and W. H. Bunnelle, *Tetrahedron Letters*, **28**, 6261 (1987); *idem*, *Org. Syn.*, submitted 1988.
[2] T. V. Lee, J. R. Porter, and F. S. Roden, *Tetrahedron Letters*, **29**, 5009 (1988).
[3] R. Oi and S. Tanenaka, *Chem. Letters*, 1115 (1988).
[4] I. Amer and H. Alper, *Am. Soc.*, **111**, 927 (1989).

Cerium(IV) sulfate, $Ce(SO_4)_2$.

Malonyl radicals.[1] Thiophenes and furans undergo substitution at C_2 when treated with a dialkyl malonate and ceric sulfate at 25°. The reaction is considered to involve oxidative generation of malonyl radicals by the cerium salt. Yields are 40–85%.

$$\text{2-CH}_3\text{-C}_4\text{H}_3\text{Z} + CH_2(COOCH_3)_2 \xrightarrow[75-85\%]{Ce(SO_4)_2,\ CH_3OH/H_2O} \text{5-CH}_3\text{-C}_4\text{H}_2\text{Z-2-}CH(COOCH_3)_2$$

Z=O, S

[1] L. M. Weinstock, E. Corley, N. L. Abramson, A. O. King, and S. Karady, *Heterocycles*, **27**, 2627 (1988).

Cesium carbonate.

Wittig–Horner reactions.[1] The reaction of 1,2-O-isoproylidene-D-glyceroylmethylphosphonate (**1**) with aldehydes fails when $N(CH_2CH_3)_3$-LiCl is used as base and gives poor yields when NaH is used. However, reactions with Cs_2CO_3 in isopropyl alcohol proceed readily at 0°. This Wittig–Horner reaction was used for a synthesis of D-*erythro*-C_{18}-sphingosine (**2**).

CH_3 CH_3 O O O O $P(OCH_3)_2$ + $C_{13}H_{27}CHO$ $\xrightarrow[85\%]{Cs_2CO_3,\ (CH_3)_2CHOH}$

1

H_3C CH_3 O O O $C_{13}H_{27}$ $\xrightarrow{\text{several steps}}$ HO NH_2 OH $C_{13}H_{27}$

2

Anionic polycyclization to a steroid.[2] When treated with Cs_2CO_3, 2-methoxycarbonylcyclohexenone (**1**) and the Nazarov reagent **2** undergo a Michael reaction followed by aldol condensation to give a tetracyclic adduct (**3**), which undergoes decarboalkoxylation (TsOH) to give the steroid derivative **4** in 47% overall yield.

O E + CH_3 O O O $COOC(CH_3)_3$ $\xrightarrow{Cs_2CO_3,\ CHCl_3,\ 25°}$ O E H CH_3 O H OH O H $COOC(CH_3)_3$

1, E = CO_2CH_3 **2** **3**

47% overall ↓ TsOH, C_6H_6, Δ

O E H CH_3 O H OH O H

4

Alkylation of cesium phenolates.[3] Alkylation of hydroxylated 9,10-anthraquinones can be accomplished fairly readily by alkylation of cesium phenolates.

HO O Cl / HO O Cl (1,4-dihydroxy-5,8-dichloroanthraquinone) $\xrightarrow{(CH_3O)_2SO_2,\ (CH_3)_2C=O}$ CH_3O O Cl / CH_3O O Cl

K_2CO_3, 64 hr. 84%
Cs_2CO_3, 15 hr. 94%

[1] T. Yamanoi, T. Akiyama, E. Ishida, H. Abe, M. Amemiya, and T. Inazu, *Chem. Letters*, 335 (1989).
[2] J.-F. Lavallée and P. Deslongchamps, *Tetrahedron Letters*, **29**, 6033 (1988).
[3] R. T. Winters, A. D. Sercel, and H. D. H. Showalter, *Synthesis*, 712 (1988).

Cesium fluoride.

2-Aza-1,3-dienes.[1] A new route to these useful substrates for [4 + 2]- and [2 + 2]cycloadditions involves protodesilylation of N-1-triethylsilylallylimines such as **1**, available in several steps from propargylamine in about 30% yield (equation I). Desilylation of **1** with CsF and 18-crown-6 in THF containing a trace of water

$$\text{(I)}\quad H_2NCH_2C\equiv CH \xrightarrow[\sim 30\%]{6\ \text{steps}} H_2N\underset{}{\overset{Si(C_2H_5)_3}{C}}HCH=CH_2 \xrightarrow{C_6H_5CHO} C_6H_5CH=N\underset{Si(C_2H_5)_3}{C}HCH=CH_2$$

1, (E, 100%)

80% ↓ CsF, 18-crown-6, THF (H_2O)

$C_6H_5CH=N-CH=CH-CH_3$ (H)

2, (E, E/E, Z) = 6 : 1

results in (E,E)- and (E,Z)-**2** in the ratio of 6:1. As the concentration of water is increased the (E,E)/(E,Z) ratio approaches unity.

Crotylization of RCHO.[2] (E)- and (Z)-Crotyltrifluorosilanes add to aldehydes in the presence of CsF to give *anti*- and *syn*-β-methylhomoallyl alcohols, respectively, with high diastereoselectivity.

$$RCHO + CH_3CH{=}CHSiF_3 \xrightarrow[68-96\%]{CsF} R\text{-}CH(OH)\text{-}CH(CH_3)\text{-}CH{=}CH_2$$

(98–99 : 2–1)

$$RCHO + CH_3CH{=}CHSiF_3 \xrightarrow[77-96\%]{CsF} R\text{-}CH(OH)\text{-}CH(CH_3)\text{-}CH{=}CH_2$$

(90–99 : 10–1)

[1] S.-F. Chen, E. Ho, and P. S. Mariano, *Tetrahedron*, **44**, 7013 (1988).
[2] M. Kira, T. Hino, and H. Sakurai, *Tetrahedron Letters*, **30**, 1099 (1989).

Cesium fluoride–18-Crown-6.

Fluorodestannylation. Bu_4NF is the most reactive source of F^-, but is difficult to obtain anhydrous. For fluorodestannylation, Harpp and Gingras[1] prefer CsF with a catalytic amount of 18-crown-6. This reagent combines with diorganotin oxides, sulfides, or selenides to liberate the corresponding anion, which can be alkylated to give ethers, sulfides, or selenides.

$$(Bu_3Sn)_2O + 2BzlBr \xrightarrow[71\%]{F^-} (Bzl)_2O + 2Bu_3SnF$$

$$[(CH_3)_3Sn]_2S + 2CH_3(CH_2)_5Br \xrightarrow[99\%]{F^-} [(CH_3(CH_2)_5]_2S + 2(CH_3)_3SnF$$

18-Crown-6 is also an effective solid–liquid phase-transfer catalyst for conversion of benzyl bromide to benzyl fluoride with CsF.[2]

[1] D. N. Harpp and M. Gingras, *Am. Soc.*, **110**, 7737 (1988).
[2] M. Gingras and D. N. Harpp, *Tetrahedron Letters*, **29**, 4669 (1988).

Cesium fluoroxysulfate, $CsSO_3OF$ (**1**), **10**, 84–85. *Caution*: This hypofluorite is explosive.

Addition to alkenes. Although this reagent has been studied mainly for electrophilic fluorination,[1] it does add to alkenes to give *cis*-selective *vic*-fluoro alkyl sulfates with some preference for *anti*-Markovnikov addition. The adducts can be converted into fluoro ethoxysulfates by reaction with $(C_2H_5)_3O^+BF_4^-$.[2]

$$BuCH{=}CH_2 \xrightarrow[62-74\%]{1,\ C_2H_5OAc,\ 20^\circ} BuCH(OSO_3^-Cs^+){-}CH_2F + BuCHF{-}CH_2OSO_3^-Cs^+ \quad 1:2$$

$$\xrightarrow[90\%]{(C_2H_5)_3O^+BF_4^-} BuCH(OSO_2OC_2H_5){-}CH_2F + BuCHF{-}CH_2OSO_2OC_2H_5$$

Cyclohexene + 1 ⟶ 2-fluorocyclohexyl $OSO_3^-Cs^+$ (*cis*-2) + *trans*-2

	cis-2	*trans*-2
CH_3CN, 0°	52%	30%
C_2H_5OAc	30%	20%

[1] E. H. Appelman, L. J. Basile, and R. C. Thompson, *Am. Soc.*, **101**, 3384 (1979).
[2] N. S. Zefirov, V. V. Zhdankin, A. S. Koz'min, A. A. Fainzilberg, A. A. Gakh, B. I. Ugrak, and S. V. Romaniko, *Tetrahedron*, **44**, 6505 (1988).

Cetyltrimethylammonium bromide (CTAB).

Williamson synthesis of phenyl alkyl and dialkyl ethers.[1] Phenols react with alkyl halides in 20% aqueous NaOH containing 1 equiv. of this surfactant at 80° to form phenolic ethers in 85–97% yield. There is no reaction in the absence of CTAB. This procedure is not useful for preparation of dialkyl ethers from alcohols and alkyl halides. Instead, the alkyl chloride, alcohol, a trace of water, and CTAB are heated in THF at 70° with NaOH (2 equiv.).

$$C_6H_5OH + C_2H_5Br \xrightarrow[93\%]{CTAB,\ NaOH,\ H_2O} C_6H_5OC_2H_5$$

$$(CH_3)_2CHOH + n\text{-}C_5H_{11}Cl \xrightarrow[91\%]{CTAB,\ THF,\ NaOH} (CH_3)_2CHOC_5H_{11}\text{-}n$$

[1] B. Juršić, *Tetrahedron*, **44**, 6677 (1988).

Cetyltrimethylammonium chromate, $n\text{-}C_{16}H_{31}\overset{+}{N}H(CH_3)_3CrO_3^-$ (**1**). This oxidant is obtained as a yellow–orange solid by reaction of CrO_3 with cetyltrimethylammonium bromide in water at 25°. It can be used to oxidize alcohols to aldehydes or ketones in an aqueous solution. However, it is not necessary to isolate the salt since essentially the same yields are obtained by oxidations with CrO_3 in H_2O/$CHCl_3$ with cetyltrimethylammonium bromide as surfactant.[1]

[1] B. Juršić, *Synthesis*, 868 (1988).

Chloramine-T (1).

2-Isoxazolines.[1] Reaction of aldoximes with chloramine-T generates nitrile oxides, which can be trapped by alkenes to give 2-isoxazolines.

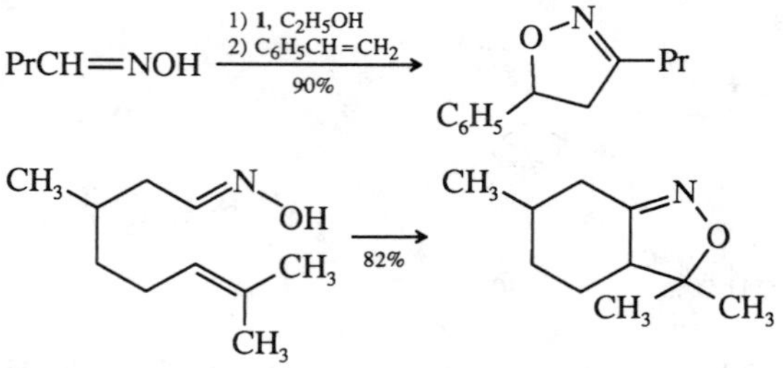

$$\text{PrCH}{=}\text{NOH} \xrightarrow[90\%]{1)\ \mathbf{1},\ C_2H_5OH;\ 2)\ C_6H_5CH{=}CH_2}$$

$$\xrightarrow{82\%}$$

[1] A. Hassner and K. M. L. Rai, *Synthesis*, 57 (1989).

1-Chloroalkyl phenyl sulfoxides, $C_6H_5S(O)CHClR$ **(1).**

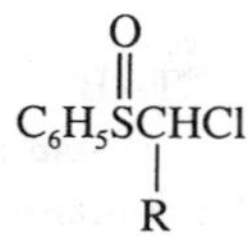

Cyclic ethers.[1] A new route to 2-acyl cyclic ethers is based on an intramolecular reaction of α,β-epoxy sulfoxides with a nucleophile as outlined in equation I. Thus reaction of the α,β-epoxy sulfoxide **2** with pyridinium *p*-toluenesulfonate

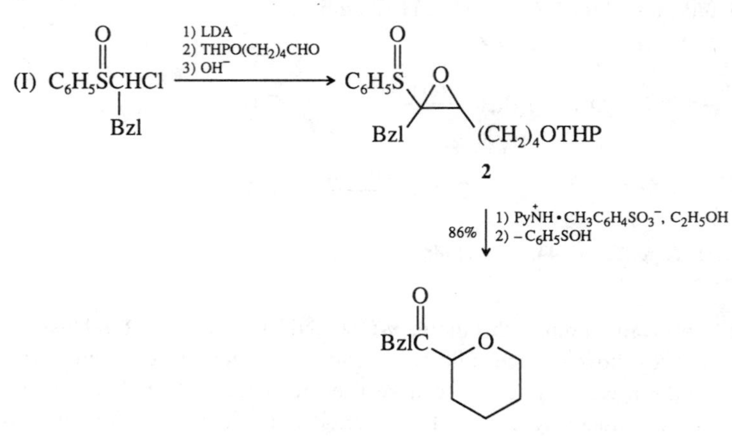

$$\text{(I)}\ C_6H_5S(O)CHCl(Bzl) \xrightarrow{1)\ LDA;\ 2)\ THPO(CH_2)_4CHO;\ 3)\ OH^-} \mathbf{2}\ \xrightarrow[86\%]{1)\ Py\overset{+}{N}H\cdot CH_3C_6H_4SO_3^-,\ C_2H_5OH;\ 2)\ -C_6H_5SOH} \mathbf{3}$$

in ethanol liberates a free hydroxy group which cleaves the epoxy group with formation of the cyclic ether **3**. A similar strategy can be used to prepare 3-oxo cyclic ethers (equation II).

$$\text{(II)}\quad C_6H_5\overset{\overset{O}{\|}}{S}\underset{\underset{(CH_2)_3OTHP}{|}}{CHCl} + C_6H_5CHO \longrightarrow \text{epoxide } [C_6H_5S(O), THPO(CH_2)_3, C_6H_5] \xrightarrow[60\%]{Py\overset{+}{N}H \cdot CH_3C_6H_4SO_3^-,\ C_2H_5OH} \text{2-phenyltetrahydropyran-3-one}$$

[1] T. Satoh, K. Iwamoto, A. Sugimoto, and K. Yamakawa, *Bull. Chem. Soc. Japan*, **61**, 2109 (1988).

μ-Chlorobis(cyclopentadienyl)(dimethylaluminum)-μ-methylenetitanium (Tebbe reagent).

Addition to nitriles.[1] Reaction of $Cp_2Ti{=}CH_2$, readily formed from the titanacyclobutane **1**, with pivalonitrile forms the [2 + 2]cycloadduct **a**, which on further reaction with a nitrile forms eventually a new metallacycle **2**. This product on acidolysis provides the β-keto enamine **3**.

$$Cp_2Ti\square{-}C(CH_3)_3\ (\mathbf{1}) \longrightarrow [Cp_2Ti{=}CH_2] \xrightarrow{(CH_3)_3CCN} [Cp_2Ti{-}CH_2{-}C(C(CH_3)_3){=}N]\ (\mathbf{a}) \xrightarrow{(CH_3)_3CCN} Cp_2Ti(N{=}C(C(CH_3)_3){-}CH{=}C(C(CH_3)_3){-}NH)\ (\mathbf{2})$$

$$\mathbf{2} \xrightarrow{1)\ HCl;\ 2)\ H_2O} Cp_2TiCl_2 + O{=}C(C(CH_3)_3){-}CH{=}C(C(CH_3)_3){-}NH_2\ (\mathbf{3})$$

[1] K. M. Doxsee and J. B. Farahi, *Am. Soc.*, **110**, 7239 (1988).

Chlorobis(cyclopentadienyl)hydridozirconium (Schwartz reagent, **1**).

Cross-coupling of alkynes.[1] A route to isomerically pure 1,3-dienes involves hydrozirconation of an alkyne followed by conversion to the alkyne complex **3**, which couples with a second alkyne to give a metallacyclopentadiene **4**. Treatment of **4** with dilute acid provides the diene **5**, usually as a single isomer (>97%). Overall yields are in the range 53–80%.

$R^1C{\equiv}CR^2 \xrightarrow{Cp_2ZrHCl}$ [$Cp_2Zr(Cl)C(R^1){=}CHR^2$] $\xrightarrow{CH_3Li}$ $Cp_2Zr(CH_3)C(R^1){=}CHR^2$ (**2**) $\xrightarrow{-CH_4}$ $Cp_2Zr(R^1C{\equiv}CR^2)$ (**3**) $\xrightarrow{R^3C{\equiv}CR^4}$ **4** $\xrightarrow{H_3O^+}$ **5**

Butenolides.[2] Addition of the Schwartz reagent to a protected propargylic alcohol (**2**) followed by carbonylation provides an acyl zirconocene complex (**3**). This is not isolated but treated *in situ* with iodine to provide an intermediate (**a**) that cyclizes to a 3,5-disubstituted butenolide (**4**). Optically active substrates undergo this sequence with no loss of optical purity.

$CH_3C{\equiv}C{-}CH(OSiR_3)CH_2CH_3$ **2** (90% ee), R = C_6H_5 $\xrightarrow{1)\ \mathbf{1};\ 2)\ CO}$ **3** $\xrightarrow{I_2}$ [**a**] $\xrightarrow{55\%}$ **4** (90% ee)

[1] S. L. Buchwald and R. B. Nielsen, *Am. Soc.*, **111**, 2870 (1989).
[2] S. L. Buchwald, Q. Fang, and S. M. King, *Tetrahedron Letters*, **29**, 3445 (1988).

Chlorobis(cyclopentadienyl)methylzirconium, $Cp_2Zr(CH_3)Cl$ (**1**).

Pyrroles.[1] This zirconocene reacts in THF with a lithio trimethylsilyl primary amine, $RCH_2N(Li)Si(CH_3)_3$, to form **2**, which is not isolated but allowed to react with various alkynes to form azametallacyclopentenes **3**. These metallacycles, which need not be isolated, react with carbon monoxide to afford pyrroles **4**. Yields are

$$\mathbf{1} + RCH_2N(Li)Si(CH_3)_3 \xrightarrow[-CH_4]{THF,\ -78^\circ} \mathbf{2} \xrightarrow{R^1C\equiv CR^2} \mathbf{3} \xrightarrow[30-70\%\ overall]{CO,\ THF,\ NH_4Cl\ or\ 4\AA MS} \mathbf{4}$$

highest when the reaction is carried out in THF in the presence of NH_4Cl (3 equiv.) at 25° or of 4-Å molecular sieves. The report includes a rationale for formation of pyrroles rather than pyrrolinones. Even acetylene can be used in this synthesis to provide 2-substituted pyrroles. 1-Alkynes react to form 2,4-disubstituted pyrroles. The yield is low in a reaction with a 1-trimethylsilyl-1-alkyne because of steric factors.

[1] S. L. Buckwald, M. W. Wannamaker, and B. T. Watson, *Am. Soc.*, **111**, 776 (1989).

Chlorobis(cyclopentadienyl)titanium(III), Cp_2TiCl. Preparation.[1]

Cyclization of epoxy alkenes.[2] The reaction of δ,ε-unsaturated epoxides with 2 equiv. of Cp_2TiCl in THF results in cyclization to cyclopentanemethanols. The reaction is compatible with keto and nitrile groups but not aldehyde groups.

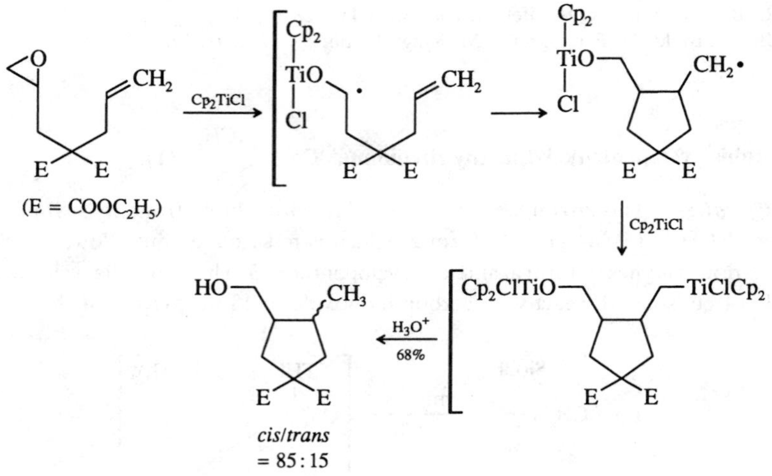

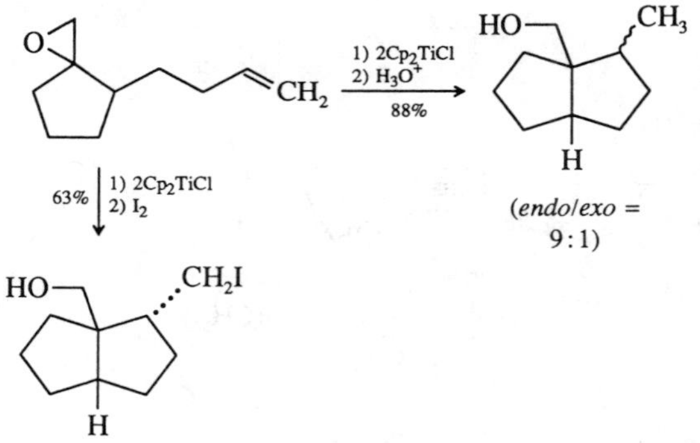

O CH$_2$ E E ($E = COOC_2H_5$) $\xrightarrow{Cp_2TiCl}$ [Cp$_2$ TiO Cl · CH$_2$ E E] $\longrightarrow$ Cp$_2$ TiO Cl CH$_2$· E E

$\downarrow$ Cp$_2$TiCl

[Cp$_2$ClTiO TiClCp$_2$ E E] $\xrightarrow[68\%]{H_3O^+}$ HO CH$_3$ E E

cis/trans = 85:15

O CH$_2$ $\xrightarrow[88\%]{1)\ 2Cp_2TiCl\quad 2)\ H_3O^+}$ HO CH$_3$ H

(*endo/exo* = 9:1)

63% $\downarrow$ 1) 2Cp$_2$TiCl 2) I$_2$

HO CH$_2$I H

[1] L. E. Manzer, *Inorg. Synth.*, **21**, 84 (1982).
[2] W. A. Nugent and T. V. RajanBabu, *Am. Soc.*, **110**, 8561 (1988).

Chlorodicarbonylrhodium(I) dimer.

Carbonylation of aziridines (**12**, 112).[1] This ring expansion of aziridines to β-lactams catalyzed by $[(CO)_2RhCl]_2$ is successful only when the ring is substituted by an aryl group and when the nitrogen is substituted by a *t*-butyl group. The

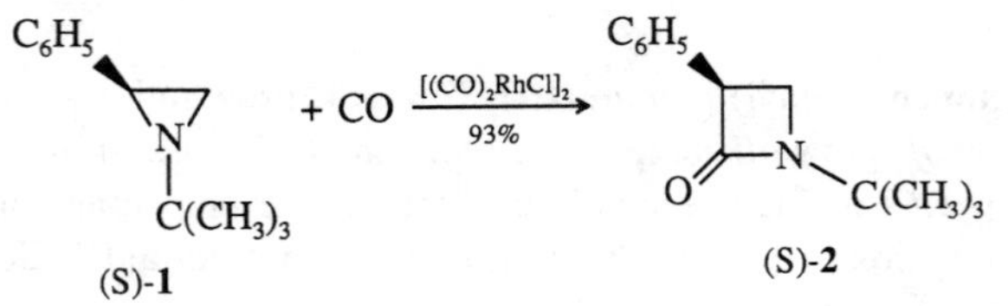

C$_6$H$_5$ N C(CH$_3$)$_3$ (S)-**1** + CO $\xrightarrow[93\%]{[(CO)_2RhCl]_2}$ C$_6$H$_5$ O N C(CH$_3$)$_3$ (S)-**2**

expansion of the optically pure aziridine **1** results in the pure (S)-β-lactam **2** in 93% yield. Carbonylation of racemic **1** catalyzed by either $[(CO)_2RhCl]_2$ or $[CODRhCl]_2$ in the presence of *l*-menthol results in optically pure (S)-**2** in 25–30% chemical yields. This asymmetric induction results from kinetic resolution.

[1] S. Calet, F. Urso, and H. Alper, *Am. Soc.*, **111**, 931 (1989).

Chlorodiethylborane, $ClB(C_2H_5)_2$ (**1**).

Allylic diethylboranes.[1] These boranes (**2**) can be prepared from 1-methylcycloalkenes and 2-alkenes by metallation with trimethylsilylmethylpotassium[2] followed by reaction with **1**. The products react with acetaldehyde to form homoallylic alcohols (**3**), which can be converted into α,β- and β,γ-unsaturated ketones.

1-Methylcyclohexene $\xrightarrow[\text{2) 1, ether}]{\text{1) } KCH_2Si(CH_3)_3}$ 1-cyclohexenyl-$CH_2B(C_2H_5)_2$ (**2**) $\xrightarrow[\text{72\% overall}]{\text{1) } CH_3CHO \text{; 2) } OH^-}$ 2-methylenecyclohexyl-$CH(OH)CH_3$ (**3**) $\xrightarrow{\text{PDC/DMF, 70\%}}$ 2-methylenecyclohexyl methyl ketone $\xrightarrow{Al_2O_3,\ 76\%}$ 2-methyl-1-cyclohexenyl methyl ketone

[1] M. Zaidlewicz, *J. Organomet. Chem.*, **293**, 139 (1985); *Synthesis*, 701 (1988).
[2] Generated from $Hg[CH_2Si(CH_3)_3]_2$ and potassium.

α-Chloroethyl chloroformate, $ClC(=O)OCHClCH_3$, **12**,112–113.

Review. Use of chloroformates (carbonochloridates) for N-alkylation (von Braun reaction) has been reviewed.[1] Cyclization/dealkylation is also a useful reaction.

Examples:

$$(C_2H_5)_2N(CH_2)_2C(C_6H_5)(COOH)CH_2CH_2N(CH_3)_2 \xrightarrow{1,\ SOCl_2}$$

$$\text{1-methyl-3-phenyl-3-}[(C_2H_5)_2NCH_2CH_2]\text{-2-pyrrolidinone} + \text{1-ethyl-3-phenyl-3-}[(CH_3)_2N(CH_2)_2]\text{-2-pyrrolidinone} \quad (3:1)$$

$$(C_2H_5)_2N(CH_2)_4COOH \xrightarrow{SOCl_2} \text{1-ethyl-2-piperidinone}$$

J. H. Cooley and E. J. Evain, *Synthesis*, 1 (1988).

Chloroform.

Reimer–Tiemann formylation of phenols.[1] Yields in this reaction can be increased by use of solid, powdered NaOH and addition of water (2 equiv. per phenol).

$$C_6H_5OH \xrightarrow{CHCl_3,\ H_2O,\ NaOH\ (solid)} \text{2-}OHC\text{-}C_6H_4OH\ \mathbf{1}\ (61\%) + OHC\text{-}C_6H_4\text{-}OH\ (para)\ \mathbf{2}\ (14\%)$$

[1] A. Thoer, G. Denis, M. Delmas, and A. Gaset, *Syn. Comm.*, **18**, 2095 (1988).

3-Chloro-1-hydroxytetrabutyldistannoxane, $ClSn(Bu)_2\text{-}O\text{-}Sn(Bu)_2OH$ (**1**). The reagent is obtained by partial hydrolysis of dibutyltin dichloride. It exists as a dimer.

Transesterification.[1] This distannoxane is an efficient catalyst for tranesterification of methyl or ethyl esters with a variety of alcohols or phenols in refluxing toluene in high yield. Since the conditions are neutral, a wide variety of functional groups is tolerated and racemization or isomerization is not observed.

[1] J. Otera, T. Yano, A. Kawabata, and H. Nozaki, *Tetrahedron Letters*, **27**, 2383 (1986).

Chloromethyllithium.

Homologation of boracyclanes.[1] Chloromethyllithium is an efficient reagent for one-carbon homologation of boracyclanes. Thus B-methoxyboracycloheptane

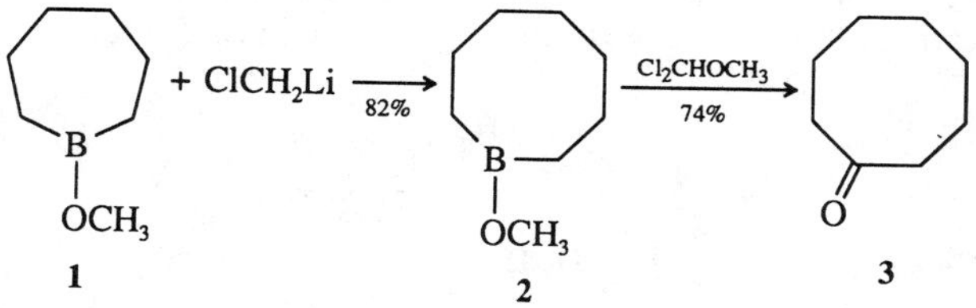

B
OCH_3
1
+ $ClCH_2Li$ → (82%)
B
OCH_3
2
Cl_2CHOCH_3 → (74%)
O
3

(**1**) is converted to boraoctane **2** in 82% yield. This homologation can be extended to synthesis of 12-membered boracyclanes. Reaction of the products with α,α-dichloromethyl methyl ether (**5**, 201–202) provides the corresponding cycloalkanones (**3**).

[1] H. C. Brown, A. S. Phadke, and M. V. Rangaishenvi, *Am. Soc.*, **110**, 6263 (1988).

2-Chloro-1-methylpyridinium iodide (**1**), **7**, 110; **8**, 95–96.

***Immunosuppressant* (−)-*FK*-506 (2).** This 23-membered lactone–lactam resembles cyclosporin A, a macrocyclic polypeptide, in that they both inhibit immune responses to transplantation, but FK-506 seems to be more active. The total synthesis of (−)-FK-506 has been achieved by a process research group at Merck.[1] The cyclization to a carboxamide was carried out with Mukaiyama's reagent (**1**) under high dilution in 81% yield. The most unusual feature of **2** is the presence of a three-keto sequence at C_8, C_9, and C_{10} (hemiketal).

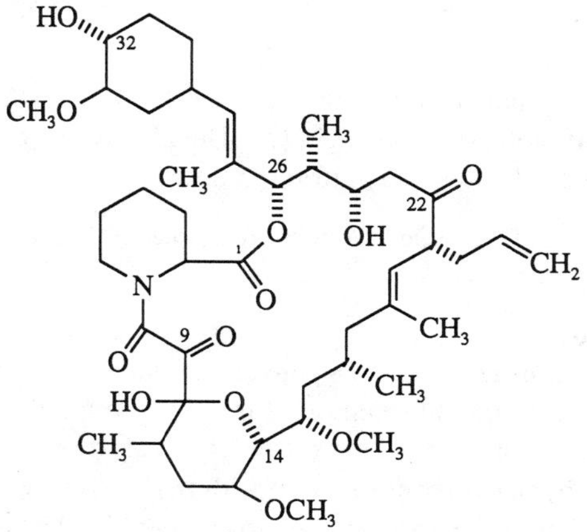

HO
32
CH_3O
CH_3
CH_3
26
22
O
O
OH
1
CH_2
N
O
O
9
O
CH_3
CH_3
HO
O
OCH_3
CH_3
14
OCH_3
2

[1] T. K. Jones, S. G. Mills, R. A. Reamer, D. Askin, R. Desmond, R. P. Volante, and I. Shinkai, *Am. Soc.*, **111**, 1157 (1989).

m-Chloroperbenzoic acid.

Oxidative contraction of glycals.[1] Reaction of 3,4-dihydro-2*H*-pyran (**1**) with benzeneselenenyl chloride and then with methanol provides the adduct **2**, which is oxidized by *m*-chloroperbenzoic acid to the ring-contracted aldehyde **3** (cf., **12**, 120). Application of this sequence to the tribenzyl-D-glucal (**4**) furnishes the adduct **5**, which undergoes similar oxidation to the aldehyde **6**, a convenient precursor to the C-nucleoside antibiotic showdomycin (**7**). The aldehyde **6** can also be prepared by oxidation of **4** with thallium(III) nitrate in acetonitrile.

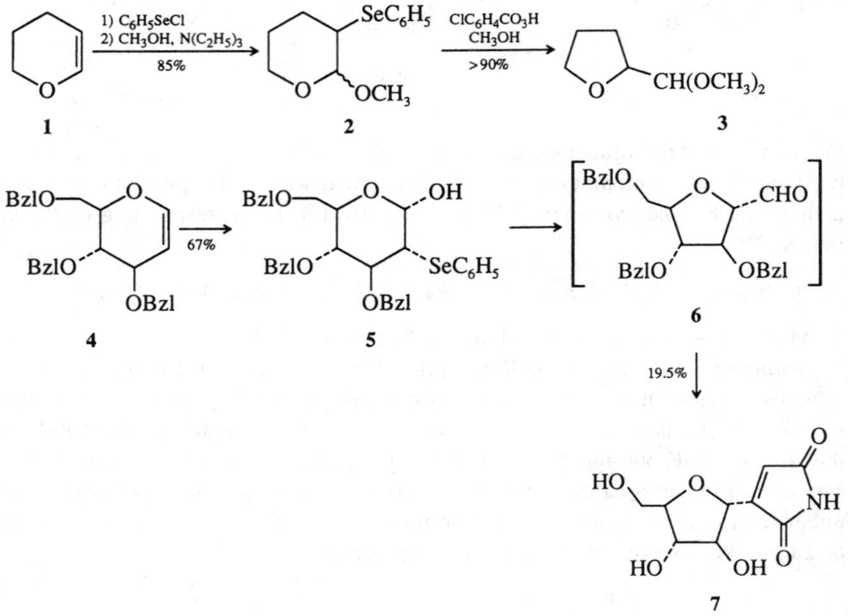

[1] A. Kaye, S. Neidle, and C. B. Reese, *Tetrahedron Letters*, **29**, 2711 (1988).

N-Chlorosuccinimide.

Chlorination of aldoximes. NCS converts aryl aldoximes to hydroxamic acid chlorides without significant chlorination of the aryl group. This reaction has been used for a novel synthesis of nitrile oxides. Thus reaction of salicylaldoxime (**1**) with NCS followed by dehydrochlorination with pyridine generates a nitrile oxide, which is trapped by styrene to give the isoxazoline **2**.[1] The N–O bond can be cleaved by catalytic hydrogenation to **3**, which is converted into the chalcone **4** on elimination of water. This product can be converted by classical methods to the flavanone **5** and the flavone **6**. An analogous route can be used to synthesize 2-

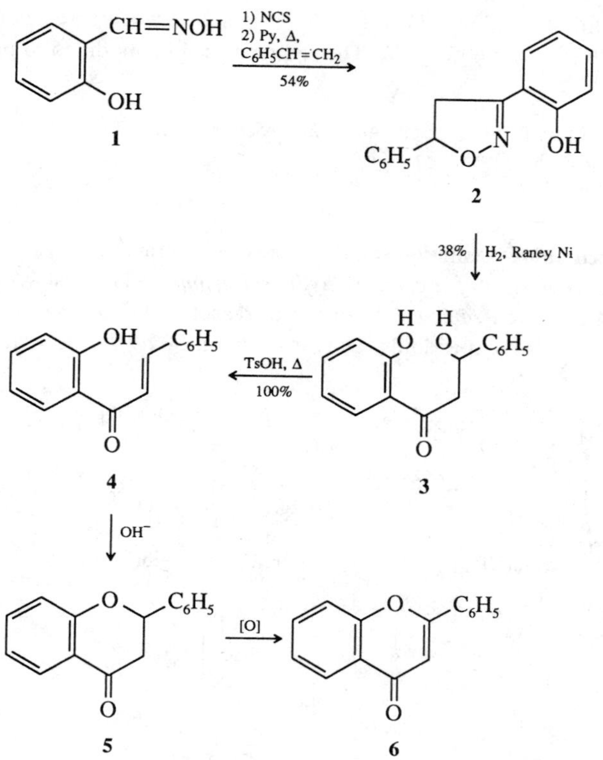

1) NCS
2) Py, Δ, $C_6H_5CH{=}CH_2$
54%

CH=NOH
OH
1

C_6H_5 O N OH
2

38% H_2, Raney Ni

OH C_6H_5
TsOH, Δ
100%
O
4

H H
O O C_6H_5
O
3

OH^-

O C_6H_5
O
5

[O]

O C_6H_5
O
6

alkyl-4-aminoquinolines.[2] Thus chlorination of *o*-nitrobenzaldoxime (**7**) followed by reaction with $N(C_2H_5)_3$ and trapping with phenylacetylene results in the isoxazole

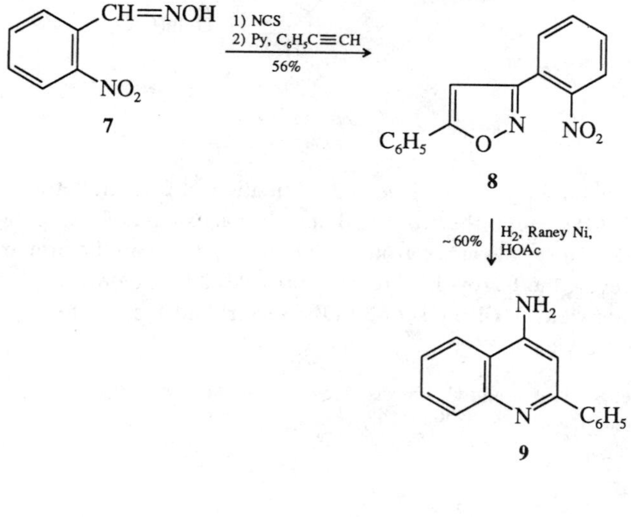

CH=NOH
NO_2
7

1) NCS
2) Py, $C_6H_5C{\equiv}CH$
56%

C_6H_5 O N NO_2
8

~60% H_2, Raney Ni, HOAc

NH_2
N C_6H_5
9

8 in 56% yield. On prolonged catalytic hydrogenation **8** can be converted into the isoquinoline **9** in about 60% yield. The sequence can be modified to provide quinazolines.

[1] I. Thomsen and K. B. G. Torssell, *Acta Chem. Scand*, **B42**, 303 (1988).
[2] *Idem*, *ibid.*, **B42**, 309 (1988).

N-Chlorosuccinimide–Dimethyl sulfide (Corey–Kim reagent).

Selective oxidation of a primary hydroxyl group.[1] The first synthesis of the secoiridoid **4** has been achieved from the tetraacetate (**1**) of secologanin, the biological precursor of secoiridoids. Dihydroxylation of **1** with OsO_4 and potassium

1 (R = Ac) → (OsO_4, $KClO_3$, Dioxane, H_2O; 45%) → **2** → (NCS • $S(CH_3)_2$, $C_6H_5CH_3$, CH_2Cl_2, $-25 \rightarrow 0°$; 29%) → **3**, R = Ac; **4**, R = H

chlorate[2] gives the bicyclic hemiacetal **2**. Oxidation of **2** to the tetraacetate (**3**) of secogalioside (**4**) presents the problem that the usual oxidants, PCC or those derived from DMSO, oxidize the lactol group of **2** more readily than the primary hydroxyl group. However, the Corey–Kim reagent provides **3** as the only product, albeit in only moderate yield. Hydrolysis of **3** to the secoiridoid **4** is effected with K_2CO_3 in methanol (60% yield).

[1] L. F. Tietze, S. Henke, and C. Bärtels, *Tetrahedron*, **44**, 7145 (1988).
[2] K. A. Hofmann, *Ber.*, **45**, 3329 (1972).

Chlorotrimethylsilane.

Diastereoselective* 1,4-*addition of cuprates to enones. 5-Trimethylcyclohexenone (**1**) has been resolved by kinetic resolution via the adduct with *p*-toluenethiol by cinchonidine. It undergoes highly diastereoselective addition with cuprates in the presence of $ClSi(CH_3)_3$ and HMPT to furnish only the *trans*-1,4-adducts in 88–95% yield.[1]

This stereoselective Michael reaction provides a synthesis of (−)-vetivone (**3**) from a cyclohexenone (**1**).[2]

$(CH_3)_3Si$ — (R) − (−)−**1**, α_D −9.5° → (CH_3MgBr, CuBr, $S(CH_3)_2$, $ClSi(CH_3)_3$; 94%) → silyl enol ether ($OSi(CH_3)_3$, CH_3) → (1) $Br(CH_2)_2$/$BrCH_2$ $>C=C(CH_3)_2$, $ZnBr_2$ (41%); 2) $NaOCH_3$ (49%)) → (+)−**2** → **3**, α_D −47°

[1] M. Asaoka, K. Shima, and H. Takei, *Tetrahedron Letters*, **28**, 5669 (1987).
[2] M. Asaoka, K. Takenouchi, and H. Takei, *Chem. Letters*, 1225 (1988).

Chlorotrimethylsilane–Selenium dioxide.

ROH → RCl.[1] Chlorotrimethylsilane itself does not convert alcohols into alkyl chlorides, but this conversion proceeds readily in the presence of SeO_2. The actual reagent is believed to be selenium(IV) oxychloride, $SeOCl_2$, which can be isolated in 74% yield from the reaction of SeO_2 with $ClSi(CH_3)_3$.

[1] J. G. Lee and K. K. Kang, *J. Org.*, **53**, 3634 (1988).

Chlorotrimethylsilane–Zinc chloride.

Intramolecular tandem Michael reaction.[1] Angular triquinane sesquiterpenes such as pentalenene (**5**) have been synthesized from the bis(enone) **1** by an intramolecular tandem Michael reaction as the key step for conversion to the tricyclic system. Lithium amides, even bulky ones, are not useful for this step, nor is dimethyl-*t*-butylsilyl triflate. Annelation can be achieved by reaction of **1** with $ClSi(CH_3)_3$, zinc chloride, and $N(C_2H_5)_3$ at 160° in either toluene or CH_2Cl_2. Use of either solvent gives rise to a mixture of **2** and **3**, but yields are higher in toluene. The ketones **2** and **3** are converted in three steps in about 70% overall yield to the keto esters **4** and then into pentalenene **5**.

$\xrightarrow[N(C_2H_5)_3,\ THF,\ 160°]{ClSi(CH_3)_3,\ ZnCl_2}$

1 **2** (65%) **3** (35%)

4 **5**

[1] M. Ihara, M. Katogi, and K. Fukumoto, *J.C.S. Perkin I*, 2963 (1988).

Chlorotris(triphenylphosphine)rhodium (1).

Decarbonylation of sugars.[1] Aldoses undergo decarbonylation when treated with this organometallic in N-methylpyrrolidin-2-one (NMP) at 130°. Thus glucose is converted into arabinitol in 88% yield (equation I).

(I) $\xrightarrow[88\%]{1,\ NMP}$ $+ ClRh(CO)[P(C_6H_5)_3]_2$

A similar reaction with ketoses is more complex. Thus fructose is converted into furfuryl alcohol (79% yield).

$\xrightarrow[-3H_2O]{1,\ NMP}$ $\xrightarrow[79\%]{}$

Hydroboration of alkenes.[2] Wilkinson's catalyst is effective for catalysis of hydroboration with catecholborane (CB), which can differ from the regioselectivity and stereochemistry of uncatalyzed hydroboration with 9-BBN. In the case of

(I)

9-BBN 86% 83:2 5:10

CB, cat. 84% 18:1 72:9

(II)

9-BBN 83% 1:1

CB, cat. 93% 9:1

cyclohexenol, the regioselectivity is reversed in the catalyzed and noncatalyzed reactions, but the diastereoselectivity is comparable. In the hydroborations with

(III) Bu ... $OSi\text{-}t\text{-}BuMe_2$, CH_2, CH_3 ⟶ Bu ... $OSiR_3$, CH_3, OH + Bu ... $OSiR_3$, CH_3, OH

9-BBN 94% 11:89

CB, cat. 80% 96:4

the exocyclic cycloalkenes (equation II), the catalyzed reaction show high *syn*-diastereoselectivity, as does the catalyzed reaction of 1,1-disubstituted allylic alcohols (equation III). The catalyzed hydroboration of monosubstituted 1-alkenes results in the primary alcohol (>99.5% at −40°).

Cyclization of* 1,6-*enynes[3] (cf. **13**, 91; **14**, 299). Cyclization of these enynes catalyzed by palladium or nickel complexes generally leads to five-membered ring products. However, cyclization catalyzed by Wilkinson's catalyst generally leads to methylene-2-cyclohexenes. Substitution on either of the terminal groups suppresses this cyclization, which probably involves insertion of Rh(I) in the C—H bond of the terminal alkyne.

Example:

$(C_2H_5OOC)_2C(CH_2C{\equiv}CH)(CH_2CH{=}CH_2)$ $\xrightarrow[73\%]{Rh(I),\ CH_3CN,\ 80^\circ}$ $(C_2H_5OOC)_2C$ (ring) $=CH_2$

[1] M. A. Andrews and S. A. Klaeren, *J.C.S. Chem. Comm.*, 1266 (1988).
[2] D. A. Evans, G. C. Fu, and A. H. Hoveyda, *Am. Soc.*, **110**, 6917 (1988).
[3] R. Grigg, P. Stevenson, and T. Worakun, *Tetrahedron*, **44**, 4967 (1988).

Chromic acid.

Jones reagent **(CrO_3-H^+) *in acetone*** (**1**, 142). Jones reagent can oxidatively cleave *sec-tert*-1,2-glycols in high yield.[1] An α-hydroxy ketone may be an intermediate (equation I). Note that an epoxide can also be cleaved to a keto acid.[1]

(I)

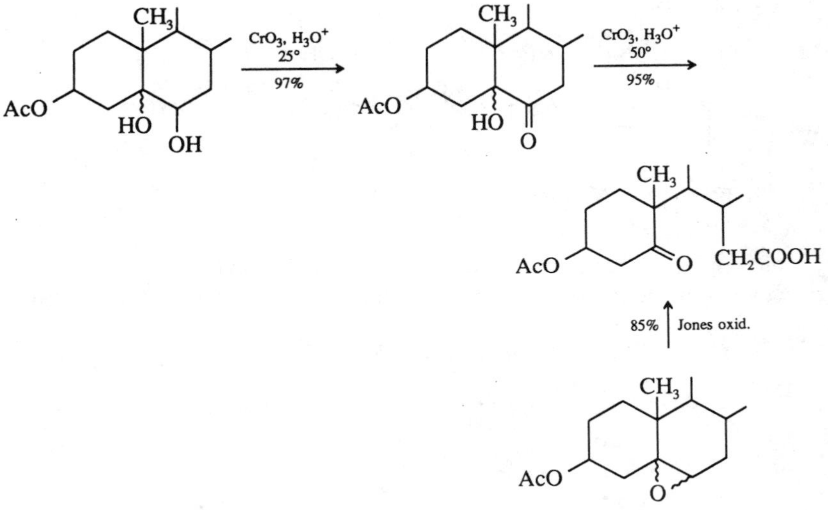

Another example:

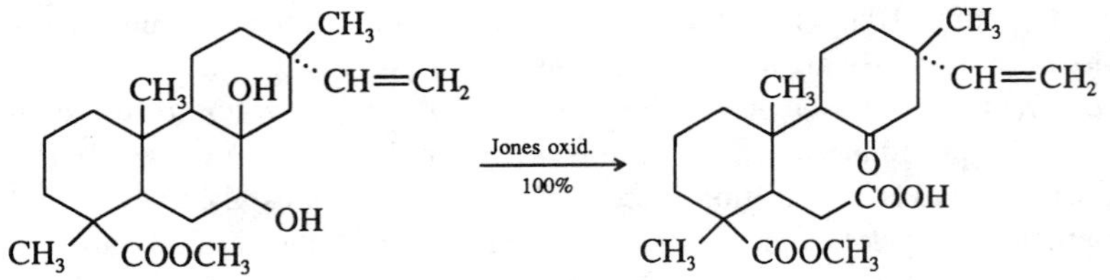

[1] R. de A. Epifanio, W. Camargo, and A. C. Pinto, *Tetrahedron Letters*, **29**, 6403 (1988).

Chromic acid–Chlorotrimethylsilane.

α-Chloro ketones. Both *cis*- and *trans*-disubstituted alkenes are oxidized by a combination of these two reagents dissolved in CCl_4 to α-chloro ketones in satisfactory yield. Unsymmetrical disubstituted alkenes afford both possible α-chloro ketones in about equal amounts. Yields for oxidation of trisubstituted alkenes are only fair. This oxidation is generally useless in the case of 1-alkenes. The report suggests that the actual oxidant is a polyoxochromium dichloride, $Cl(CrO_2)_nCl$.[1]

$$\text{cyclohexene} \xrightarrow[83\%]{CrO_3,\ ClSi(CH_3)_3,\ CCl_4,\ 25°} \text{2-chlorocyclohexanone}$$

$$CH_3(CH_2)_4CH{=}CHCH_3 \longrightarrow \underset{(40\%)}{CH_3(CH_2)_4C(=O)CHClCH_3} + \underset{(40\%)}{CH_3(CH_2)_4CHClC(=O)CH_3}$$

$$(C_6H_5)_2C{=}CH_2 \xrightarrow[91\%]{} (C_6H_5)_2C{=}O$$

[1] J. G. Lee and D. S. Ha, *Tetrahedron Letters*, **30**, 193 (1989).

Chromium carbene reagents.

[(*Dimethylamino*)*methylene*]*chromium pentacarbonyl* (1).[1] This complex reacts with diphenylacetylene at 80° in THF to form *in situ* a complexed enaminoketene (**a**). This ketene (**a**) reacts with imines to form bicyclic lactams, usually as a single stereoisomer.

$$\underset{\mathbf{1}}{(CO)_5Cr{=}C(N(CH_3)_2)H} + C_6H_5C{\equiv}CC_6H_5 \xrightarrow[-CO]{THF,\ 80°} \underset{\mathbf{a}}{\left[O{=}C{=}C(C_6H_5){-}C(C_6H_5){=}CHN(CH_3)_2 \cdot Cr(CO)_3\right]}$$

$$\mathbf{a} \xrightarrow[20-75\%]{R^1N{=}CHR^2} \text{bicyclic lactam: N-}R^1\text{, C=O, CH}R^2\text{; ring-fusion }C_6H_5, C_6H_5\text{; cyclopropane C with }(CH_3)_2N, H$$

Cyclobutanones.[2] The carbene **1** when irradiated undergoes [2+2]cycloaddition with electron-rich alkenes to give, after decomplexation (O_2), cyclobutanones in ~60–85% yield. The regio- and stereochemistry of the products correspond to that observed with ethoxyketene, which suggests that a ketene intermediate is involved. An intramolecular version of this cycloaddition is also efficient.

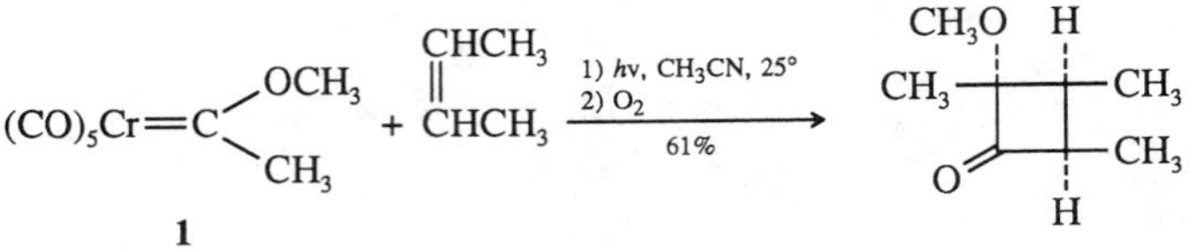

$(CO)_5Cr{=}C(OCH_3)CH_3$ (**1**) + $CHCH_3{=}CHCH_3$ → 1) $h\nu$, CH_3CN, 25°; 2) O_2; 61%

$(CO)_5Cr{=}C(O(CH_2)_3CH{=}CH_2)CH_3$ → 1) $h\nu$, CO, CH_3CN; 2) oxid.; 86%

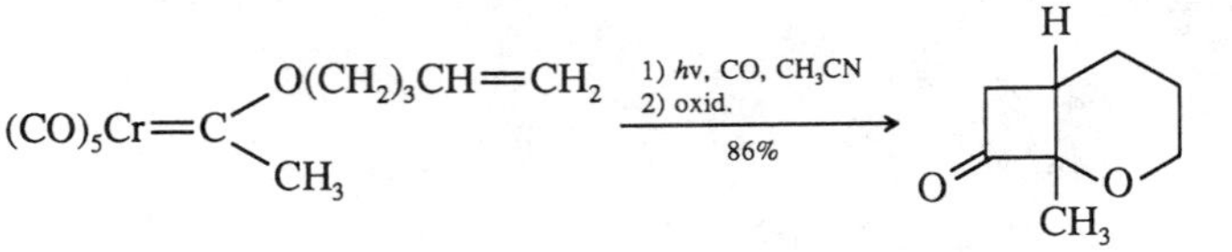

Furanocoumarins.[3] The angular furanocoumarin sphondin (**3**) has been synthesized in four steps from $Cr(CO)_6$ via the carbene **1** in 20% overall yield (Scheme I).

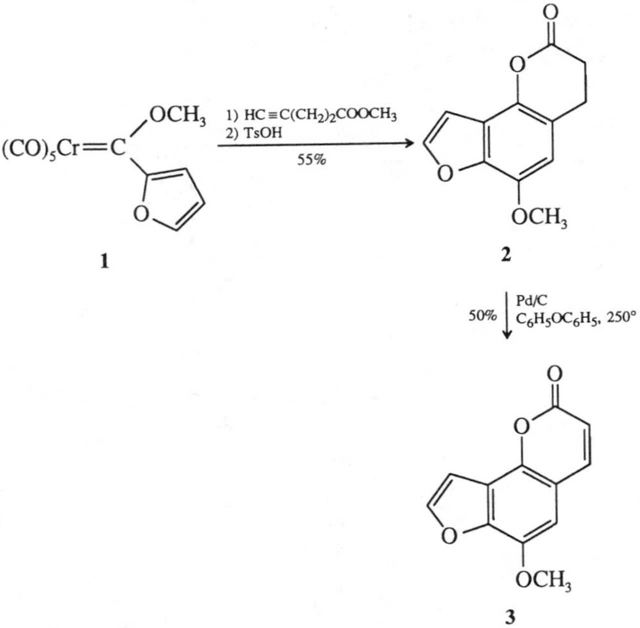

1 → 1) $HC{\equiv}C(CH_2)_2COOCH_3$; 2) TsOH; 55% → **2**

2 → Pd/C, $C_6H_5OC_6H_5$, 250°; 50% → **3**

Scheme I

[2+2]*Cycloadditions of alkynyl carbenes.*[4] Alkynyl carbene complexes of chromium and tungsten undergo [2+2]cycloaddition at room temperature with a wide range of enol ethers.

Examples:

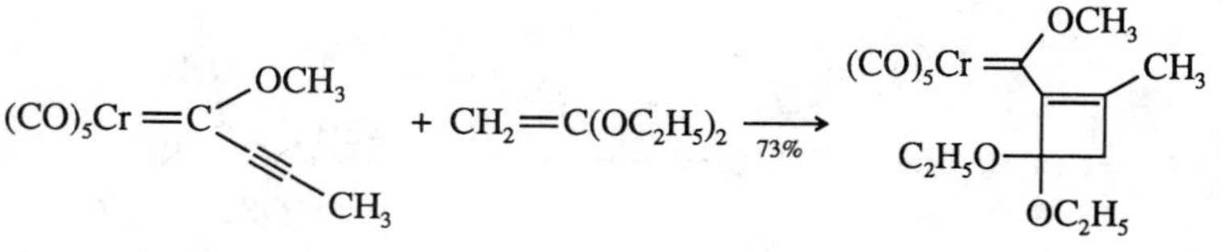

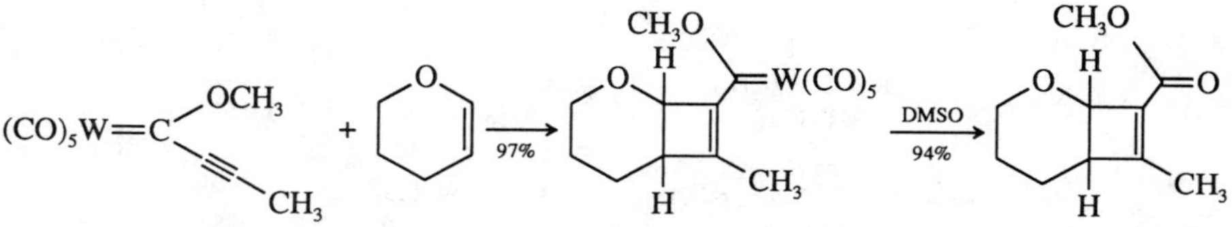

$(CO)_5Cr{=}C(OCH_3)C{\equiv}CCH_3 + CH_2{=}C(OC_2H_5)_2 \xrightarrow{73\%}$

$(CO)_5W{=}C(OCH_3)C{\equiv}CCH_3$ + dihydropyran $\xrightarrow{97\%}$ $\xrightarrow[94\%]{DMSO}$

[1] L. S. Hegedus and D. B. Miller, Jr., *J. Org.*, **54**, 1241 (1989).
[2] M. A. Sierra and L. S. Hegedus, *Am. Soc.*, **111**, 2335 (1989).
[3] W. D. Wulff, J. S. McCallum, and F.-A. Kunng, *ibid.*, **110**, 7419 (1988).
[4] K. L. Faron and W. D. Wulff, *ibid.*, **110**, 8727 (1988).

Chromium(II) chloride.

Intramolecular Nozaki reaction. The addition of a vinyl–chromium compound to an aldehyde (**12**, 137; **14**, 96) is useful for macrocyclization, particularly if the $CrCl_2$ reagent is activated by a catalytic amount of $Ni(acac)_2$. Under these conditions, the vinyl iodide with a ω-formyl group (**1**) cyclizes to (+)-brefeldin C (**2**) and the 4-epimer as a 1:4 mixture in 60% yield.[1]

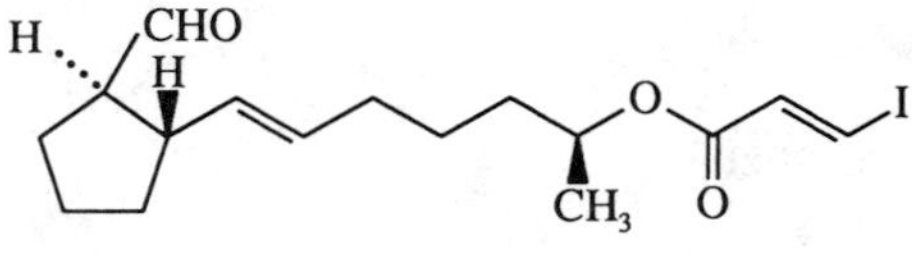

1

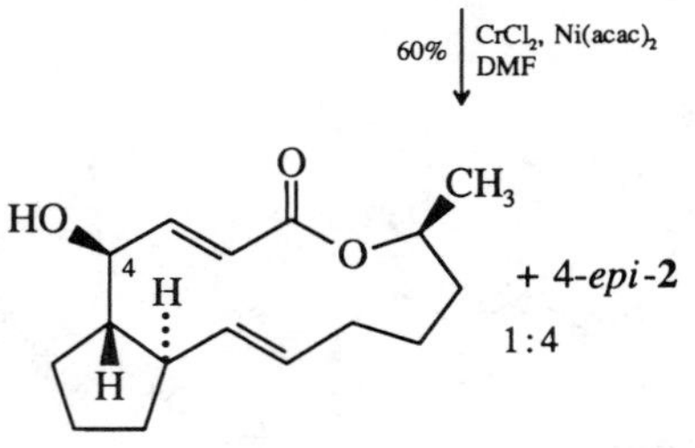

60% $CrCl_2$, $Ni(acac)_2$, DMF

2 + 4-*epi*-**2**

1:4

***syn*-3,4-Butene-1,2-diols.**[2] Reduction *in situ* of an acrolein dialkyl acetal with $CrCl_2$ in THF provides a γ-alkoxy substituted allylic chromium reagent that adds to an aldehyde to afford 3,4-butene-1,2-diols. The reaction rate and the diastereoselectivity are increased by addition of $ISi(CH_3)_3$. Highest diastereoselectivity obtains at $-40°$, but then the reaction is too slow for practicable use.

$$CH_2{=}CHCH(OBzl)_2 + RCHO \xrightarrow[\text{THF, } -30°]{CrCl_2,\ ISi(CH_3)_3} CH_2{=}CH\text{–}CH(OBzl)\text{–}CH(OH)\text{–}R$$

R	Yield	*syn/anti*
R = C_6H_5	98%	88:12
R = $C_6H_5(CH_2)_2$	99%	88:12
R = $C(CH_3)_3$	91%	33:67

[1] S. L. Schreiber and H. V. Meyers, *Am. Soc.*, **110**, 5198 (1988).
[2] K. Takai, K. Hitta, and K. Utimoto, *Tetrahedron Letters*, **29**, 5263 (1988).

Chromium(II) chloride–Nickel(II) chloride.

C-Sucrose.[1] A key step in a synthesis of this saccharide involves as the first step the coupling of a vinyl iodide with an aldehyde mediated by $CrCl_2$ and a trace of $NiCl_2$ (**14**, 97–98). Thus coupling of the vinyl iodide **1** with the aldehyde **2**, derived from D-arabinose, provides an allylic alcohol, which on Mitsunobu inversion provides the *threo* alcohol **3** as the major product. This is converted to C-sucrose

BzlO, OBzl, O, BzlO, OBzl, CH_2, I — **1** + CH_3, CH_3, O, O, CHO, OBzl — **2**

1) $CrCl_2$, $NiCl_2$
2) inversion

BzlO, OBzl, O, BzlO, OBzl, CH_2, 3′, OH, OBzl, O, O, CH_3, CH_3 — **3**

1) $ClC_6H_4CO_3H$
2) CSA
3) H_2, $Pd(OH)_2$
30% overall

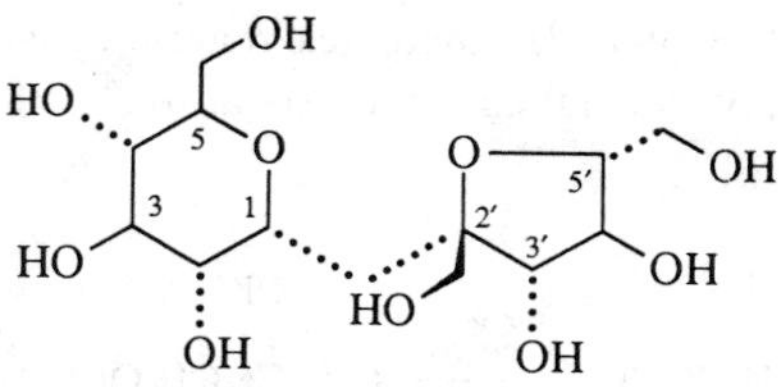

4, α_D + 21.6°

(**4**) in 30% overall yield by stereoselective epoxidation and treatment with camphorsulfonic acid (CSA), which cleaves the acetonide and the epoxide group with cyclization to a triol, which is a protected derivative of **4**. Stereoselective epoxidation of **3** and its *erythro* isomer can provide the four $C_{2'}$ and $C_{3'}$ stereoisomers of **4**.

[1] U. C. Dyer and Y. Kishi, *J. Org.*, **53**, 3383 (1988).

Cobalt(II) chloride.

RCOSi(CH₃)₃ → RCSSi(CH₃)₃.[1] Acylsilanes are converted into thioacylsilanes by reaction with bis(trimethylsilyl) sulfide catalyzed by $CoCl_2 \cdot 6H_2O$ (60–90% yield). The reaction is also applicable to aldehydes, which are converted into thioaldehydes, which can be trapped by dienes.

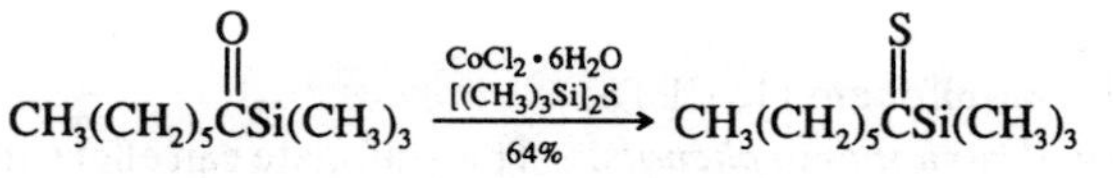

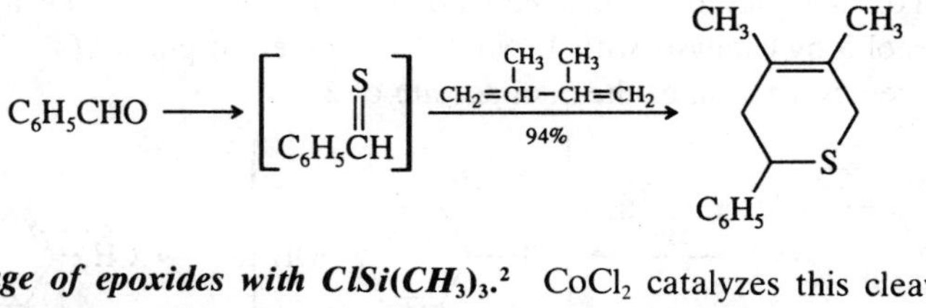

$$CH_3(CH_2)_5C(=O)Si(CH_3)_3 \xrightarrow[64\%]{CoCl_2 \cdot 6H_2O,\ [(CH_3)_3Si]_2S} CH_3(CH_2)_5C(=S)Si(CH_3)_3$$

$$C_6H_5CHO \longrightarrow [C_6H_5CH{=}S] \xrightarrow[94\%]{CH_2{=}C(CH_3){-}C(CH_3){=}CH_2}$$

Cleavage of epoxides with ClSi(CH₃)₃.[2] $CoCl_2$ catalyzes this cleavage with high selectivity to provide primary chlorides when possible.

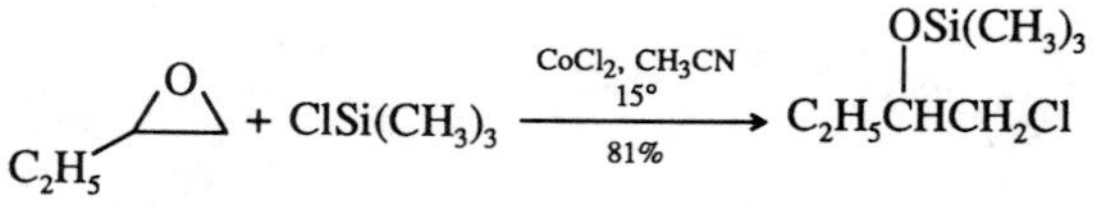

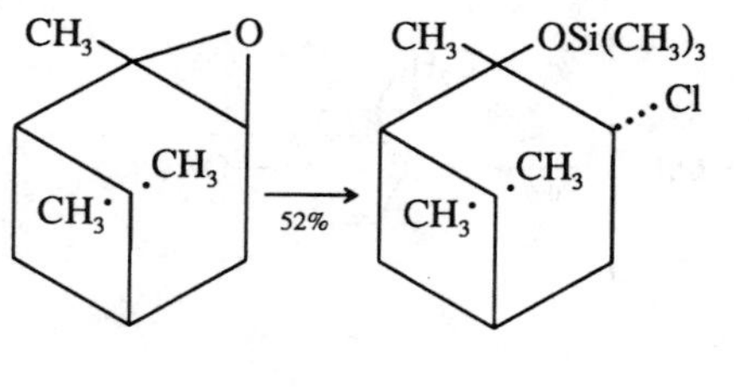

$$C_2H_5\text{-epoxide} + ClSi(CH_3)_3 \xrightarrow[81\%]{CoCl_2,\ CH_3CN,\ 15°} C_2H_5CH(OSi(CH_3)_3)CH_2Cl$$

$$\xrightarrow{52\%}$$

vic-Chloro esters.[3] Cobalt(II) mediates regioselective cleavage of oxiranes with acyl chlorides to give vicinal chloro esters at 0°.

$$C_2H_5\text{-oxirane} + C_6H_5COCl \xrightarrow[87\%]{CoCl_2,\ CH_3CN,\ 0°} C_2H_5CH(Cl)CH_2OC(=O)C_6H_5 + C_2H_5CH(OC(=O)C_6H_5)CH_2Cl \quad (4:1)$$

$$C_6H_5OCH_2\text{-oxirane} + CH_3CH{=}CHC(=O)Cl \xrightarrow[80\%]{} ClCH_2CH(CH_2OC_6H_5)OC(=O)CH{=}CHCH_3$$

$$\text{cyclohexene oxide} + C_6H_5COCl \xrightarrow[87\%]{} \textit{trans}\text{-2-chlorocyclohexyl } OC(=O)C_6H_5$$

[1] A. Ricci, A. Degl'Innocenti, A. Capperucci, and G. Reginato, *J. Org.*, **54**, 19 (1989).
[2] J. Iqbal and M. A. Kahn, *Chem. Letters*, 1157 (1988).
[3] J. Iqbal, M. A. Khan, and R. R. Srivastava, *Tetrahedron Letters*, **29**, 4985 (1988).

Collidine *p*-toluenesulfonate (1, CPTS).

Cyclization of enynones to phenols. This sulfonate can effect an acid-catalyzed cyclization of enynones to phenols (equation I). Thus the enynone **2** can be cyclized to the phenol **3** by catalysis with **1**. Note that a different phenol (**4**) obtains from acid-catalyzed cyclization of the enol acetate of **2**.

(I) $$\mathbf{2}\ ((C_2H_5)_2C{=}CHC(=O)C{\equiv}CCH_3) \xrightarrow{1,\ C_6H_5CH_3,\ 250°} [\text{dienol intermediate}] \xrightarrow[82\%]{} \mathbf{3}$$

$$\mathbf{2} \xrightarrow{CH_3CO_2C(CH_3){=}CH_2,\ TsOH} [\text{enol acetate}] \xrightarrow[80\%]{} \mathbf{4}$$

For cyclization of these substrates to methylenecyclopentenones *see* 4-*t*-butylcatechol.

[1] P. A. Jacobi and J. I. Kravitz, *Tetrahedron Letters*, **29**, 6873 (1988).

Copper, Cu(0)*, **9**, 122.

A highly active form of copper slurry (Cu*) can be prepared by reduction of $CuI \cdot P(C_2H_5)_3$ with lithium naphthalenide. This Cu(0) converts aryl, alkynyl, and vinyl halides to organocopper reagents at sufficiently low temperatures to prevent self-coupling (Ullmann reactions). However, reaction with primary alkyl, allyl, and benzyl iodides at 25° gives homocoupled products in high yield. The arylcopper compounds cross-couple with acid chlorides or with alkyl halides in moderate to high yield.[1]

This Cu* species on reaction with epoxyalkyl halides forms an epoxyalkylcopper reagent. This undergoes an intramolecular cyclization via epoxy cleavage. The regioselectivity of reaction of Cu* with the substrate **1** can be controlled by the

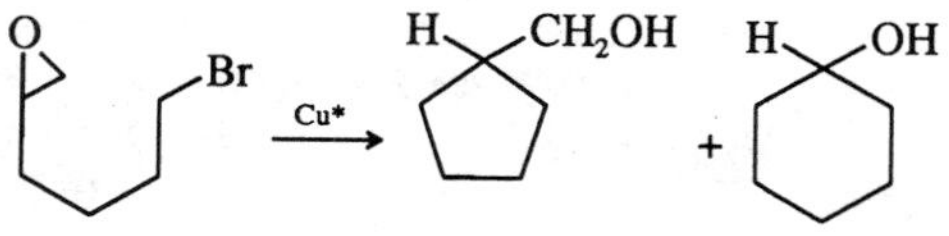

1		(*exo*) : (*endo*)
THF, –78°	56%	1:6
$C_6H_5CH_3$	45%	1:0

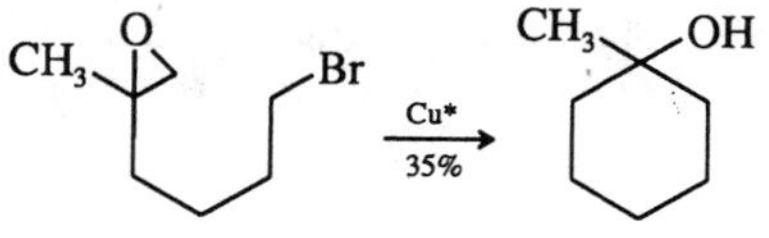

2 → (Cu*, 35%) → **3**

solvent, but the substrate **2** cyclizes to 1-methylcyclohexanol (**3**) regardless of the solvent. This reaction can also effect cyclization to three-membered rings.[2]

[1] G. W. Ebert and R. D. Rieke, *J. Org.*, **53**, 4482 (1988).
[2] T.-C. Wu and R. D. Rieke, *Tetrahedron Letters*, **29**, 6753 (1988); R. D. Rieke, R. M. Wehmeyer, T.-C. Wu, and G. W. Ebert, *Tetrahedron*, **45**, 443 (1989).

Copper(II) acetate–Anthracene.

Knoevenagel condensation; methylidenemalonates. In the presence of $Cu(OAc)_2$ and KOH, di-*t*-butyl malonate undergoes a Knoevenagel condensation with paraformaldehyde to give di-*t*-butyl methylidenemalonate.[1] This method is not useful for the lower analogs because of anionic polymerization of the products. A more general route to methylidenemalonates involves a Knoevenagel condensation of a dialkyl malonate with paraformaldehyde in the presence of anthracene, which traps the product by a Diels–Alder reaction. The [4+2]cycloadduct (**1**), when heated at 200–250°, undergoes a retro-Diels–Alder reaction to give the methylidenemalonate (**2**) in 30–80% yield.[2]

$$(ROOC)_2CH_2 + (CH_2O)_n + \text{anthracene} \xrightarrow[40-80\%]{Cu(OAc)_2,\ HOAc,\ Xylene,\ 100^\circ} \mathbf{1} \xrightarrow[30-80\%]{200-250^\circ} CH_2{=}C(COOR)_2\ (\mathbf{2})$$

[1] P. Ballesteros, B. W. Roberts, and J. Wong, *J. Org.*, **48**, 3603 (1983).
[2] J.-L. De Keyser, C. J. C. De Cock, J. H. Poupaert, and P. Dumont, *ibid.*, **53**, 4859 (1988).

Copper(II) bromide.

Oxidation of allylstannanes.[1] Oxidation of these compounds with $CuBr_2$ in the presence of water, alcohols, or sodium acetate provides allylic alcohols, ethers, or acetates. Allylic amines can be obtained by oxidation in THF in the presence of an amine.

$$(CH_3)_2C{=}CHCH(SnBr_3)Bu \xrightarrow{CuBr_2} CH_3C(CH_3){=}CHCH(Nu)Bu + (CH_3)_2C(Nu)CH{=}CHBu$$

Nu	Yield	Ratio
$Nu = OC_2H_5$	73%	83:17
$Nu = NH(CH_2)_2C_6H_5$	73%	14:86

[1] T. Takeda, T. Inoue, and T. Fujiwara, *Chem. Letters*, 985 (1988).

Copper(I) bromide–Dimethyl sulfide, $CuBr{\cdot}S(CH_3)_2$.

S_N2'-Allylation of R_2Zn or RZnCl.[1] Cinnamyl chloride reacts with a variety of organozinc reagents in the presence of this Cu(I) reagent (5 mole %) at 20° to give the S_N2'-products of allylation. The regioselectivity is comparable to that of $RCu{-}BF_3$ at $-70°$.

$$C_6H_5CH{=}CHCH_2Cl + CH_3ZnCl \longrightarrow C_6H_5CH(CH_3)CH{=}CH_2 + C_6H_5CH{=}CHCH_2CH_3$$

Cu(I)	67%	92:8
Ni(II)	65%	8:92

Allylation catalyzed by $NiCl_2$ complexed with a phosphine involves an S_N2 reaction with equally high selectivity.

[1] K. Sekiya and E. Nakamura, *Tetrahedron Letters*, **29**, 5155 (1988).

Copper(I) chloride.

Alkyl carbamates. In the presence of CuCl (1 equiv.) alkyl isocyanates react in DMF at 25° with alcohols, even tertiary ones, to form alkyl carbamates in 45–96% yield.

$$ROH + C_6H_5CH(R)NCO \xrightarrow[45-98\%]{CuCl,\ DMF,\ 25°} C_6H_5CH(R)NHC(=O)OR$$

$(R = H, CH_3)$

[1] M. E. Duggan and J. S. Imagire, *Synthesis*, 131 (1989).

Copper(II) nitrate–Montmorillonite (Claycop), **12**, 231.

Nitration and oxidation.[1] Clay-supported $Cu(NO_3)_2$, unlike clayfen (**12**, 231), is shelf-stable for months. Like clayfen, it is a convenient source of NO_2^+ and can cleave thioacetals or selenoacetals to the carbonyl compound at 25° in high yield. It effects aromatization of 1,4-dihydropyridines in 80–92% yield. In the presence of acetic anhydride, it can effect nitration even of halobenzenes at 25° with marked *para*-preference, which can be enhanced by use of lower temperatures.

[1] P. Laszlo and A. Cornélis, *Aldrichimica Acta*, **21**, 97 (1988).

Copper(I) thiocyanate, CuSCN.

Arene thiocyanation.[1] Thiocyanation of arenes with metal thiocyanates usually results in both arenethiocyanates and the more stable areneisothiocyanates. The reaction can be effected with CuSCN supported on charcoal. Thus iodo- and bromoarenes react with this supported reagent (excess) in *t*-butylbenzene at 150° to give the corresponding thiocyanates in 70–80% yield.

[1] J. H. Clark, C. W. Jones, C. V. A. Duke, and J. M. Miller, *J.C.S. Chem. Comm.*, 81 (1989).

Cyano(methylthio)methyltrimethylsilane, $(CH_3)_3SiCH(CN)SCH_3$ (**1**). The reagent is available by reaction of methylthioacetonitrile, CH_3SCH_2CN, with trimethylsilyl triflate and $N(C_2H_5)_3$ in ether at 0–5° (80–85% yield).

α-Cyano vinyl sulfides.[1] The lithium anion of **1** undergoes a Peterson reaction with aldehydes or ketones to give these cyano substituted sulfides as E/Z mixtures in 60–95% yield.

$$\mathbf{1} + (C_6H_5)(CH_3)C{=}O \xrightarrow[61\%]{BuLi} (C_6H_5)(CH_3)C{=}C(CN)(SCH_3) + (CH_3)_3SiOLi$$

E/Z = 5 : 1

[1] D. I. Han and D. Y. Oh, *Syn. Comm.*, **18**, 2111 (1988).

2-Cyanopyridinium chlorochromate (1), 13, 87.

α-Chloro ketones. This chlorochromate (5–11 equiv.) converts alkenes to α-chloroketones in 40–95% yield. The workup is facilitated by use of Celite as adsorbent. In rare cases allylic oxidation competes with oxychlorination, and epoxides are occasionally formed as by-products.[1]

$$PrCH{=}CHPr \xrightarrow[96\%]{1,\ CH_2Cl_2,\ 25°} PrC(=O){-}CHClPr$$

$$\text{norbornene} \xrightarrow{53\%} \text{3-chloro-2-norbornanone}$$

(*exo*)

[1] A. F. Guerrero, H. Kim, and M. F. Schlecht, *Tetrahedron Letters*, **29**, 6707 (1988).

Cyanotrimethylsilane, $(CH_3)_3SiCN$. This silane is present in solution in part as the isocyanide isomer, $(CH_3)_3SiNC$.[1] And indeed it is more useful than NaCN for cyanation of the iron complex **1**[2] to provide **2** and **3**, intermediates in a projected synthesis of tridachione.

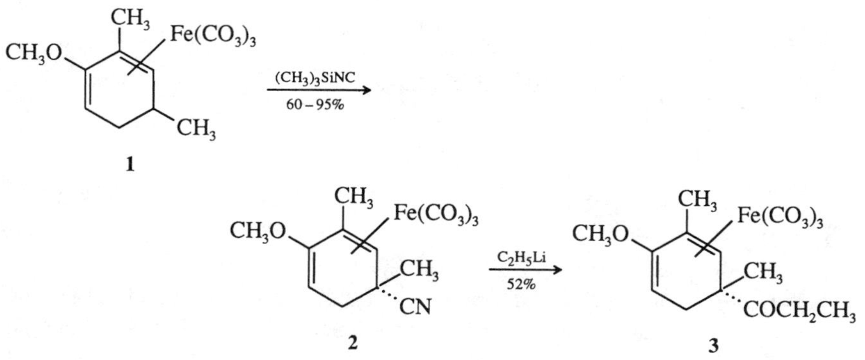

$$\mathbf{1} \xrightarrow[60-95\%]{(CH_3)_3SiNC} \mathbf{2} \xrightarrow[52\%]{C_2H_5Li} \mathbf{3}$$

Asymmetric hydrocyanation of aldehydes.[3] This reaction can be effected by reaction of aliphatic or aromatic aldehydes with cyanotrimethylsilane and an optically active reagent (**1**) derived from (2R,3R)-tartaric acid, and dichlorodiisopropoxytitanium(IV). The actual chiral reagent may be **2**, shown by ^1H NMR to be

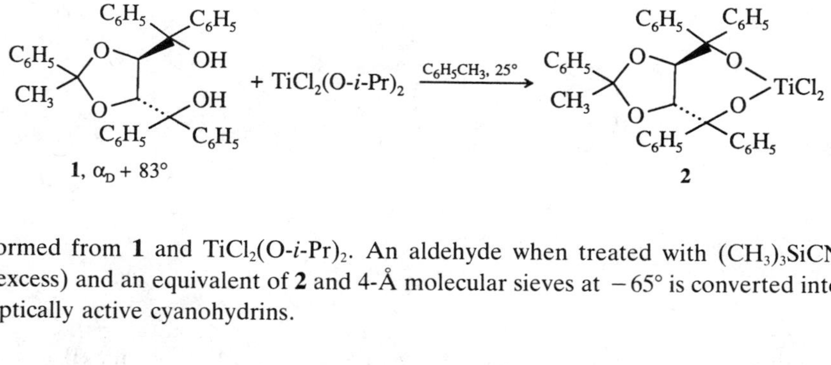

$$\mathbf{1},\ \alpha_D + 83° + TiCl_2(O\text{-}i\text{-}Pr)_2 \xrightarrow{C_6H_5CH_3,\ 25°} \mathbf{2}$$

formed from **1** and $TiCl_2(O\text{-}i\text{-}Pr)_2$. An aldehyde when treated with $(CH_3)_3SiCN$ (excess) and an equivalent of **2** and 4-Å molecular sieves at −65° is converted into optically active cyanohydrins.

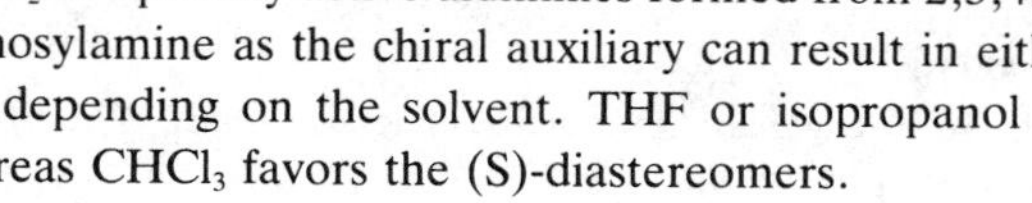

$$R(Ar)CHO + CNSi(CH_3)_3 \xrightarrow[65-90\%]{\mathbf{2}} R(HO)CH(CN)$$

(73–96% ee)

Asymmetric Strecker amino acid synthesis.[4] Addition of cyanotrimethylsilane catalyzed by $ZnCl_2$ to optically active aldimines formed from 2,3,4,6-tetrapivaloyl-β-D-galactopyranosylamine as the chiral auxiliary can result in either (R)- or (S)-α-aminonitriles, depending on the solvent. THF or isopropanol favors (R)-diastereomers, whereas $CHCl_3$ favors the (S)-diastereomers.

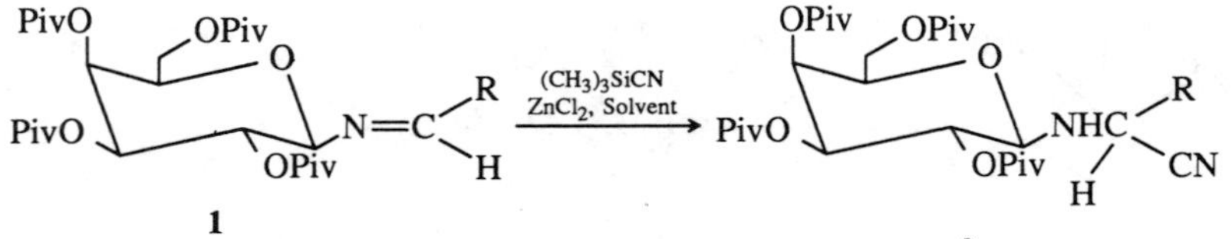

$(CH_3)_2CHOH$ (R/S) = 7–13 : 1
$CHCl_3$ (S/R) = 3–9 : 1

2,4-Dialkyl-1-naphthols.[5] The *p*-quinols **1**, formed by addition of R′Li to the monoadduct of 1,4-naphthoquinone with cyanotrimethylsilane (**5**, 722), can undergo a second addition of R^1Li or R^2Li to provide the diol **2**. Reduction of **2** with HI

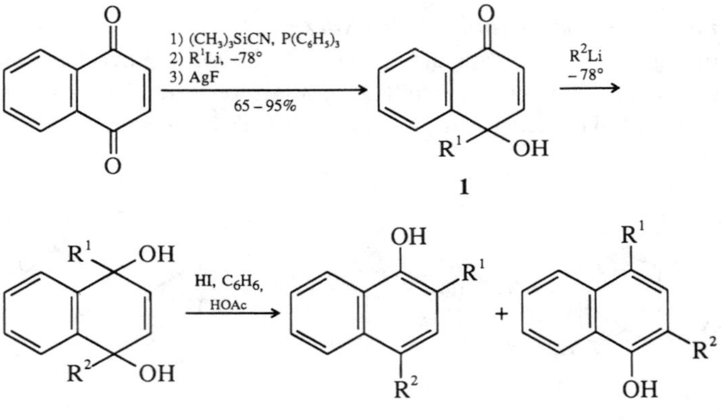

2		**3**	**4**
$R^1 = R^2 = CH_3$	89%		
$R^1 = Bu$, $R^2 = CH_3$		43%	7%
$R^1 = C_6H_5$, $R^2 = Bu$		54%	9%

in THF gives a 1,4-dialkylnaphthalene as expected, but reduction with HI under more acidic conditions results in a dienone–phenol rearrangement to a 2,4-dialkyl-1-naphthol, **3** or **4**. When $R^1 = R^2$, a single product is obtained in 75–95% yield. But when $R^1 \neq R^2$, two different naphthols, **3** or **4**, can be formed, depending on the migratory aptitudes of R^1 and R^2. In any case, the naphthols are separable by chromatography.

[1] R. P. Alexander, T. D. James, and G. R. Stephenson, *J.C.S. Dalton*, 2013 (1987).
[2] R. P. Alexander, C. Morley, and G. R. Stephenson, *J.C.S. Perkin I*, 2069 (1988).
[3] H. Minamikawa, S. Hayakawa, T. Yamada, N. Iwasawa, and K. Narasaka, *Bull. Chem. Sc. Japan*, **61**, 4379 (1988).
[4] H. Kunz, W. Sager, W. Pfrengle, and D. Schanzenback, *Tetrahedron Letters*, **29**, 4397 (1988).
[5] J. A. Dodge and A. R. Chamberlin, *ibid.*, **29**, 4827 (1988).

Cyanuric chloride.

Lactonization of ε-hydroxy acids. The ε-hydroxy acid **1** is converted into the oxazepine **2** in 55% yield by cyanuric chloride and pyridine as base in 55% yield. The combination of cyanuric chloride and triethylamine has been recommended

1	$(CNCl)_3$, CH_3CN → 2 (2S, 3R)
+ $N(C_2H_5)_3$	27%
+ Py	55%
2,6-Lutidine	<5%

for macrocyclic lactonization (**10**, 114–115), but this method is less useful in the case of **1**. In fact, the present lactonization is remarkably sensitive to the base.

[1] R. T. Brown and M. J. Ford, *Syn. Comm.*, **18** 1801 (1988).

Cyclic sulfates of 1,2-diols.

Unlike epoxides, these five-membered heterocyclics have received scant attention from organic chemists. But the recent catalytic asymmetric dihydroxylation of alkenes (**14**, 237–239), which is now widely applicable (this volume), and the ready access to optically active natural 1,2-diols has led to study of these compounds, including a convenient method for synthesis. They are now generally available by reaction of a 1,2-diol with thionyl chloride to form a cyclic sulfite of a 1,2-diol, which is then oxidized in the same flask by the Sharpless catalytic RuO_4 system, as shown in equation I.[1]

(I) R = n-C_8H_{17}; $SOCl_2$, CH_2Cl_2 → [cyclic sulfite]; $NaIO_4$, $RuCl_3$, CH_3CN, H_2O, 92% → cyclic sulfate

Cyclic sulfates can even be prepared from diols containing acid-sensitive groups (acetonide, silyloxy) by reaction with thionyl chloride and $N(C_2H_5)_3$ followed by oxidation of the isolated sulfites with RuO_4 (catalytic). After reactions with a nucleophile, the resulting sulfate esters can be hydrolyzed by water (0.5–1.0 equiv.) in THF catalyzed by H_2SO_4. The use of a minimal amount of water is crucial for chemoselectivity.[2]

$SOCl_2$, $N(C_2H_5)_2$, CH_2Cl_2 ; $RuCl_3$, $NaIO_4$, 0° ; LiN_3, 25° ; H_3O^+, 73–91% overall

Reactions with nucleophiles. Many of these reactions have been carried out with cyclic sulfate **1**, available in 90–93% yield from diisopropyl (+)-tartrate. This sulfate is reduced at pH 4–5 by sodium cyanoborohydride in THF to diisopropyl (R)-malate in 55% (equation II).

(II) 1) $NaBH_3CN$ 2) H_3O^+, 55%; O_2S(O)(O) COO-*i*-Pr, COO-*i*-Pr → *i*-PrOOC–CH(OH)–CH_2–COO-*i*-Pr

1

The reaction of **1** with sodium azide at 0° provides β-azidomalates, which are useful precursors to α-amino-β-hydroxy acids (equation III).

(III) **1** $\xrightarrow[81\%]{NaN_3,\ 0\text{–}25°,\ (CH_3)_2C{=}O,\ H_2O}$ ROOC–CH(OH)–CH(N_3)–COOR

Since α,β-unsaturated esters undergo catalytic dihydroxylation in high yield, a reaction of **2** with a number of nucleophiles has been investigated (equation IV). In all reported cases, the nucleophile reacts exclusively at the α-position. In contrast, reaction of epoxides with these nucleophiles shows no such regioselectivity.[1]

(IV) **2** (n-$C_{15}H_{31}$, $COOCH_3$, O_2S) $\xrightarrow{Nu:}$ $C_{15}H_{31}$-n–CH(OH)–CH(Nu)–CH_3OOC

Nu = $NaBH_4$(90%); $C_6H_5CO_2NH_4$(83%); NH_4SCN(90%); Bu_4NF(63%)

[1] Y. Gao and K. B. Sharpless, *Am. Soc.*, **110**, 7538 (1988).
[2] B. M. Kim and K. B. Sharpless, *Tetrahedron Letters*, **30**, 655 (1989).

β-Cyclodextrin (CD).

Asymmetric synthesis of $Ar(CH_2)_nS^*(=O)CHCl_2$.[1] Oxidation (NaOCl) of arenethioacetates, $Ar(CH_2)_nSCH_2COOR$, complexed with β-cycloodextrin results in optically active α,α-dichlorosulfoxides in 4–54% ee in 70–83% chemical yield. The highest optical yield is obtained in the oxidation of $C_6H_5SCH_2COOBu$.

Photochemical processes of CD complexes.[2] Differences in photochemical reactions conducted in solution and in CD complexes have been reviewed. For example, photo-Fries rearrangement of phenyl esters in solvents results in a mixture of *o*- and *p*-phenolic ketones via a radical reaction. Rearrangement of the same encapsulated ester results in exclusive rearrangement to the *ortho*-position (equation I).

(I) $C_6H_5OC(=O)C_6H_5 \xrightarrow{h\nu}$ *o*-$HOC_6H_4COC_6H_5$ + *p*-$HOC_6H_4COC_6H_5$

CH_3OH	1.6 : 1
β–CD	99 : 1

Similar selective photo-Claisen rearrangement obtains with *m*-alkoxyphenyl allyl ethers to provide only one of the two possible *ortho*-isomers.

[1] K. Rama Rao and P. B. Saltur, *J.C.S. Chem. Comm.*, 342 (1989).
[2] V. Ramamurthy and D. F. Eaton, *Acc. Chem. Res.*, **21**, 300 (1988).

(Cyclooctadiene)(1,6-cyclooctatriene)ruthenium, Ru(COD)COT).

Silylation of alkenes. The complex promotes silylation of alkenes to provide allylsilanes as the major product.

$H_2C{=}C(CH_3)C_2H_5$ or $CH_3C(CH_3){=}CHCH_3 + HSi(C_2H_5)_3 \xrightarrow[59-61\%]{Ru(COD)(COT)} CH_3C(CH_3){=}CHCH_2Si(C_2H_5)_3$

$CH_3(CH_2)_3CH{=}CH_2 + HSi(C_2H_5)_3 \xrightarrow[81\%]{} CH_3(CH_2)_2CH{=}CHCH_2Si(C_2H_5)_3$

E/Z = 65 : 35

[1] Y. Hori, T. Mitsudo, and Y. Watanabe, *Bull. Chem. Soc. Japan*, **61**, 3011 (1988).

Cyclopropane.

The use of cyclopropanes in synthesis has been reviewed (293 references).[1]

[1] H. N. C. Wong, M.-Y. Hon, C.-W. Tse, Y.-C. Yip, J. Tanko, and T. Hudlicky, *Chem. Rev.*, **89**, 165 (1989).

Cyclopropyldiphenylsulfonium tetrafluoroborate,

$$\triangleright\!-\overset{+}{S}(C_6H_5)_2\ \overset{-}{B}F_4$$ (**1**), **4**, 213–214.

Cationic cyclization with vinylcyclopropyl silyl ethers.[1] These ethers can serve as terminators in a cationic cyclization involving acetals. Thus reaction of the keto acetal **2** with **1** and then with a base and $ClSi(CH_3)_3$ provides a vinylcyclopropyl silyl ether **3**, which undergoes facile cyclization to an eight-membered ring (**4**) in the presence of trimethylsilyl triflate.

$CH_3C(=O)(CH_2)_5CH(OCH_3)_2$ (**2**) —[1) **1**, KOH, DMSO; 2) $LiN(C_2H_5)_2$, $ClSi(CH_3)_3$]→ **3** —[$(CH_3)_3SiOTf$, Py, CH_2Cl_2; 64%]→ **4** (9:1)

This strategy can also be used to obtain fused-ring systems such as **5**.

OTMS, CH_2, $CH_2CH(OCH_3)_2$ —[$(CH_3)_3SiOTf$, 91%]→ H, O, H, OCH_3 **5** (7:1)

[1] B. M. Trost and D. C. Lee, *Am. Soc.*, **110**, 6556 (1988).

D

1,4-Diazabicyclo[2.2.2]octane (DABCO).

Reaction of ethyl 2,3-butadienoate with aldehydes.[1] This ester (**1**) reacts with propionaldehyde in ether at 23° in the presence of DABCO to provide **2** in 42% yield, formed by replacement of the α-proton of **1**. This condensation can also be effected with butyllithium, but then **2** is accompanied by **3**, formed via a dianion of **1**.

$$CH_2{=}C{=}CHCOOC_2H_5 \xrightarrow{CH_3CH_2CHO}$$

1
+ DABCO
+ BuLi

$$CH_2{=}C{=}C(CHOHC_2H_5)COOC_2H_5 + C_2H_5CH(OH)CH{=}C{=}C(CHOHC_2H_5)COOC_2H_5$$

	2	**3**
+ DABCO	42%	–
+ BuLi	58%	6%

α-Methylene-β-hydroxy-γ-alkoxyketones.[2] These compounds can be prepared by an aldol-type reaction of alkyl vinyl ketones with α-alkoxy aldehydes mediated by DABCO or 1-azabicyclo[2.2.2]octan-3-ol. The reaction is *anti*-selective.

Example:

$$CH_3CH(OCH_2OCH_3)CHO + CH_2{=}CHCOCH_3 \xrightarrow[55\%]{DABCO} CH_3CH(OCH_2OCH_3)CH(OH)C({=}CH_2)COCH_3$$

(*anti/syn* = 71 : 29)

[1] S. Tsuboi, S. Takatsuka, and M. Utaka, *Chem. Letters*, 2003 (1988).
[2] S. E. Drewes, T. Manickum, and G. H. P. Roos, *Syn. Comm.*, **18**, 1065 (1988).

1,8-Diazabicyclo[5.4.0]-7-undecene (DBU, **1**).

O-and S-Akylation. This base facilitates esterification of carboxylic acids[1] and S-alkylation of *p*-toluenesulfinic acids.[2]

$$p\text{-}CH_3C_6H_4SO_2H + RX + DBU \xrightarrow[80-92\%]{CH_3CN} p\text{-}CH_3C_6H_4SO_2R$$

[1] D. Mal, *Syn. Comm.*, **16**, 331 (1986).
[2] G. Biswas and D. Mal, *J. Chem. Research* (*S*), 308 (1988).

Diazidodiisopropoxytitanium, $(N_3)_2Ti(O\text{-}i\text{-}Pr)_2$ (**1**), **13**, 217. Preparation.[1] The reagent can also be prepared *in situ* from $Ti(O\text{-}i\text{-}Pr)_4$ and 2 equiv. of $(CH_3)_3SiN_3$ in refluxing benzene.

α-Amino acids.[2] Treatment of 2,3-epoxy alcohols with **1** affords 3-azido 1,2-diols, which can be converted to amino acids by RuO_4 oxidation to α-azido carboxylic acids followed by catalytic hydrogenation.

C_6H_5(epoxide)CH_2OH —**1**, 76%→ $C_6H_5CH(N_3)CH(OH)CH_2OH$ (100:1) —$RuCl_3$, H_5IO_6, 85%→ $C_6H_5CH(N_3)COOH$ —H_2, Pd/C, HCl, 76%→ $C_6H_5CH(NH_2 \cdot HCl)COOH$

[1] R. Choukroun and D. Gervais, *J.C.S. Dalton*, 1800 (1980).
[2] M. Caron, P. R. Carlier, and K. B. Sharpless, *J. Org.*, **53**, 5185 (1988).

Di(benzene)chromium(0). Preparation.[1]

Dehydrogenation catalysts.[2] Reduction of di(benzene)chromium with potassium sand in DME generates a species that dehydrogenates 1,4-cyclohexadiene to benzene. The catalyst prepared from benzene(1,3-cyclohexadiene)iron(0) promotes dehydrogenation and disproportionation to cyclohexene and benzene.

$(C_6H_6)_2Cr$ + 1,4-cyclohexadiene —K, DME, 25°→ benzene

[1] E. O. Fischer, *Inorg. Syn.*, **6**, 132 (1960).
[2] G. Fochi, *Organometallics*, **7**, 2255 (1988).

L-N,N-Dibenzyl-α-amino aldehydes, (1).

Preparation:

$RCH(NH_2)COOH$ —BzlBr→ $RCH(NBzl_2)COOBzl$ —1) $LiAlH_4$; 2) $(COCl)_2$/DMSO→ $RCH(NBzl_2)CHO$ (L-**1**)

β-Amino alcohols.[1] Alkyllithium and Grignard reagents add to **1** to provide β-amino alcohols (**2** and **3**), with the latter predominating. Addition of Lewis acids, CH_3TiCl_3, $SnCl_4$, can favor the chelation-controlled product **2**.

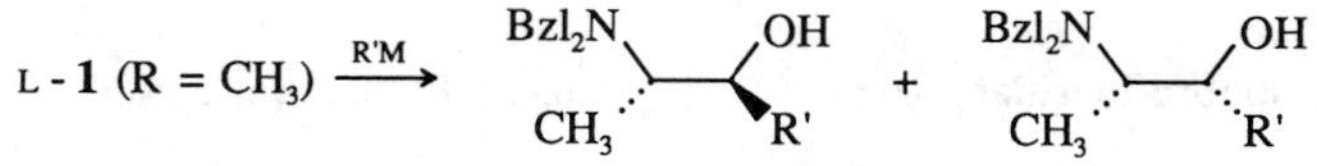

L-**1** ($R = CH_3$) $\xrightarrow{R'M}$ Bzl_2N, OH, CH_3, R' (**2**) + Bzl_2N, OH, CH_3, R' (**3**)

R'M	Yield	2 : 3
CH_3MgI	87%	5:95
CH_3Li	91%	9:91
CH_3TiCl_3	82%	94:6
$(CH_3)_2CuLi$	80%	25:75
$CH_2{=}C(OLi)(OCH_3)$	82%	5:95
$CH_2{=}C(OC_6H_5)(OSi(CH_3)_3)$ •$TiCl_4$	25%	95:5

[1] M. T. Reetz, M. W. Drewes, and A. Schmitz, *Angew. Chem. Int. Ed.*, **26**, 1141 (1987).

Dibenzyl azodicarboxylate (1).

2-Amino glycosides.[1] Furanoid or pyranoid glycals (**2**) when irradiated with this azodicarboxylate undergo a [4+2]cycloaddition to form a dihydrooxadiazine (**3**) in 70–80% yield. The adducts on treatment with methanol (TsOH) open with inversion at C_1 to give hydrazines (**4**), which furnish 2-amino glycosides (**5**) on hydrogenolysis and deprotection.

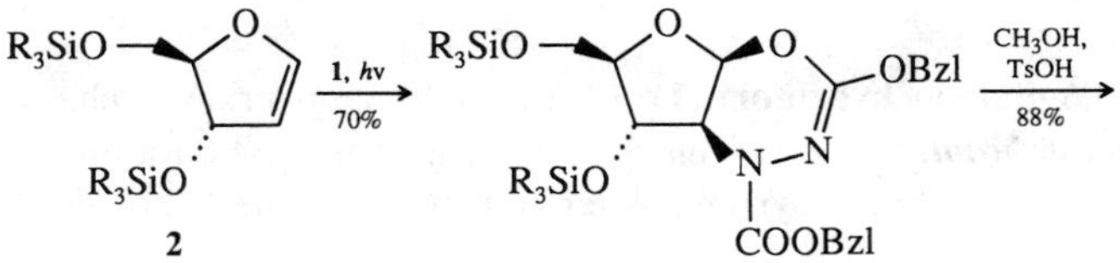

R_3SiO, O, R_3SiO (**2**) $\xrightarrow{1,\ h\nu,\ 70\%}$ R_3SiO, O, O, OBzl, N–N, R_3SiO, COOBzl (**3**) $\xrightarrow{CH_3OH,\ TsOH,\ 88\%}$

R_3SiO, O, OCH_3, R_3SiO, N—NHCOOBzl, COOBzl (**4**) $\xrightarrow{1)\ H_2,\ Raney\ Ni;\ 2)\ Bu_4NF;\ 73\%}$ HO, O, OCH_3, HO, NH_2 (**5**)

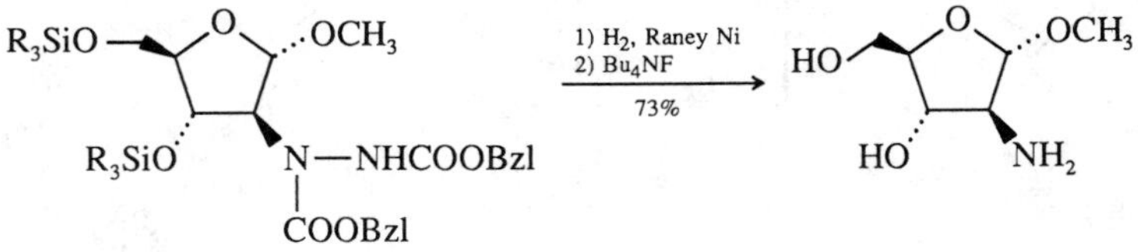

Actually, the adducts are cleaved by various primary or secondary alcohols, including various carbohydrate alcohols, and thus can be used to obtain amino-polysaccharides.

[1] B. J. Fitzsimmons, Y. Leblanc, and J. Rokach, *Am. Soc.*, **109**, 285 (1987); B. J. Fitzsimmons, Y. Leblanc, N. Chan, and J. Rokach, *ibid.*, **110**, 5229 (1988); Y. Leblanc, B. J. Fitzsimmons, J. P. Springer, and J. Rokach, *ibid.*, **111**, 2995 (1989).

Dibenzyl phosphorofluoridate, $(BzlO)_2P(=O)F$ (**1**), b.p. ~150°/0.2 mm.

$$(BzlO)_2P(=O){-}OH + \text{1-methyl-2-fluoropyridinium } TsO^- \xrightarrow[90\%]{N(C_2H_5)_3} \mathbf{1}$$

$$[(BzlO)_2P(=O)]_2O \xrightarrow[80\%]{CsF,\ CH_3CN} \mathbf{1}$$

Phosphorylation.[1] Dibenzyl phosphorochloridate is widely used for phosphorylation of alcohols, but is more difficult to prepare than the fluoridate. Phenol is very readily phosphorylated by **1** and CsF in CH_3CN, and indeed a phenolic hydroxyl group is selectively attacked even in the presence of a primary hydroxyl group. Primary and anomeric alcohols are preferentially phosphorylated in the presence of a secondary hydroxyl group. Alcohols that are insoluble in acetonitrile can be phosphorylated by **1** and DBU in place of CsF.

[1] Y. Watanabe, N. Hyodo, and S. Ozaki, *Tetrahedron Letters*, **29**, 5763 (1988).

1,3-Dibromo-3,5-dimethylhydantoin (1), 12, 158–159. Supplier: Aldrich.

Regioselective bromination.[1] This reagent is superior to NBS for bromination of pyrrole to provide 2-bromopyrrole. After conversion to the N-boc derivative,

$$\text{pyrrole} + \mathbf{1} \xrightarrow{AIBN,\ THF} [\text{2-bromopyrrole}] \xrightarrow[85\%]{(CH_3)_3COCO)_2O,\ DMAP} \text{N-Boc-2-bromopyrrole}$$

$$\xrightarrow[70\%]{1)\ BuLi;\ 2)\ ClSi(CH_3)_3,\ THF,\ -78 \rightarrow 0^\circ} \text{N-Boc-2-}Si(CH_3)_3\text{-pyrrole}$$

lithium–halogen exchange provides N-boc-2-lithiopyrrole for subsequent reaction with electrophiles.

[1] W. Chen, E. K. Stephenson, and M. P. Cava, *Org. Syn.*, submitted (1989).

3,5-Di-*t*-butyl-1,2-benzoquinone, 3, 78–79.

$R_2CHNH_2 \rightarrow R_2C{=}O$.[1] Although this conversion was originally reported in 1969, it has not been used widely. A recent overview reports yields of 83–93% from both acyclic and cyclic amines. A further advantage is that the by-product, 2-amino-4,6-di-*t*-butylphenol, can be converted to the benzoquinone by electrochemical or dichromate oxidation in 55–65% yield.

[1] R. F. X. Klein, L. M. Bargas, and V. Horak, *J. Org.*, **53**, 5994 (1988).

Dibutylboryl trifluoromethanesulfonate (1).

Boron enolates of thioglycolates. Esters do not form boron enolates because of the low acidity of the α-protons. However, methyl phenylthioacetate (**2**) forms a boron enolate on treatment with Hunig's base and dibutylboryl triflate, and this enolate undergoes aldol reactions with aldehydes with high *syn*-diastereoselectivity.[1]

$$CH_3CHO + \underset{\mathbf{2}}{CH_2(SC_6H_5)COOCH_3} \xrightarrow[74\%]{\mathbf{1},\ \text{base, ether, } -78^\circ} CH_3CH(OH)CH(SC_6H_5)COOCH_3 + CH_3CH(OH)CH(SC_6H_5)COOCH_3 \quad (96:4)$$

[1] Y. Sugano and S. Naruto, *Chem. Pharm. Bull.*, **36**, 4619 (1988).

Di-*t*-butyl dicarbonate (t-Boc_2O).

NHCbo → NHBoc. This interconversion can be effected in high yield by hydrogenation catalyzed by Pd–C in CH_3OH in the presence of excess t-Boc_2O. It can also be effected by reaction with $(C_2H_5)_3SiH$ catalyzed by $Pd(OAc)_2$ and triethylamine in the presence of excess t-Boc_2O in C_2H_5OH.

[1] M. Sakaitani, K. Hori, and Y. Ohfune, *Tetrahedron Letters*, **29**, 2983 (1988).

2,6-Di-*t*-butyl-4-methylphenol.

Conjugate addition to α,β-unsaturated esters. Esters (BHT) of this phenol are too hindered to form ester enolates,[1] but α,β-unsaturated BHT esters do undergo conjugate addition of organolithium reagents,[2] as do esters (BHA) of 2,6-di-*t*-butyl-4-methoxyphenols, which are easily cleaved by CAN oxidation (**13**, 94–95). The observation that BHT esters are reduced by sodium in liquid NH_3 to the corresponding alcohol provides a synthesis of angularly fused triquinane pentalenene (**7**). Thus conjugate addition of **1** to **2** followed by *in situ* methylation furnishes **3**, which is reductively cleaved to the alcohol **4**. This is converted by known reactions to the enyne **5**, which undergoes cyclization when heated with $Co_2(CO)_8$ (**12**, 164) to a mixture of two triquinanes **6**, in which the product with the desired configuration at C_1 and C_9 predominates. Remaining steps to the pentalene **7** are known reactions.

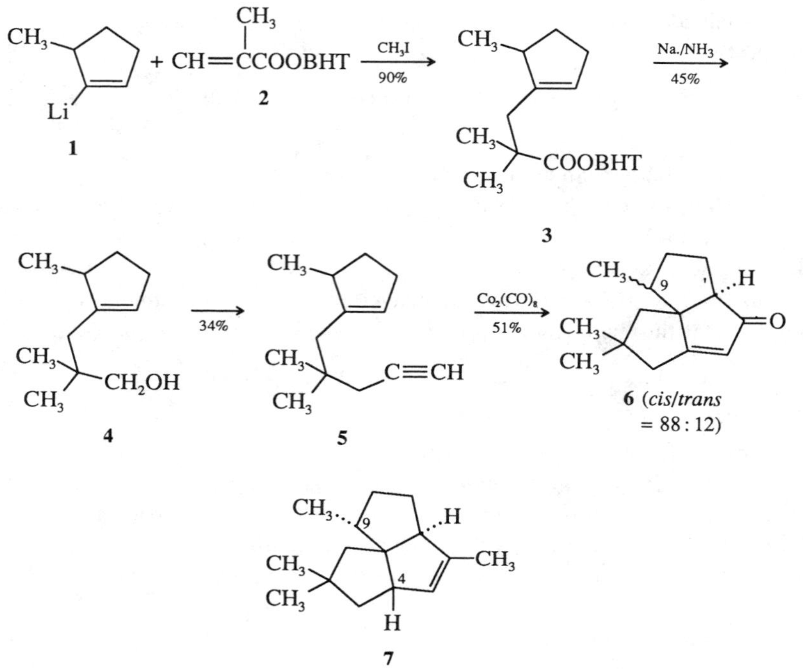

[1] R. Häner, T. Laube, and D. Seebach, *Am. Soc.*, **107**, 5396 (1985).
[2] N. E. Schore and E. G. Rowley, *ibid.*, **110**, 5224 (1988).

2,2-Dibutyl-2-stanna-1,3-dithiane, Bu_2Sn (S–S ring) (**1**).

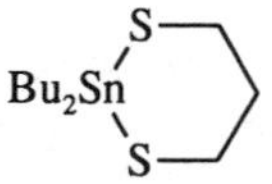

The reagent is prepared[1] in about 50% yield by reaction of propane-1,3-dithiol with Bu_2SnCl_2 and $N(C_2H_5)_3$.

Reaction with RCHO and acetals.[2] In the presence of dibutyltin bis(trifluoromethanesulfonate), **2**,[3] this reagent converts aldehydes or their dialkyl acetals or ethylene acetals into 2-alkyl-1,3-dithianes (70–85% yield). However, the reagent differentiates between aldehydes and acetals as shown in the examples. An aliphatic aldehyde reacts more readily than an ethylene acetal, whereas the selectivity is reversed in the case of an aromatic aldehyde. The reagent can also distinguish between a straight-chain aldehyde and a branched-chain aldehyde.

$$BuCHO + n\text{-}C_7H_{15}CH(OCH_2CH_2O) \xrightarrow{1,2} BuCH(SCH_2CH_2CH_2S) + n\text{-}C_7H_{15}CH(SCH_2CH_2CH_2S)$$

93:7

$$C_{10}H_7CHO + C_6H_5CH(OCH_2CH_2O) \longrightarrow C_{10}H_7CH(SCH_2CH_2CH_2S) + C_6H_5CH(SCH_2CH_2CH_2S)$$

3:97

$$BuCHO + (CH_3)_3CCHO \longrightarrow BuCH(SCH_2CH_2CH_2S) + (CH_3)_3CCH(SCH_2CH_2CH_2S)$$

80:20

[1] M. Wieber and M. Schmidt, *J. Organomet. Chem.*, **1**, 336 (1964).
[2] T. Sato, E. Yoshida, T. Kobayashi, J. Otera, and H. Nozaki, *Tetrahedron Letters*, **29**, 3971 (1988).
[3] M. Schmeisser, P. Sartori, and B. Lippsmeier, *Ber.*, **103**, 868 (1970).

3,4-Di-*t*-butylthiophene 1,1-dioxide (1).
Preparation:

$$(CH_3)_3CCOCH_2SCH_2COC(CH_3)_3 \xrightarrow[53\%]{TiCl_4,\ Zn,\ THF,\ 0^\circ} \text{3,4-dihydroxy-3,4-di-}t\text{-butyltetrahydrothiophene} \xrightarrow[83\%]{1)\ TsOH;\ 2)\ ClC_6H_4CO_3H} \mathbf{1}$$

1

1,2-Di-*t*-butylbenzenes.[1] This reagent undergoes a Diels–Alder reaction with phenyl vinyl sulfoxide, an acetylene equivalent (**8**, 399), with loss of benzenesulfenic acid to give *o*-di-*t*-butylbenzene in 77% yield.

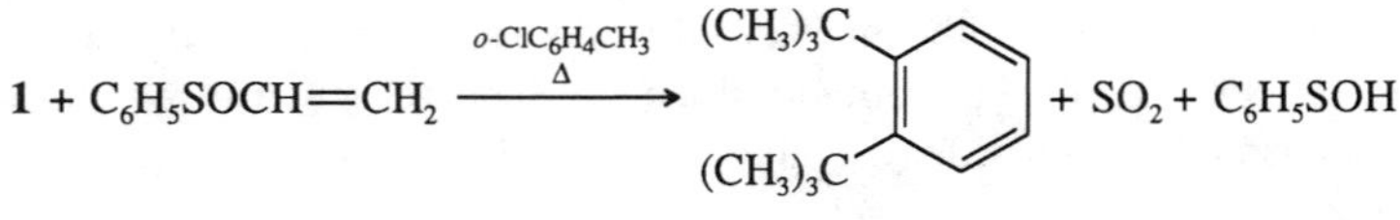

2 (77%)

Diels–Alder reactions of **1** with a variety of alkynes provides a route to various substituted *o*-di-*t*-butylbenzenes.

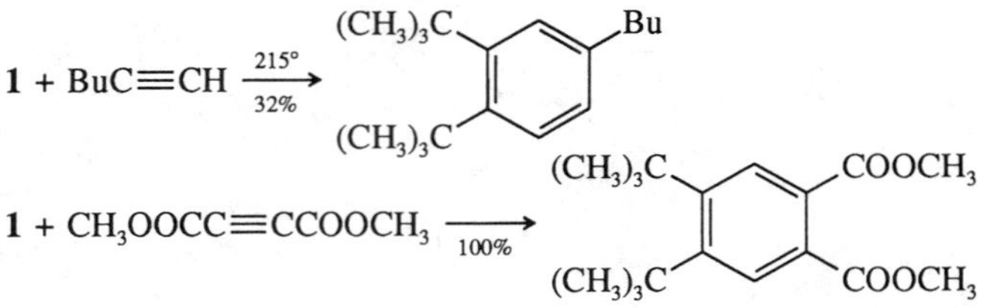

[1] J. Nakayama, S. Yamaoka, T. Nakanishi, and M. Hoshino, *Am. Soc.*, **110**, 6598 (1988).

Dibutyltin oxide.

Intramolecular Williamson reaction.[1] The final steps in a total synthesis of octosyl acid A (**1**) from ribose involve a cyclization of **2** to **3** via a 2′,3′-stannylene (**a**).

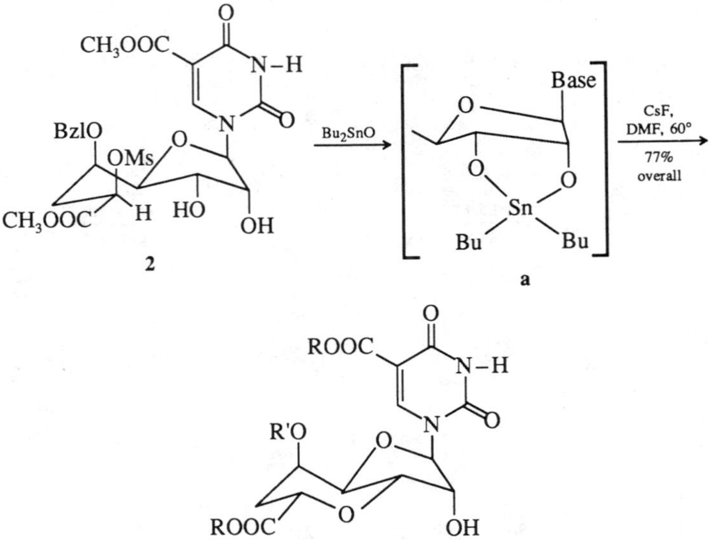

3, R = CH_3, R′ = Bzl
1, R = R′ = H

[1] S. J. Danishefsky, R. Hungate, and G. Schulte, *Am. Soc.*, **110**, 7434 (1988).

Dicarbonyl(cyclopentadienyl)[(dimethylsulfonium)methyl]iron tetrafluoroborate, $Cp(CO)_2FeCH_2\overset{+}{S}(CH_3)_2BF_4^-$ (**1**), air-stable, stable to storage (**13**, 98–99).

$$[(CO)_2FeCp]_2 \xrightarrow[70\%]{1)\ Na;\ 2)\ ClCH_2SCH_3;\ 3)\ CH_3I;\ 4)\ NaBF_4} \mathbf{1}$$

Cyclopropanation.[1] This stable methylene transfer reagent effects cyclopropanation of alkenes in generally high yield. It can effect selective cyclopropanation of 1,1-disubstituted double bonds.

CH3, COOCH3, CH3, CH2 $\xrightarrow{1}$ CH3, COOCH3, CH3 (86%) + $Cp(CO)_2FeS(CH_3)_2{}^+BF_4^-$

$$C_6H_5(CH_3)C{=}CH_2 \xrightarrow[96\%]{} C_6H_5(CH_3)\text{-cyclopropane}$$

[1] M. N. Mattson, E. J. O'Connor, and P. Helquist, *Org. Syn.*, submitted (1989).

Di-μ-carbonylhexacarbonyldicobalt, $Co_2(CO)_8$.

Alkyne–Co₂(CO)₆ complexes, $(RC{\equiv}CR)Co_2(CO)_6$.[1] These complexes (**11**, 162–163; **12**, 163–165; **14**, 117) can be prepared *in situ* and used for synthesis of cyclopentenones in moderate yield.

$$CoBr_2 + Zn + CO \xrightarrow{THF} \left[Co_2(CO)_8 \xrightarrow{HC\equiv CC_6H_{13}} HC{\equiv}CC_6H_{13}\cdot Co_2(CO)_6\right]$$

a

a + (norbornene) $\xrightarrow[88\%]{65°}$ (tricyclic cyclopentenone, O, C_6H_{13})

Cobalt-based annelation of silyl enol ethers (**10**, 129). A potential route to psuedoguaianolides involves the reaction of the ethoxy-substituted cation complex

1 with a silyl enol ether such as **2** followed by demetallation. A mixture of four products is obtained, all with complete *exo*-selectivity at C_2 adjacent to the carbonyl group. Two of the products are epimeric at C_7, and two are epimeric at $C_{1'}$.

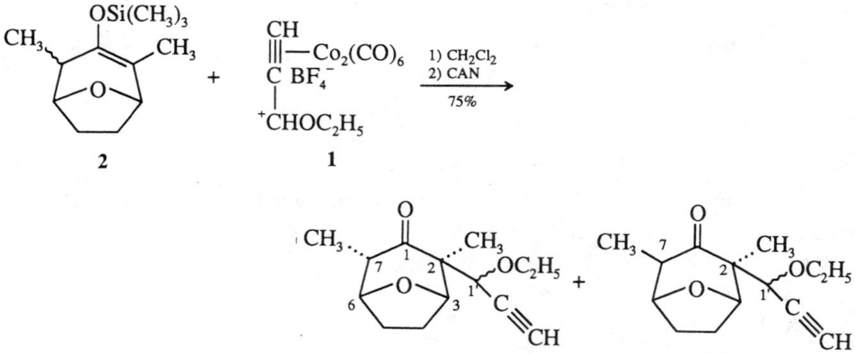

Siloxymethylation.[3] Glycosyl acetates undergo C-siloxymethylation on reaction with a hydrosilane and CO catalyzed by $Co_2(CO)_8$. Only the anomeric acetoxy group undergoes this reaction, which takes place from the side opposite to that of the acetoxy group at C_3. The reactive reagent is probably a silylcobalt carbonyl, $R_3SiCo(CO)_4$.

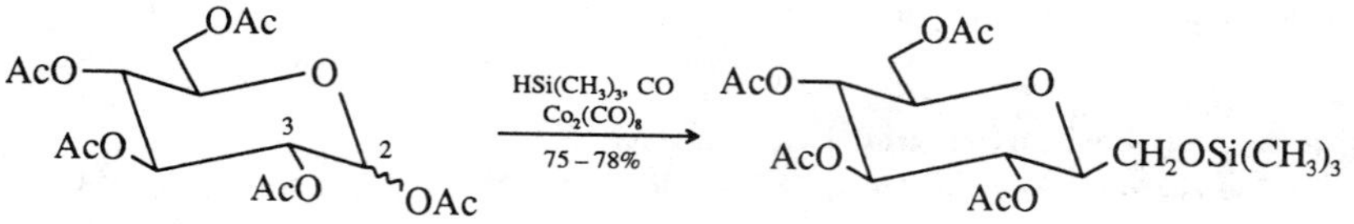

[1] A. Devasagayaraj and M. Periasamy, *Tetrahedron Letters*, **30**, 595 (1989).
[2] A. M. Montaña, K. M. Nicholas, and M. A. Khan, *J. Org.*, **53**, 5193 (1988).
[3] N. Chatani, T. Ikeda, T. Sano, N. Sonoda, H. Kurosawa, Y. Kawasaki, and S. Murai, *ibid.*, **53**, 3387 (1988).

Di-μ-carbonylhexacarbonyldicobalt–Methyl iodide.

Carbonylation of vinyl halides. These two reagents react to form tetracarbonylmethylcobalt, $CH_3Co(CO)_4$. In the presence of hydroxide ion, this organometallic and carbon monoxide convert vinyl bromides or chlorides into the corresponding carboxylic acids.

H H (E) or (Z) C_6H_5 Br $\xrightarrow{CH_3Co(CO)_4}$ [H H C_6H_5 $Co(CO)_3COOH$ CH_3] $\xrightarrow[94\%]{CO, Ca(OH)_2}$ C_6H_5 H H COOH (E)

[1] M. Miura, K. Okuro, A. Hattori, and M. Nomura, *J.C.S. Perkin I*, 73 (1989).

Dichlorobis(cyclopentadienyl)hafnium–Silver perchlorate (1).

O-Aryl glycoside → C-aryl glycoside.[1] The reaction of a glycosyl fluoride with phenol or 2-naphthol at −78° in CH_2Cl_2 in the presence of a Lewis acid results in O-glycosidation. As the temperature increases to 0° the O-glycosidation rearranges to a C-glycoside. Cp_2HfCl_2–$AgClO_4$ (or Cp_2ZrCl_2–$AgClO_4$) is particularly effective for this rearrangement. Use of other Lewis acids permits isolation of both O- and C-glycosides (equation I). In the case of glycosidation with β-naphthol in the presence of the hafnium activator, only the C-glycoside is isolated (equation II).

(I) CH_3O O F CH_3O OCH_3 + C_6H_5OH $\xrightarrow[-78 \rightarrow 5°]{BF_3 \cdot O(C_2H_5)_2}$

CH_3O O OC_6H_5 CH_3O OCH_3 (45%; α/β = 5:1) + CH_3O O OH CH_3O OCH_3 (28%)

(II)

$Cp_2HfCl_2 - AgClO_4$, CH_2Cl_2, $-78 \rightarrow -5°$, 92%

($\beta/\alpha = 9:1$)

[1] T. Matsumoto, M. Katsuki, and K. Suzuki, *Tetrahedron Letters*, **29**, 6935 (1988).

Dichlorobis(cyclopentadienyl)titanium–Magnesium

Reduction of Cp_2TiCl_2 with 1 equiv. of Mg in THF results in a green substance (**1**), which may be a THF adduct of Cp_2Ti.

Deoxygenation. This substance reduces epoxides to alkenes in 85–95% yield, generally with retention of configuration, particularly in reduction of *trans*-epoxides. This reduction proceeds more readily than that of carbonyl or ester groups, but α,β-enals are coupled reductively by **1** at 25° to diallylic ethers.

[1] R. Schobert, *Angew. Chem. Int. Ed.*, **27**, 855 (1988).

Dichlorobis(cyclopentadienyl)zirconium, Cp_2ZrCl_2 (1).

2,6-Dialkylphenols. These phenols are obtained by a cross-aromatization of cyclohexanones with aldehydes catalyzed by this zirconium complex. Cp_2TiCl_2 and Cp_2HfCl_2 are also effective catalysts.[1]

$$\text{cyclohexanone} + 2\,RCHO \xrightarrow{\mathbf{1},\ 150°} \text{2,6-}(RCH_2)_2C_6H_3OH$$

$R = C_2H_5$ 69%
$R = CH(CH_3)_2$ 16%

$$\text{3-methylcyclohexanone} + 2CH_3(CH_2)_3CHO \xrightarrow[40\%]{} \text{2,6-bis}(CH_3(CH_2)_4)\text{-3-methylphenol}$$

$$\text{cyclohexanone} + CH_3(CH_2)_3CHO \xrightarrow{Cp_2ZrH_2} \text{2-}[=CH(CH_2)_3CH_3]\text{cyclohexanone}$$

$$\downarrow CH_3CH_2CHO,\ \mathbf{1}$$

$$\text{2-}CH_3(CH_2)_2\text{-6-}(CH_2)_4CH_3\text{-phenol (OH)}$$

Glycosidation.[2] The ready access of glycosyl fluorides, thanks to DAST (**12**, 183) and pyridinium poly(hydrogen fluoride) (**12**, 419), has led to a search for suitable activators for use of these fluorides in glycosidation (**12**, 466; **13**, 302). Although the metallocenes of Zr and Hf are inert as such, in the presence of $AgClO_4$ (1 equiv.), they effect rapid glycosidation of a typical glycosyl fluoride at $-20°$ to 0° with a variety of alcohols in generally high yields. The stereoselectivity can be controlled to some extent by the metal and the sugar and to a greater extent by the solvent. A further advantage of this system is that even highly hindered alcohols can undergo this glycosidation.

[1] T. Nakano, H. Shirai, H. Tamagawa, Y. Ishii, and M. Ogawa, *J. Org.*, **53**, 5181 (1988).
[2] K. Suzuki, H. Maeta, T. Matsumoto, and G. Tsuchihashi, *Tetrahedron Letters*, **29**, 3567, 3571 (1988).

Dichlorobis(diphenylphosphinomethane)platinum(II), $PtCl_2$(dppm), **1**.

Thioethers; thioacetals.[1] In the presence of this catalyst and a weak base, thiols react with alkyl iodides to form thioethers. Under the same conditions thiols condense with 1,1-dihalides to form thioacetals.

$$HS(CH_2)_3SH + 2C_2H_5I \xrightarrow[92\%]{\mathbf{1},\ Na_2CO_3,\ \text{acetone}} C_2H_5S(CH_2)_3SC_2H_5$$

$$\downarrow CH_2I_2,\ \mathbf{1},\ Na_2CO_3\ (70\%)$$

$$\text{1,3-dithiane}$$

[1] P. C. Bulman Page, S. S. Klair, M. P. Brown, M. M. Harding, C. S. Smith, S. J. Maginn, and S. Mulley, *Tetrahedron Letters*, **29**, 4477 (1988).

Dichloro(1,4-bisdiphenylphosphinopropane)nickel, $Cl_2Ni(dppp)$ (**1**).

Coupling of alkenylsulfoximines with R_2Zn. The chiral alkenyl sulfoximines **2**, obtained by reaction of a ketone with enantiomerically pure (R)-Li-$CH_2SO(NMe_2)C_6H_5$ (**3**, 105–106; **5**, 231) in the presence of a metallic salt ($MgBr_2$, LiBr, $ZnCl_2$) undergo highly diastereoselective coupling with R_2Zn or Ar_2Zn in the presence of this nickel catalyst. Coupling of **2** with $[RO(CH_2)_4]_2Zn$ is equally diastereoselective, but the yield is lower because of reduction by the dialkylzinc.

$$\mathbf{2}\ (C_6H_5(CH_3)_2N\text{-}S(=O)\text{-}CH{=}CR^1R^2) + (C_6H_5)_2Zn \xrightarrow[\sim 85\%]{1,\ MgBr_2,\ ether} C_6H_5CH{=}CR^1R^2\ (\sim 99\%\ de)$$

[1] I. Erdelmeier and H.-J. Gais, *Am. Soc.*, **111**, 1125 (1989).

Dichloro[bis(1,2-diphenylphosphino)ethane]nickel, $NiCl_2(dppe)$.

Cross-coupling of RMgX with alkadienyl sulfides.[1] This coupling is the final step in a new route to (E,Z)- or (Z,E)-1,3-dienes. The first step involves addition of (Z)-dialkenylcuprates[2] to phenylthioacetylene to provide alkadienyl sulfides. The coupling proceeds with retention of configuration.

$$\underset{\mathbf{1}}{C_6H_5SC{\equiv}CH} + (C_2H_5CH\overset{Z}{=}CH)_2CuLi \xrightarrow[85\%]{} \underset{\mathbf{2}}{C_2H_5CH\overset{Z}{=}CHCH\overset{E}{=}CHSC_6H_5}$$

$$\mathbf{2} + EEO(CH_2)_6MgBr \xrightarrow[78\%]{NiCl_2(dppe),\ 25^\circ} \underset{(7E,9Z,\ 97\%)}{C_2H_5CH{=}CHCH{=}CH(CH_2)_6OEE}$$

[1] V. Fiandanese, G. Marchese, F. Naso, L. Ronzini, and D. Rotunno, *Tetrahedron Letters*, **30**, 243 (1989).
[2] M. Gardette, N. Jabri, A. Alexakis, and J. F. Normant, *Tetrahedron*, **40**, 2741 (1984).

Dichloro{1,1¹-[bis(diphenylphosphine)]ferrocene}palladium, $PdCl_2(dppf)$.

Coupling of B-alkyl-9-BBN with ArX and vinyl halides.[1] This coupling can be effected with $PdCl_2(ddpf)$ as catalyst and a base, NaOH, K_2CO_3, or K_3PO_4 (3 equiv.). A number of B-alkylboranes are known to undergo this coupling, but couplings with B-alkyl-9-BBN have been examined most thoroughly.

$$CH_3(CH_2)_5CH{=}CH_2 \xrightarrow[70-90\%]{\text{1) 9-BBN; 2) ArBr, NaOH, THF, 65°; Pd(II)}} Ar(CH_2)_7CH_3$$

$$CH_3(CH_2)_5CH{=}CH_2 \xrightarrow[85\%]{\text{1) 9-BBN; 2) (E)-}C_6H_5CH{=}CHBr\text{, Pd(II)}} (E)\text{-}C_6H_5CH{=}CH(CH_2)_7CH_3$$

This coupling can also effect cyclization of bromoalkenes:

$$o\text{-}BrC_6H_4CH_2CH{=}CH_2 \xrightarrow[86\%]{\text{1) 9-BBN, THF, 0°; 2) Pd(II), NaOH}} \text{indan}$$

$$C_6H_5CHBrCH{=}CHCH{=}CH_2 \xrightarrow{67\%} \text{1-}C_6H_5\text{-cyclopentene}$$

[1] N. Miyaura, T. Ishiyama, H. Sasaki, M. Ishikawa, M. Satoh, and A. Suzuki, *Am. Soc.*, **111**, 314 (1989).

Dichlorobis(triphenylphosphine)palladium(II).

Cyclopentenones. Negishi and co-workers[1] have extended their carbonylation of dienyl iodides (**13**, 103–104) to carbonylation of unsaturated allylic halides as a route to cyclopentenones. Thus carbonylation of the dienyl chloride **2** catalyzed by $Cl_2Pd[P(C_6H_5)_3]_2$ (**1**) results in the cyclopentenone **3**, probably via the intermediate **a**.

$$\mathbf{2} \xrightarrow{CO,\ N(C_2H_5)_3,\ CH_3OH;\ \mathbf{1},\ CH_3CN/C_6H_6,\ 100°} [\mathbf{a}] \xrightarrow{90\%} \mathbf{3}$$

(**2**: n-C_5H_{11}(H)C=C(CH_2Cl)$CH_2CH{=}CH_2$; **a**: 3-(n-C_5H_{11}CH=)-cyclopentanone bearing CH_2COOCH_3; **3**: n-C_6H_{13}-cyclopentenone bearing CH_2COOCH_3)

Lower yields obtain in carbonylation of dienyl chlorides such as **4**.

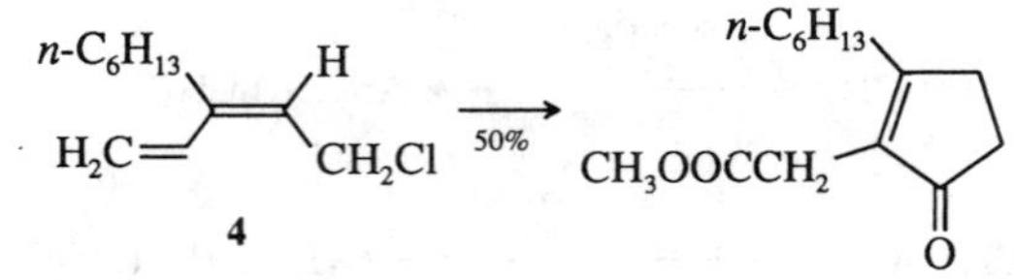

Coupling of $R^1Br(Cl)$ with R^2SnBu_3.[2] Alkyl halides couple with 2-(tributylstannyl-4,4-dimethyl-2-oxazoline (**2**) in the presence of this Pd catalyst (equation I). Surprisingly, aroyl chlorides react with **2** in the absence of a catalyst to give bis(N-aroyl-4,4-dimethyl-2-oxazolinylidenes) in high yield (equation II).

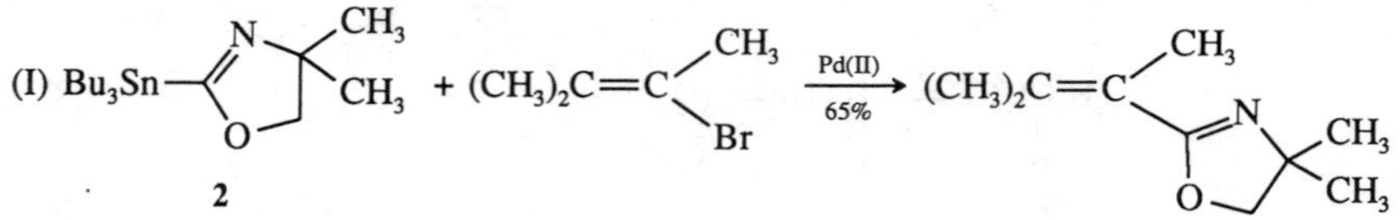

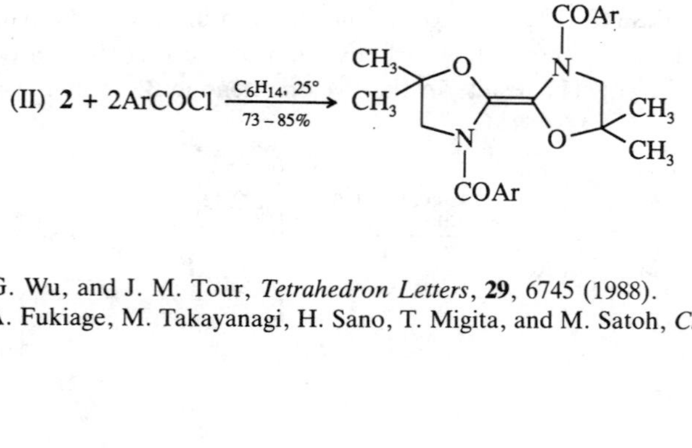

[1] E. Negishi, G. Wu, and J. M. Tour, *Tetrahedron Letters*, **29**, 6745 (1988).
[2] M. Kosugi, A. Fukiage, M. Takayanagi, H. Sano, T. Migita, and M. Satoh, *Chem. Letters*, 1351 (1988).

Dichloro(*p*-cymene)triphenylphosphineruthenium(II).

Isopropenyl esters of N-protected amino acids. In the presence of the Ru(II) catalyst, Cbo- or Boc-amino acids add to propyne at 100° to provide isopropenyl esters in 55–85% yield with no racemization. Unprotected amino acids do not add to the alkyne. These esters react with primary amines at room temperature in ethyl acetate to afford amides in 75–90% yield.

$$\text{Cbo-NHCH}_2\text{COOH} + \text{HC}\equiv\text{CCH}_3 \xrightarrow[76\%]{\text{Ru(II)}} \text{Cbo-NHCH}_2\overset{\text{O}}{\overset{\|}{\text{C}}}\text{O}\underset{\underset{\text{CH}_2}{\|}}{\text{C}}\text{CH}_3$$

$$\downarrow 91\%\ \ C_6H_5CH_2NH_2,\ AcOC_2H_5$$

$$\text{Cbo-NHCH}_2\overset{\text{O}}{\overset{\|}{\text{C}}}\text{—NHCH}_2\text{C}_6\text{H}_5$$

[1] C. Ruppin, P. H. Dixneuf, and S. Lecolier, *Tetrahedron Letters*, **29**, 5365 (1988).

Dichlorodicyanobenzoquinone (DDQ).

Oxidation of chiral esters of phenylacetic acids.[1] These esters can be converted into acetates of mandelic esters by oxidation with DDQ in acetic acid. The reaction is diastereoselective when carried out on esters of chiral alcohols, of which 8-phenylmenthol is the most useful. The presence of substituents on the phenyl group has slight effect on the diastereoselectivity, which depends on formation of a donor–acceptor complex between the substrate and the quinone with removal of a hydrogen atom. The acetoxy group then enters from the opposite, more bulky face.

$$\underset{(R^* = \text{8-phenylmenthyl})}{RC_6H_4CH_2COOR^*} \xrightarrow[60-90\%]{\text{DDQ, HOAc, } 20^\circ} \underset{(62-67\%\ \text{de})}{RC_6H_4\overset{\overset{\text{OAc}}{|}}{\overset{*}{\text{C}}}\text{HCOOR}^*}$$

Aryl ketones or aldehydes.[2] Reaction of arylalkanes with DDQ in aqueous acetic acid can result in carbonyl products. Thus reaction of **1** with DDQ in HOAc–H_2O results in the ketone **2**, believed to be formed by acetylation at the benzylic position followed by acid-catalyzed hydrolysis to the corresponding alcohol, which is oxidized to a ketone by DDQ.

1 $\xrightarrow{\text{DDQ, HOAc, } H_2O,\ \Delta}$ [**a** (OAc)] $\xrightarrow[76\%]{1)\ H_3O^+\ \ 2)\ \text{DDQ}}$ **2**

This oxidation also oxidizes methylarenes to aryl aldehydes.

$$\text{6-}CH_3O\text{-2-methylnaphthalene} \xrightarrow{90\%} \text{6-}CH_3O\text{-2-naphthaldehyde (CHO)}$$

Alkylaryl azides.[3] Reaction of an alkylarene with DDQ and $N_3Si(CH_3)_3$ in $CHCl_3$ at 25° results in replacement of a benzylic hydrogen by an azide group to give an alkylaryl azide in 50–95% yield. Hydrazoic acid can be used in place of azidotrimethylsilane, but this reaction is generally slower.

$$p\text{-}CH_3OC_6H_4CH(C_2H_5)CH_2OH \xrightarrow[81\%]{DDQ,\ N_3Si(CH_3)_3,\ CHCl_3} p\text{-}CH_3OC_6H_4C(C_2H_5)(N_3)CH_2OH + DDHQ$$

Benzylic cyanation.[4] The combination of DDQ and $Si(CH_3)_3CN$ (**1**) in CH_2Cl_2 or CH_3CN effects benzylic cyanation of alkylarenes.

$$p\text{-}CH_3OC_6H_4C_2H_5 + \mathbf{1} \xrightarrow[69\%]{CH_3CN} p\text{-}CH_3OC_6H_4CH(CN)CH_3 + \text{(NC)}_2\text{Cl}_2C_6(OH)(OSi(CH_3)_3)$$

[1] A. Guy, A. Lemor, D. Imbert, and M. Lemaire, *Tetrahedron Letters*, **30**, 327 (1989).
[2] H. Lee and R. G. Harvey, *J. Org.*, **53**, 4587 (1988).
[3] A. Guy, A. Lemor, J. Doussot, and M. Lemaire, *Synthesis*, 900 (1988).
[4] M. Lemaire, J. Doussot, and A. Guy, *Chem. Letters*, 1581 (1988).

Dichloroketene.

Butyrolactones.[1] Reaction of dichloroketene with an N-protected 3-sulfinylindole provides a synthesis of an indoline-fused butyrolactone in 35–40% yield

$$\text{(I)}\quad \text{1-}C_6H_5SO_2\text{-3-(RS(=O))indole} \xrightarrow{Cl_2C=C=O} \text{SR, H, Cl, Cl furoindolinone (N-}C_6H_5SO_2\text{)} \xrightarrow[35-40\%\ \text{overall}]{Al(Hg)} \text{SR, H furoindolinone (N-}C_6H_5SO_2\text{)}$$

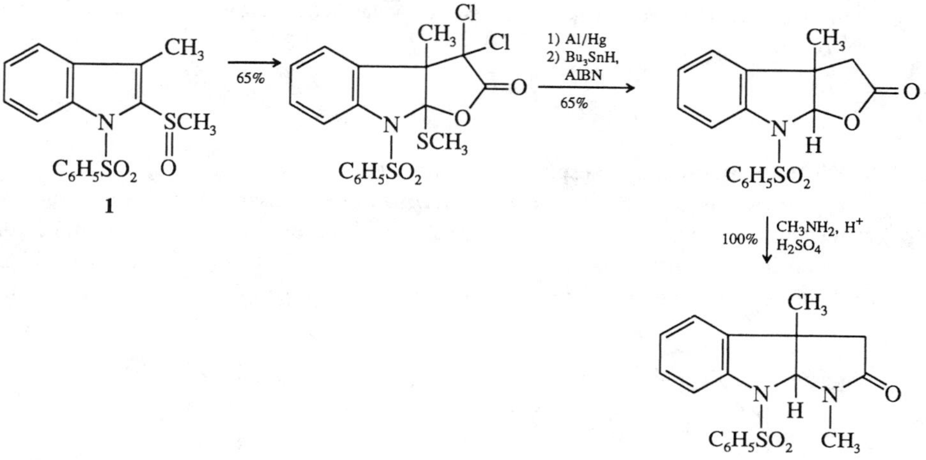

(equation I). A similar reaction of the 2-sulfinylindole **1** provides an intermediate to the system (**2**) of physostigmine alkaloids.

[1] J. P. Marino, M.-W. Kim, and R. Lawrence, *J. Org.*, **54**, 1782 (1989).

1,1-Dichloro-1-phenyl-2,2,2-trimethyldisilane, $C_6H_5Cl_2SiSi(CH_3)_3$ (**1**). The reagent is prepared by reaction of $(CH_3)_3Si\text{-}Si(C_6H_5)_3$ with hydrogen chloride and $AlCl_3$.[1]

1,4-Disilylation of enones.[2] In the presence of $Pd[P(C_6H_5)_3]_4$, this disilane undergoes 1,4-addition to α,β-enones to give γ-(phenyldichlorosilyl) silyl enol ethers, which can be converted into lithium enolates by exchange with methyllithium. The reaction can provide β-hydroxy ketones. The Michael addition is enantioselective when catalyzed by $Cl_2Pd[(+)\text{-BINAP}]$ (**12**, 53–57).

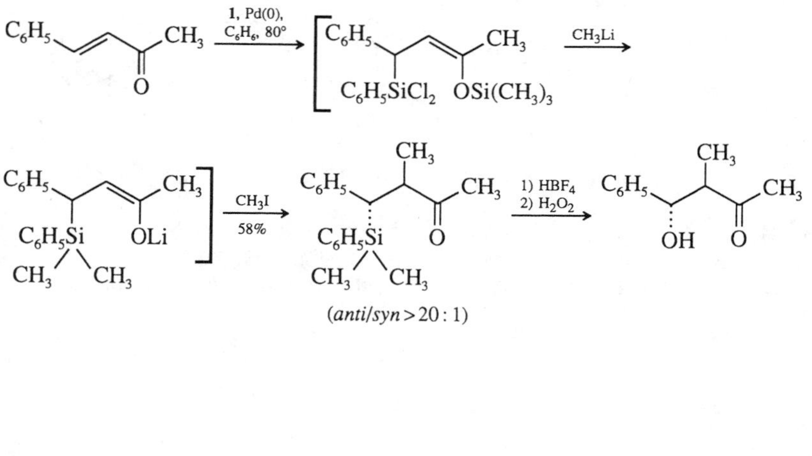

(*anti/syn* > 20 : 1)

[1] E. Hengge, G. Bauer, E. Brandstätter, and G. Kollmann, *Monatsh. Chem.*, **106**, 887 (1975).
[2] T. Hayashi, Y. Matsumoto, and Y. Ito, *Tetrahedron Letters*, **29**, 4147 (1988); *idem.*, *Am. Soc.*, **110**, 5579 (1988).

Dichlorotris(triphenylphosphine)ruthenium (1).

Intramolecular Kharasch[1]-type cyclizations.[2] Unsaturated α,α-dichloro esters or carboxylic acids undergo cyclization in the presence of various transition metal catalysts, particularly $Cl_2Ru[P(C_6H_5)_3]_3$ or $Cl_2Fe[P(OC_2H_5)_3]_3$.[3] Thus ethyl 2,2-dichloro-6-heptenoate (**2**) in the presence of **1** cyclizes rapidly to the *cis*-1,2-disub-

(I) **2** → (1, 155°; 60–70%) **3** + (3–5%)

7 hrs. 3(*cis*/*trans* = 1 : 1)
16 hrs. (" = 1 : 4)

(II) **4** → (1, 155°; 94%) **5** + (4%)

stituted cyclopentane, and epimerizes slowly to the *trans*-isomer. The corresponding acid (**4**) cyclizes to the *cis*-bicyclic lactone (**5**) under the same conditions. In both cases only a trace of cyclohexyl products are formed.

This radical cyclization can provide bridged and fused cyclic systems.

Example:

→ (1, 155°) (62%) + (12%)

[1] C. Walling and E. S. Huyser, *Org. React.*, **13**, 91 (1963).
[2] T. K. Hayes, R. Villani, and S. M. Weinreb, *Am. Soc.*, **110**, 5533 (1988).
[3] S. D. Ittel, A. D. English, C. A. Tolman, and J. P. Jesson, *Inorg. Chim. Acta*, **33**, 101 (1979).

9,10-Dicyanoanthracene (DCA).

Triplex catalysis.[1] 1,3-Cyclohexadiene does not undergo cycloadditions under usual conditions although it is known to dimerize on irradiation in the presence of a sensitizer. Actually, 1,3-cyclohexadiene can undergo both [2+2]- and [4+2]cycloaddition with a silyl enol ether such as 1-trimethylsiloxy-1-phenylethylene (**1**). Similar results obtain with trimethylsilylketene acetal. These sensitized

$CH_2{=}C(C_6H_5)OSi(CH_3)_3$ (**1**) + 1,3-cyclohexadiene $\xrightarrow[80\%]{DCA,\ h\nu,\ C_6H_6,\ 25°}$ **2** (*exo*/*endo* = 1:1) + **3** (*exo*/*endo* = 1:1) [+ dimer]; 3:1

cycloadditions involve a complex (triplex) composed of the sensitizer, the diene, and the dienophile. When 1,4-dicyanonaphthalene is used as the sensitizer in the reaction with **1**, the ratio of [4+2]adducts decreases relative to the [2+2]adducts; essentially only [2+2]adducts form when benzophenone is the sensitizer.

Photocyclization of enones substituted by an α-silylamino group. Irradiation (>320 nm) of these substrates with DCA as sensitizer results in N-heterocycles.

$(CH_3)_3SiCH_2N(Bzl)CH_2CH_2$-cyclohexenone $\xrightarrow[78\%]{h\nu,\ DCA,\ CH_3OH\ (N_2)}$ spiro N-Bzl pyrrolidine ketone

4-[2-(N-CH_3-N-$CH_2Si(CH_3)_3$)aminoethyl]cyclohexenone $\longrightarrow$ N-CH_3 octahydroisoquinolinone (71%) + N-CH_3 octahydroindolone (13%)

[1] N. Akbulut and G. B. Schuster, *Tetrahedron Letters*, **29**, 5125 (1988).
[2] W. Xu, Y. T. Jeon, E. Hasegawa, U. C. Yoon, and P. E. Mariano, *Am. Soc.*, **111**, 406 (1989).

Dicyanobenzene (DCB).

Photosensitized cyclobutanation.[1] Conjugated dienes and enamides undergo photosensitized cross addition to provide cycloadducts, which can be hydrolyzed to vinyl cyclobutylamines. These secondary amines undergo a base-catalyzed rearrangement to give cyclohexenes corresponding to the Diels–Alder adducts of the diene and the enamide.

CH_3 (E,E) diene + $CH_2=CH-N(CH_3)-C(=O)CH_3$ $\xrightarrow[41\%]{h\nu,\ DCB,\ CH_3CN}$ cyclobutane (CH_3, H, $NCOCH_3$ (CH_3), H, $CH_3CH=CH$) (3.5 : 1)

$\xrightarrow[85\%]{KOH,\ 1\text{–}110^\circ,\ NH_2NH_2,\ HOCH_2CH_2OH}$ cyclobutane (CH_3, H, $NHCH_3$, H, $CH_3CH=CH$) (3.2 : 1)

$\xrightarrow[91\%]{KH,\ THF,\ \Delta}$ cyclohexene (CH_3, $NHCH_3$, CH_3) (*endo*)

[1] N. L. Bauld, B. Harirchian, D. W. Reynolds, and J. C. White, *Am. Soc.*, **110**, 8111 (1988).

Dicyclohexylborane.

RCH_2CN. These primary nitriles can be obtained in 70–98% yield by hydroboration of a terminal alkene with dicyclohexylborane followed by addition of an excess of CuCN and $Cu(OAc)_2$ and a catalytic amount of $Cu(acac)_2$, which is not essential but which markedly improves the yield. The overall process effects *anti*-Markownikoff hydrocyanation of 1-alkenes.[1]

$$R^1R^2C{=}CH_2 \xrightarrow[0^\circ]{R_2BH,\ THF} [R^1R^2CH{-}CH_2BR_2] \xrightarrow[70-98\%]{1)\ CuCN,\ 0\rightarrow 20^\circ;\ 2)\ Cu(OAc)_2,\ Cu(acac)_2} R^1R^2CHCH_2CN$$

[1] Y. Masuda, M. Hoshi, and A. Arase, *J.C.S. Chem. Comm.*, 266 (1989).

Dicyclohexylcarbodiimide–Dimethylaminopyridine (**13**, 107–108).

Macrolactonization. This version of Steglich esterification was developed specifically for lactonization of **1** to (+)-colletodiol **2**.[1]

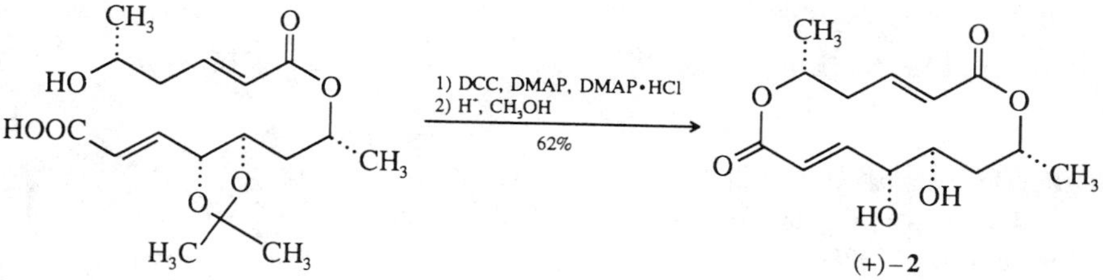

CH_3 O HO O HOOC CH_3 O O H_3C CH_3

1

1) DCC, DMAP, DMAP•HCl
2) H^+, CH_3OH
62%

CH_3 O O O O CH_3 HO OH

(+)-**2**

[1] G. E. Keck, E. P. Boden, and M. R. Wiley, *J. Org.*, **54**, 896 (1989).

Di(cyclooctadiene)nickel(0), Ni(COD)$_2$.

[4+4]*Cycloadditions.* This reaction provides an enantioselective synthesis of a cyclooctane-containing terpenoid (**5**). The optically active precursor (**3**) was obtained by reduction of the *t*-alkyl alkynyl ketone **1** with lithium aluminum hydride

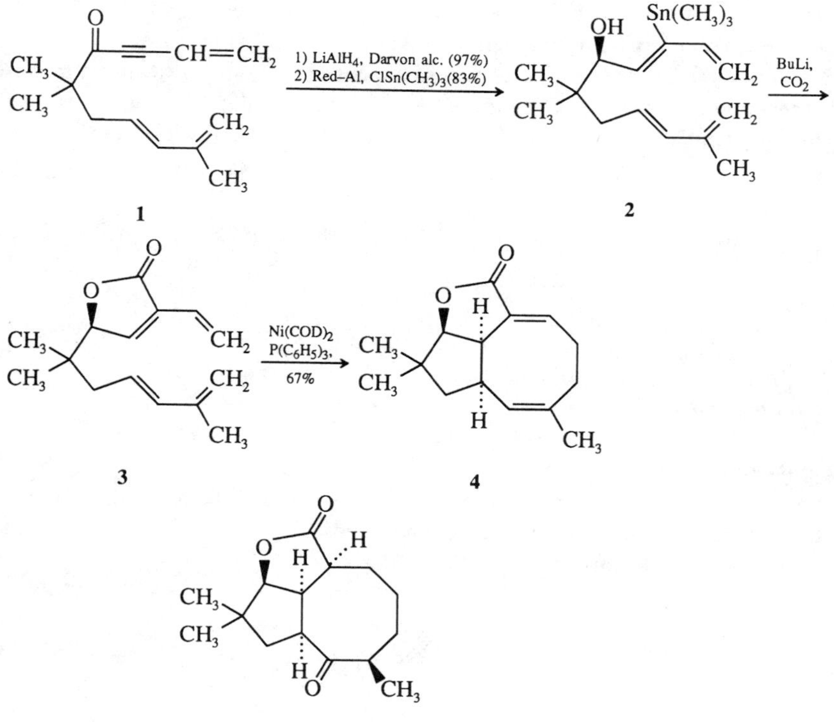

O CH=CH$_2$ CH_3 CH_3 CH_2 CH_3

1

1) $LiAlH_4$, Darvon alc. (97%)
2) Red-Al, $ClSn(CH_3)_3$ (83%)

OH $Sn(CH_3)_3$ CH_3 CH_3 CH_2 CH_2 CH_3

2

BuLi, CO_2

O O CH_3 CH_3 CH_2 CH_2 CH_3

3

$Ni(COD)_2$
$P(C_6H_5)_3$,
67%

O O H CH_3 CH_3 H CH_3

4

O O H H CH_3 CH_3 H O CH_3

5

modified by Darvon alcohol in >98% ee. The alcohol was converted to the bis-diene **3**, which undergoes nickel-catalyzed intramolecular cyclization to **4** in 67% yield. Remaining steps to **5** include conjugate reduction (copper hydride), hydroboration, and oxidation.

[1] P. A. Wender, N. C. Ihle, and C. R. D. Correia, *Am. Soc.*, **110**, 5904 (1988).

Dicyclopentadienylyttrium chloride, Cp_2YCl. Preparation.[1]

Fulvenes.[2] Cp_2YCl reacts with aldehydes or ketones in DME at 80° to form fulvenes in 60–95% yield.

$$C_2H_5C(=O)CH_3 + Cp_2YCl \xrightarrow{91\%} C_5H_4{=}C(CH_3)(C_2H_5)$$

[1] R. E. Maginn, S. Manastyrskyj, and M. Dubeck, *Am. Soc.*, **85**, 672 (1968).
[2] C. Qian and A. Qiu, *Tetrahedron Letters*, **29**, 6931 (1988).

Dicyclopentylboryl trifluoromethanesulfonate, $(c\text{-}C_5H_9)_2BOTf$ (**1**).

trans-*Glycidic esters.* The Darzens condensation[1] usually provides a mixture of *trans-* and *cis-*products. Only the *trans-*esters are obtained by condensation of the dicyclopentylboryl enolate of α-bromo-*t*-butylthioacetate with aromatic aldehydes.[2]

$$ArCHO + BrCH_2C(=O)SC(CH_3)_3 \xrightarrow[62-74\%]{1, i\text{-}Pr_2NC_2H_5} Ar\text{-}(epoxide)\text{-}C(=O)SC(CH_3)_3$$

[1] M. Ballester, *Chem. Revs.*, **55**, 283 (1955).
[2] R. P. Polniaszek and S. E. Belmont, *Syn. Comm.*, **19**, 221 (1989).

Diethyl aminomalonate.

[3 + 2]*Cycloaddition to pyrrolidines.* The reaction of diethyl aminomalonate with formaldehyde can generate an azomethine ylid (**a** ↔ **b**) which cycloadds to 1,2-dipolarophiles to form pyrrolidines.

$$H_2NCH(COOC_2H_5)_2 \xrightarrow{HCHO,\ C_6H_5CH_3} [\underset{\mathbf{a}}{CH_2{=}NCH(COOC_2H_5)_2} \rightleftharpoons$$

$$CH_2{=}\overset{+}{N}H\overset{-}{C}(COOC_2H_5)_2]$$

b

$\downarrow$ 84%, $CH_2{=}CHCOOC_2H_5$, Δ

(pyrrolidine: 4-C_2H_5OOC, 2,2-bis($COOC_2H_5$), N–H)

[1] S. Husinec, V. Savic, and A. E. A. Porter, *Tetrahedron Letters*, **29**, 6649 (1988).

Diethyl oxalate, $(C_2H_5OOC)_2$ (**1**).

Ethyl 2-oxo-3-alkenoates, $RCH{=}CHC(=O)COOC_2H_5$.[1] Vinyl Grignard reagents react with diethyl oxalate in THF/ether (1:1) at $-80°$ by 1,2-addition to only one ester group of diethyl oxalate to form the ester of 2-oxo-3-alkenoic acids.

$$CH_3CH{=}CHMgBr + \mathbf{1} \xrightarrow[73\%]{-80°} CH_3CH{=}CHC(=O){-}C(=O)OC_2H_5$$

[1] M. Rambaud, M. Bakasse, G. Duguay, and J. Villieras, *Synthesis*, 564 (1988).

4-(Diethylphosphono)-2-butenal, $(C_2H_5O)_2P(=O)CH_2CH{=}CHCHO$ (**1**, E/Z = 9:1).
Preparation:

$$(C_2H_5O)P(O)CH_2CH{=}CHCH_2OH \xrightarrow[70\%]{CrO_3,\ H_2SO_4} \mathbf{1} \xrightarrow[100\%]{C_6H_{11}NH_2,\ MS} (C_6H_5O)_2P(=O)CH_2CH{=}CHCH{=}NC_6H_{11}\text{-}c$$

2

Dienals.[1] The cyclohexylimine (**2**) of **1** reacts with carbonyl compounds in THF to give (E,E)-dienals in 65–80% yield. The reaction of **2** with α,β-enals to provide trienals proceeds in generally lower yield (25–45%).

[1] T. Rein, B. Åkermark, and P. Helquist, *Acta Chem. Scand.*, **42B**, 569 (1988).

Diethyl thioxomalonate, $S{=}C(COOC_2H_5)_2$ (**1**). This reagent can be generated *in situ* from the Bunte salt (**2**) prepared by reaction of diethyl chloromalonate with sodium thiosulfate.

$$(C_2H_5OOC)_2CHCl + Na_2S_2O_3 \longrightarrow \underset{\mathbf{2}}{(C_2H_5OOC)_2CHSSO_3^-Na^+}$$

Annelation of* 1,3-*dienes.[1] The salt **2** undergoes cycloaddition with 1,3-dienes to provide six-membered adducts which undergo ring contraction to cyclopentenes when treated with LDA or $KN[Si(CH_3)_3]_2$ and CH_3I. The ring contraction is equivalent to a [1,2]Wittig rearrangement.

Examples:

2-methyl-1,3-butadiene + 1 —(82%)→ 4-methyl-3,6-dihydro-2H-thiopyran-2,2-dicarboxylate ($COOC_2H_5$, $COOC_2H_5$) + 5-methyl isomer (3 : 1)

—($KN[Si(CH_3)_3]_2$, −64°; CH_3I; 90%)→ cyclopentene with $COOC_2H_5$, $COOC_2H_5$, CH_3, SCH_3

2-(trimethylsilyloxy)-1,3-butadiene [$(CH_3)_3SiO$] + 1 —(79%)→ tetrahydrothiopyranone with $COOC_2H_5$, $COOC_2H_5$ —(LDA, HMPT; CH_3I; 70%)→ cyclopentanone with $COOC_2H_5$, $COOC_2H_5$, SCH_3

[1] S. D. Larsen, *Am. Soc.*, **110**, 5932 (1988).

Dihalomethyllithium.

Transmetallation.[1] X_2CHLi reagents have limited use because of thermal instability even at −100°. However, transmetallation of Cl_2CHLi with $Ti(O\text{-}i\text{-}Pr)_4$ provides the ate complex $Cl_2CHTi(O\text{-}i\text{-}Pr)_4Li$ (**1**), which is stable at 0°. Moreover the ate complex, unlike the carbene, shows considerable group selectivity for aldehydes and α-amino ketones (equations I and II).

$$\text{(I)}\quad C_6H_5CHO + C_6H_5COCH_3 \xrightarrow[76-80\%]{1,\ 0^\circ} C_6H_5CH(OH)CHCl_2$$

$$\text{(II)}\quad CH_3C(=O)CH_2N(CH_3)_2 + C_2H_5COC_2H_5 \xrightarrow[71\%]{1,\ -30^\circ} CH_3C(OH)(CHCl_2)CH_2N(CH_3)_2$$

In contrast, transmetallation of X_2CHLi with $ClTi(O\text{-}i\text{-}Pr)_3$ and especially with CuCl increases the instability.

[1] T. Kauffmann, R. Fobker, and M. Wensing, *Angew. Chem. Int. Ed.*, **27**, 943 (1988).

Dihydridotetrakis(triphenylphosphine)ruthenium, $RuH_2[P(C_6H_5)_3]_4$ (**1**).

Isomerization of alkynes.[1] This complex (**1**) or $IrH_5[P(i\text{-}Pr)_3]_2$ (**2**) in combination with Bu_3P catalyzes isomerization of 2-ynoic esters to (2E,4E)-dienoic esters at 80–100° in yields generally of about 80–90%.

[1] D. Ma and X. Lu, *Tetrahedron Letters*, **30**, 843 (1989).

Dihydrogen hexachloroplatinate(IV), $H_2PtCl_6\cdot 6H_2O$, **1**, 890.

Intramolecular hydrosilylation. The hydrodimethylsilyl ether (**2**) of a homopropargyl alcohol (**1**) undergoes intramolecular hydrosilylation catalyzed by H_2PtCl_6 to give a cyclic (E)-vinylsilane (**3**). This can be converted to a β-hydroxy ketone (**4**) by hydrogen peroxide oxidation or into (Z)-3-bromo-3-decene-1-ol (**5**) by bromination/bromodesilylation with inversion of geometry.

$$n\text{-}C_6H_{13}C{\equiv}C(CH_2)_2OH \ (\mathbf{1}) \xrightarrow{HN[SiH(CH_3)_2]_2} RC{\equiv}C(CH_2)_2OSi(CH_3)_2H \ (\mathbf{2})$$

$$\mathbf{2} \xrightarrow[\sim 70\%]{H_2PtCl_6,\ CH_2Cl_2,\ 25^\circ} \mathbf{3}$$

$$\mathbf{5} \xleftarrow[78\%\ \text{overall}]{1)\ Br_2,\ 2)\ KHF_2} \mathbf{3} \xrightarrow[72\%\ \text{overall}]{H_2O_2,\ KF,\ K_2CO_3} n\text{-}C_6H_{13}CH_2C(=O)(CH_2)_2OH \ (\mathbf{4})$$

3: cyclic CH_3—Si(CH_3)—O ring with H and R on the vinyl carbon; **5**: $n\text{-}C_6H_{13}$(H)C=C(Br)$(CH_2)_2OH$

[1] K. Tamao, K. Maeda, T. Tanaka, and Y. Ito, *Tetrahedron Letters*, **29**, 6955 (1988).

2,2'-Dihydroxy-1,1'-binaphthyl (1).

Resolution of sulfoximines.[1] Alkyl aryl sulfoximines (**2**) can be resolved by complex formation with (+)- or (−)-**1** (equation I).

$$\text{(I)}\quad ArS(=O)(=NH)\text{—}R + (+)\text{-}\mathbf{1} \xrightarrow[80\%]{C_6H_6,\ 25^\circ} (-)\text{-}\mathbf{2}\ (90\%\ ee) + (+)\text{-}\mathbf{1}\text{–}(+)\text{-}\mathbf{2}\ (1:1)$$

(±)-**2**, Ar = $m\text{-}CH_3C_6H_4$, R = CH_3

$$(+)\text{-}\mathbf{1}\text{–}(+)\text{-}\mathbf{2} \xrightarrow[70\%]{1)\ \text{crystallization},\ 2)\ NaOH} (R)\text{-}(-)\text{-}\mathbf{2}\ (100\%\ ee)$$

The efficiency of the resolution is dependent on the Ar and R groups, being highest when Ar is *m*-tolyl and R is CH_3 or C_2H_5. No complex is formed when Ar is C_6H_5 and R is CH_3 or from dialkyl sulfoximines. However, some dialkyl sulfoximines can be resolved with (+)- or (−)-**3**, 1,6-di(*o*-chlorophenyl)-1,6-diphenylhexa-2,4-diyne-1,6-diol. This reagent is available commercially as Chiral Host Ace.[2]

$$C_6H_5-C(OH)(C_6H_4Cl\text{-}o)-C\equiv C-C\equiv C-C(OH)(C_6H_4Cl\text{-}o)-C_6H_5$$

3

[1] K. Mori and F. Toda, *Chem. Letters*, 1997 (1988).
[2] Ube Industries, Ltd., Tokyo, Japan.

2,2′-Dihydroxy-3,3′-bis(triphenylsilyl)1,1-binaphthyl–Trimethylaluminum, 14, 46–47. These two reagents react to form the aluminum naphthoxide **1**. The binaphthol

[Structure: 3,3′-bis($Si(C_6H_5)_3$)-1,1′-binaphthyl-2,2′-dioxy–$AlCH_3$] (**1**)

is prepared[1] from 3,3′-dibromo-1-1′-binaphthol by conversion to the bis(triphenylsilyloxy) ether, which rearranges on treatment with *t*-butyllithium to 3,3′-triphenylsilyl-1,1′binaphthol (80–95% yield). Reaction of this binaphthol with $Al(CH_3)_3$ provides the naphthoxide **1**. Both optical isomers of the starting dibromobinaphthol are known, or racemic **1** can be resolved by a kinetic resolution with a chiral ketone.

Asymmetric ene reactions.[2] The reaction of alkenes with prochiral, electron-deficient aldehydes can furnish optically active homoallylic alcohols when catalyzed by (+)- or (−)-**1** and in the presence of activated 4-Å molecular sieves. In the absence of the sieves, a stoichiometric amount of the organoaluminum complex is essential for high chemical and optical yields.

$$C_6F_5CHO + CH_2{=}C(CH_3)SC_6H_5 \xrightarrow[88\%]{\text{(R)-1, 4-Å MS, } CH_2Cl_2,\ -78^\circ} C_6F_5CH(OH)CH_2C(=CH_2)SC_6H_5$$

(88% ee)

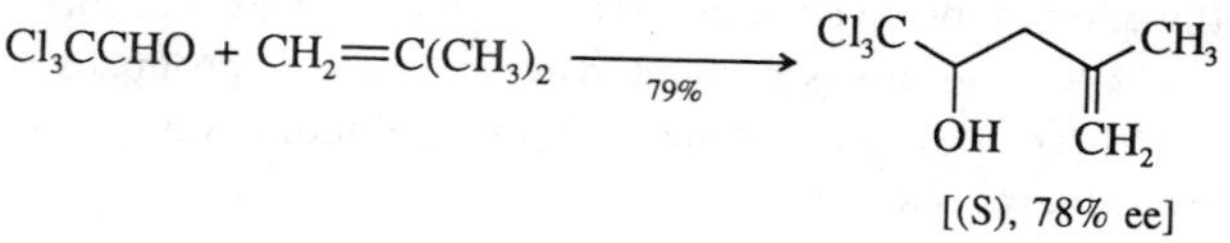

$Cl_3CCHO + CH_2{=}C(CH_3)_2$ —79%→ Cl_3C–CH(OH)–CH_2–C(=CH_2)–CH_3

[(S), 78% ee]

[1] K. Maruoka, T. Itoh, T. Shirasaka, and H. Yamamoto, *Am. Soc.*, **110**, 310 (1988).
[2] K. Maruoka, Y. Hoshino, T. Shirasaka, and H. Yamamoto, *Tetrahedron Letters*, **29**, 3967 (1988).

Diisobutylaluminum 2,6-di-*t*-butyl-4-methylphenoxide (**1**), **9**, 171.

Diastereospecific reduction of an enone.[1] A key step in a synthesis of the lichen macrolide (+)-aspicilin (**4**) is the reduction of the oxodienelactone **2** to the dienol **3** with Yamamoto's reagent (**1**) in 96% yield. Dihydroxylation of **3** with OsO_4 and pyridine followed by H_2S reduction gives the triol (**4**) in 40% yield.

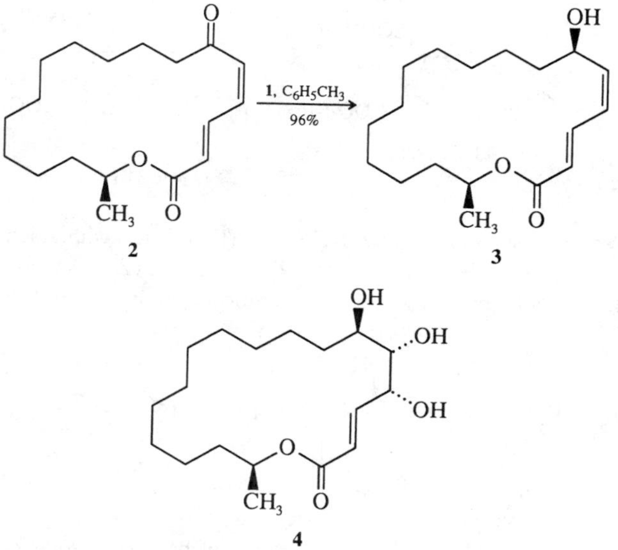

2 —1, $C_6H_5CH_3$, 96%→ **3**

4

[1] G. Quinkert, N. Heim, J. Glenneberg, U. Döller, M. Eichhorn, U.-M. Billhardt, C. Schwarz, G. Zimmermann, J. W. Bats, and G. Dürner, *Helv.*, **71**, 1719 (1988).

Diisobutylaluminum hydride (DIBAH).

Hydroalumination of ω-t-butoxyalkynes.[1] Hydroalumination of the ω-*t*-butyl ether of propargylic alcohol or of the homopropargylic alcohol followed by iodination yields exclusively the corresponding (Z)-vinyl iodides because of chelation. On the other hand, if the *t*-butoxy group is separated from the triple bond by three or more methylene groups, hydroalumination/iodination gives exclusively (E)-vinyl iodides. The *t*-butoxy group is converted to an acetoxy group by reaction with $Ac_2O/FeCl_3$ in ether in 87–93% yield. This hydroalumination is a useful reaction for synthesis of several pheromones.

$$HC{\equiv}C(CH_2)_nOC(CH_3)_3 \xrightarrow[62-74\%]{1)\ DIBAH\quad 2)\ I_2} ICH{=}CH(CH_2)_nOC(CH_3)_3$$

n = 1, 2 (Z, >99%)
n = 3–10 (E, >99%)

[1] A. Alexakis and J. M. Duffault, *Tetrahedron Letters*, **29**, 6243 (1988).

Diisopropyl crotylboronates, 1. These chiral crotylboranates are prepared in 98–99% ee from diisopropyl tartrate.

(E)-(R,R)-**1** (98% ee) (Z)-(R,R)-**1** (99% ee)

Diastereoselective reactions with RCHO.[1] (E)-**1** reacts with aldehydes to form the *anti*-adduct with high selectivity (95–99:5–1), and (Z)-**1** shows equally high selectivity for *syn*-adducts. However, the enantioselectivity is somewhat less in the

$C_9H_{19}CHO \longrightarrow$ R–CH(OH)–CH(CH_3)–CH=CH_2 + C_9H_{19}–CH(OH)–CH(CH_3)–CH=CH_2

(E)-**1**		>99 : 1	
(Z)-**1**	88% ee	1 : 99	82% ee

case of (Z)-**1**. Moreover, the enantioselectivity of both (E)- and (Z)-**1** decreases in reactions with β-alkoxy aldehydes (72–85% ee) and aromatic or α,β-unsaturated aldehydes (55–77% ee).

[1] W. R. Roush, K. Ando, D. B. Powers, R. L. Halterman, and A. D. Palkowitz, *Tetrahedron Letters*, **29**, 5579 (1988).

2,2-Dimethoxypropane–Tin(II) chloride.

Isopropylidenation.[1] Reaction of D-glucono-1,4-lactone (**1**) with 2,2-dimethoxypropane catalyzed by $SnCl_2$ in DME results in the ketal in 57% yield. Periodate cleavage of **2** leads to (R)-O-isopropylidineglyceraldehyde, an attractive chiral

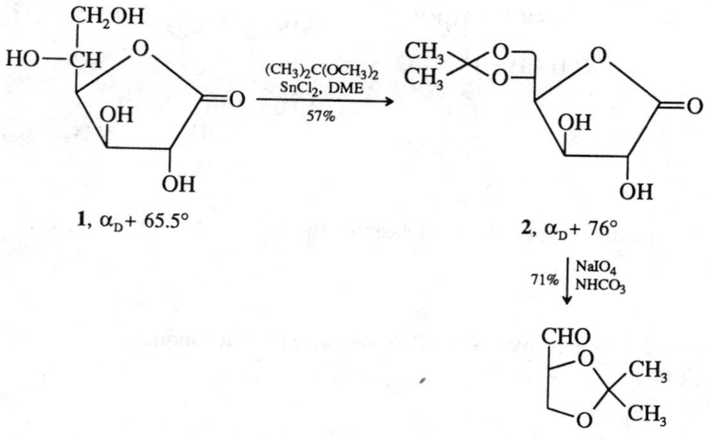

CH$_2$OH, HO—CH, O, =O, OH, OH → $(CH_3)_2C(OCH_3)_2$, $SnCl_2$, DME, 57% → CH_3, CH_3, O, O, O, =O, OH, OH

1, $\alpha_D + 65.5°$

2, $\alpha_D + 76°$

71% ↓ $NaIO_4$, $NHCO_3$

CHO, O, CH_3, O, CH_3

3, $\alpha_D + 63°$

synthon. This route via D-gluconic acid is economically more attractive than the usual route via D-mannitol.

[1] G. J. F. Chittenden, *Rec. Trav.*, **107**, 455 (1988).

1,1-Dimethoxypropene, $CH_3CH{=}C(OCH_3)_2$. Supplier: Aldrich.

2,2-*Dimethoxyoxetanes.*[1] In the presence of $ZnCl_2$, this ketal reacts with β-hydroxy aldehydes to give 2,2-dimethoxyoxetanes, which can be converted into 4-hydroxy-3-methyl-δ-lactones and 5,6-dihydro-2-pyrones. In one case, an optically active lactone was obtained in 80% ee from an optically active γ-hydroxy aldehyde.

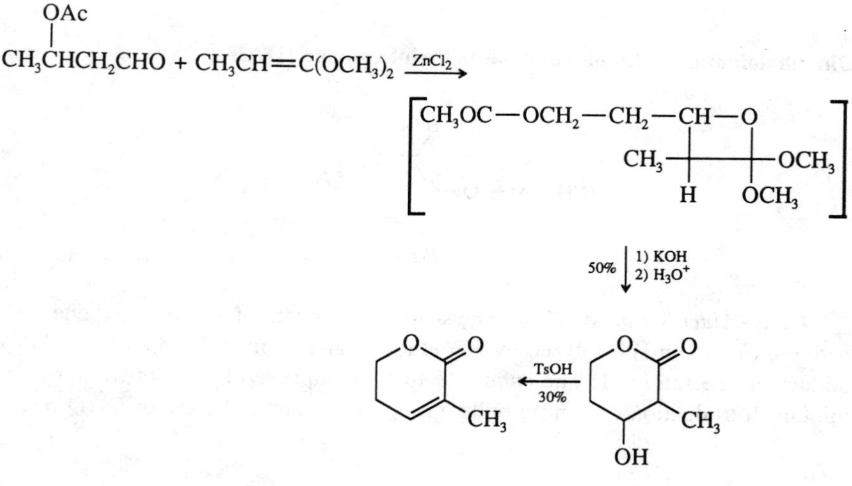

$CH_3CH(OAc)CH_2CHO + CH_3CH{=}C(OCH_3)_2 \xrightarrow{ZnCl_2}$

[$CH_3OC—OCH_2—CH_2—CH—O$, CH_3, H, OCH_3, OCH_3]

50% ↓ 1) KOH, 2) H_3O^+

O, O, CH_3 ← TsOH, 30% — O, O, CH_3, OH

$$C_6H_5CH(OCH(CH_3)OC_2H_5)-CH(CH_3)CHO \xrightarrow{61\%}$$

(*anti*)

(4 : 1)

[1] R. G. Hofstraat, J. Lange, H. W. Scheeren, and R. J. F. Nivard, *J.C.S. Perkin I*, 2315 (1988).

(R,R)-(+)-2,3-Dimethoxy-N,N,N',N'-tetramethylsuccinamide (1), m.p. 61–62°, α_D + 115°.

$CON(CH_3)_2$ / H–C–OCH_3 / CH_3O–C–H / $CON(CH_3)_2$

1

The reagent is prepared from (R,R)-(+)-diethyl tartrate by reaction with dimethylamine followed by methylation (dimethyl sulfate and 2,2-dimethoxypropane).

***Resolution of 2,2'-dihydroxy-1,1'-binaphthyl* (2).** The succinamide **1** forms an insoluble 1:1 complex with (±)-**2** at room temperature in benzene/hexane. The complex after recrystallization and chromatography on silica gel furnishes (S)-(−)-**2** of 100% ee. (R)-(+)-**2** of 100% ee can be obtained from the filtrate.

[1] F. Toda, K. Tanaka, L. Nassimbeni, and M. Niven, *Chem. Letters*, 1371 (1988).

Dimethylaluminum 2,6-di-*t*-butyl-4-methylphenoxide (DAD),

$(CH_3)_2Al-O-C_6H_2(Bu\text{-}t)_2-CH_3$ (1)

Diels–Alder catalyst. The Diels–Alder reaction of cyclopentadiene with acrolein (equation I) catalyzed by BF_3 etherate gives a mixture of *endo*- and *exo*-adducts in the ratio 9:1. The *endo*-selectivity is improved by use of trimethylaluminum. Introduction of a more bulky group on aluminum by use of DAD results

(I) cyclopentadiene + $CH_2=CH-CHO$ $\xrightarrow{C_6H_5CH_3,\ -78°}$ *endo*-2 (CHO) + *exo*-2

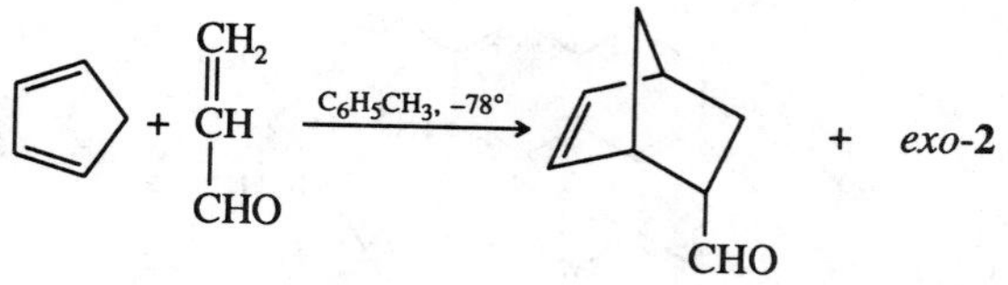

$BF_3 \cdot O(C_2H_5)_2$	46%	9:1
$(CH_3)_3Al$	57%	24:1
DAD	52%	32:1
MAD	56%	12:1

in a further increase in *endo*-selectivity. But use of the even more hindered catalyst MAD [methylaluminum bis(2,6-di-*t*-butyl-4-methylphenoxide), **13**, 203–206] lowers the *endo*-selectivity.

The usual rule of *endo*-selectivity is not observed in the cycloaddition of cyclopentadiene with methacrolein (equation II). In this reaction both DAD and MAD enhance *exo*-selectivity because of steric repulsion between the bulky catalyst and

(II) cyclopentadiene + $CH_2=C(CH_3)-CHO$ ⟶ *exo*-3 (CHO, CH_3) + *endo*-3 (CH_3, CHO) 5:1

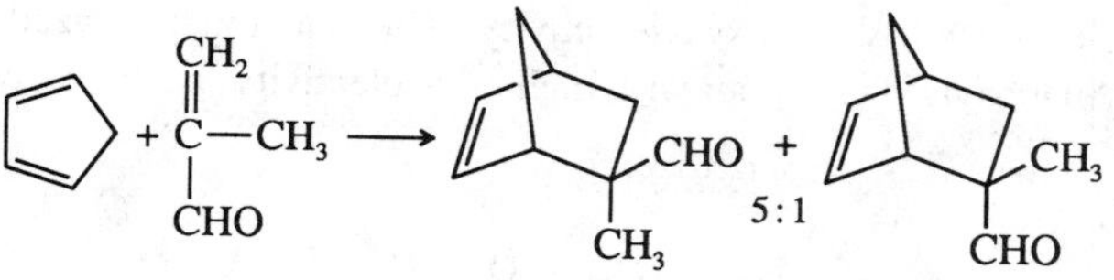

+ $(CH_3)_3Al$	64%	15:1
DAD	67%	24:1
MAD	76%	49:1

the diene, rather than by an electronic effect involving the carbonyl group of the dienophile, as in the reaction with acrolein.

An even greater steric effect of the Lewis acid catalyst is observed with the organoaluminum reagents **4–6** in the Diels–Alder reactions of an aldehyde with a diene (equation III).

(III) diene (OCH_3, CH_3, $(CH_3)_3SiO$, CH_3) $\xrightarrow{\text{1) } C_6H_5CHO\text{, cat.; 2) } CF_3COOH}$ *cis*-7 (CH_3, O, C_6H_5, CH_3) + *trans*-7 (CH_3, O, C_6H_5, CH_3)

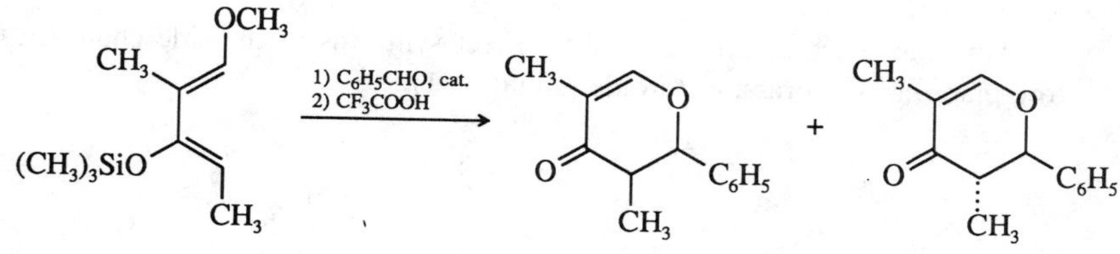

+ **4**	72%	1:1
+ **5**	91%	2:1
+ **6**	84%	12:1

4, R = CH_3
5, R_3 = Me_2-*t*-Bu
6, R = C_6H_5

[1] K. Maruoka, K. Nonoshita, and H. Yamamoto, *Syn. Comm.*, **18**, 1453 (1988).

Dimethylaluminum trifluoromethanesulfonate, $(CH_3)_2AlOTf$.

Glyoxylate ene reactions. The diastereoselectivity of the ene reaction of methyl glyoxylate (**1**) with 2-butene (**2**) can be controlled by the choice of the Lewis acid catalyst. Reactions catalyzed by $SnCl_4$ with either (E)- or (Z)-**2** proceed in quantative yield and fairly high *syn*-selectivity. The reaction catalyzed by $(CH_3)_2AlCl$ proceeds in low yield (5%) but with high *anti*-selectivity, particularly in the reaction

$CH_3CH{=}CHCH_3$ (**2**) + $HC(O)COOCH_3$ (**1**) $\xrightarrow[-78^\circ]{CH_2Cl_2}$ *anti*-**3** + *syn*-**3**

2			
E	$SnCl_4$	100%	82:18
Z	$SnCl_4$	100%	72:28
E	$(CH_3)_2AlOTf$	29%	21:79
Z	$(CH_3)_2AlOTf$	65%	9:91

with (E)-**2**. Higher yields obtain with dimethylaluminum triflate, particularly in the case of (Z)-**2**.

This ene reaction can be used to effect synthesis of the side chain present in the plant growth brassinosteroids (equation I).

(I) (2) + $HC(O)COOCH_3$ $\xrightarrow[67\%]{(CH_3)_2AlCl}$ (*anti*, 100%)

[1] K. Mikami, T.-P. Loh, and T. Nakai, *Tetrahedron Letters*, **29**, 6305 (1988).

Dimethyldioxirane, 12, 413; **13**, 120.

Arene dioxides. Arene dioxides have not been identified as metabolites of polycyclic arenes, but some identified metabolites may be formed from arene dioxides. Arene dioxides can be obtained by reaction of polyarenes containing a double bond comparable to the 9,10-double bond of phenanthrene (such as **2**) by oxidation with dimethyldioxirane (**1**), but in modest yield. The dioxide **5** cannot be obtained by this method, but can be obtained via a *trans,vic*-bromoacetate.[1]

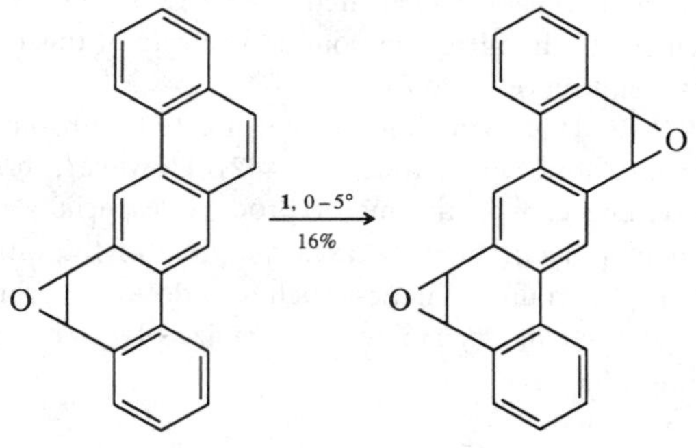

1, 0–5°, 16%

2 **3**

1) NBA LiOAc
2) NBS
3) $NaOCH_3$
60%

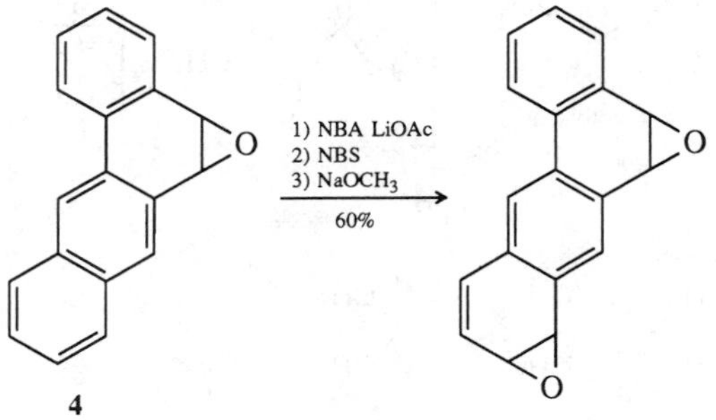

4 **5**

Aflatoxin B₁ epoxide.[2] This epoxide (**2**), which may be responsible for the carcinogenicity of aflatoxin B_1, has been prepared by reaction of aflatoxin B_1 (**1**)

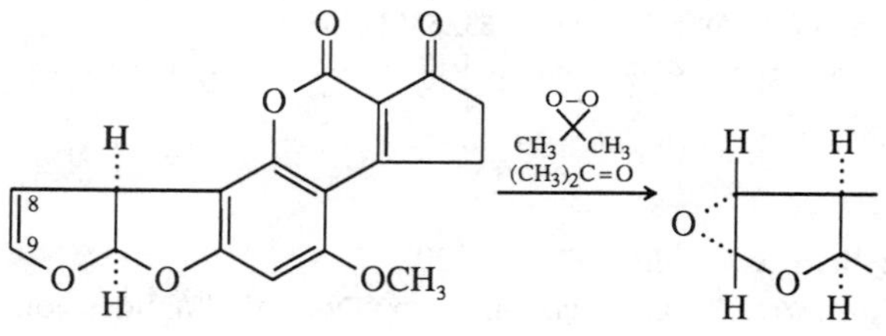

CH_3 CH_3 dioxirane, $(CH_3)_2C=O$

1 **2**

with thoroughly dried (molecular sieves or K_2CO_3) solutions of dimethyldioxirane in acetone or CH_2Cl_2. The epoxide is very sensitive to water, but can be stored at −10°. The epoxide (**2**), the first known epoxide of a tetrahydrofuran, reacts very specifically with the guanine unit of DNA to form an adduct at the N^7-position.

RNCO → RNO₂.[3] Ozone does not react with isocyanates under a variety of conditions; oxidation with $KMnO_4$ or $ClC_6H_4CO_3H$ gives complex mixtures. Surprisingly, oxidation with Oxone® in acetone affords nitro compounds in surprisingly high yields from both aromatic and primary, secondary, and tertiary aliphatic precursors. Water is an essential ingredient, probably because the oxidation proceeds through the carbamic acid, which then loses CO_2 to give an amine that is the immediate precursor to the nitro compound. The rate of this oxidation can be greatly increased by addition of Triton B.

$ArNH_2$ → $ArNO_2$.[4] This oxidation can be effected with dimethyldioxirane generated *in situ* from Oxone® in acetone (**11**, 442). However, in some cases the intermediate nitrosoarene can be the major product, especially when a limited amount of Oxone is employed. In such cases the yield of the nitroarene can be improved by addition of certain aromatics, such as indoles or furans.

Nitroxides.[5] Cyclic secondary amines are oxidized by dimethyldioxirane (**1**) to nitroxides in yields >95%.

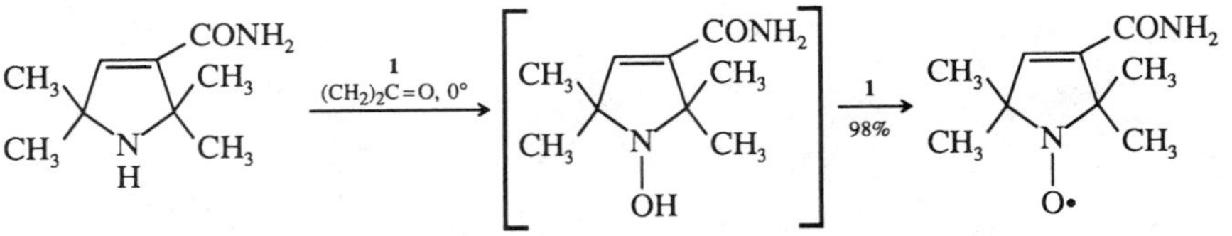

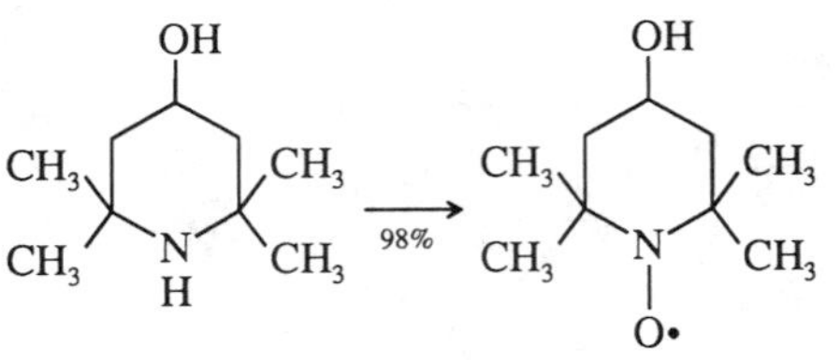

[1] S. K. Agarwal, D. R. Boyd, W. B. Jennings, R. M. McGuckin, and G. A. O. Kane, *Tetrahedron Letters*, **30**, 123 (1989).
[2] S. W. Baertschi, K. D. Raney, M. P. Stone, and T. M. Harris, *Am. Soc.*, **110**, 7929 (1988).
[3] P. E. Eaton and G. E. Wicks, *J. Org.*, **53**, 5353 (1988).
[4] D. L. Zabrowski, A. E. Moormann, and K. R. Beck, Jr., *Tetrahedron Letters*, **29**, 4501 (1988).
[5] R. W. Murray and M. Singh, *ibid.*, **29**, 4677 (1988).

N,O-Dimethylhydroxylamine, 11, 201–202.

β,δ-Diketo esters.[1] Various N-methoxy-N-methyl amides couple with the dianion of alkyl acetoacetates to form β,δ-diketo esters in 42–91% yield. These products are reduced to β,δ-dihydroxy esters by $NaBH_4$ in combination with diethyl(methoxy)borane.

$$(CH_3)_2C{=}CHC(=O)N(OCH_3)CH_3 + CH_3COCH_2COOBu\text{-}t \xrightarrow[90\%]{1)\ NaH;\ 2)\ BuLi} (CH_3)_2C{=}CHCOCH_2COCH_2COOBu\text{-}t$$

$$\xrightarrow{(C_2H_5)_2BOCH_3,\ NaBH_4,\ THF/CH_3OH} (CH_3)_2C{=}CHCH(OH)CH_2CH(OH)CH_2COOBu\text{-}t$$

syn/anti >95 : 5

[1] T. Hanamoto and T. Hiyama, *Tetrahedron Letters*, **29**, 6467 (1988).

1,3-Dimethyl-2-imidazolidinone (DMI), (1), **11**, 202.

Ullman ether synthesis.[1] *m*-Phenoxybenzyl alcohol (**3**) can be prepared by condensation of *m*-hydroxybenzyl alcohol (**2**) with chlorobenzene catalyzed by CuCl and 8-hydroxyquinoline (**4**). The yield is markedly dependent on the solvent, being highest in DMI.

$$m\text{-}HOC_6H_4CH_2OH\ (\mathbf{2}) + C_6H_5Cl \xrightarrow[150-170^\circ]{CuCl,\ \mathbf{4},\ K_2CO_3} m\text{-}C_6H_5OC_6H_4CH_2OH\ (\mathbf{3})$$

Solvent	Yield
DMF	21%
DMSO	58%
DMI	81%

[1] R. Oi, C. Shimakawa, and S. Takenaka, *Chem. Letters*, 899 (1988).

Dimethylphosphinothioyl chloride, $ClP(=S)(CH_3)_2$ (**1**). Preparation.[1]

Peptide synthesis in methanol or ethanol.[2] The dimethylphosphinothioic anhydrides prepared from N-protected amino acids with **1** are stable to water or alcohols in the absence of a base, probably because of the P=S bond. As a consequence these active esters can be used for peptide syntheses in methanol or ethanol rather than the dipolar aprotic solvents (HMPT, DMF, or DMSO) usually required. The report includes a synthesis of a heptapeptide by this expedient.

[1] M. Ueki, T. Inazu, and S. Ikeda, *Bull. Chem. Soc. Japan*, **52**, 2424 (1979).
[2] M. Ueki, T. Saito, J. Sasaya, S. Ikeda, and H. Oyamada, *ibid.*, **61**, 3653 (1988).

Dimethyl sulfoxide–Chlorotrimethylsilane.

Cleavage of α,β-epoxy ketones.[1] The reagent (**1**), possibly $(CH_3)_2\overset{+}{S}\overset{-}{Cl}$, formed from DMSO and $ClSi(CH_3)_3$, cleaves α,β-epoxy ketones to 2-chloro-3-hydroxy ketones with high regioselectivity (equation I). If the epoxide has a phenyl substituent, 3-chloro-2-hydroxy ketones are formed (equation II).

(I) [epoxy ketone with CH_3, O, $CH(CH_3)_2$] $\xrightarrow[91\%]{1,\ CH_3CN,\ 25°}$ [product with HO, Cl, CH_3, O, $CH(CH_3)_2$]

(II) [C_6H_5 epoxide with $COCH_3$] $\xrightarrow{84\%}$ C_6H_5–CH(Cl)–CH(OH)–C(=O)–CH_3

[1] F. Ghelfi, R. Grandi, and U. M. Pagnoni, *J. Chem. Res. (S)*, 200 (1988).

Dimethyl sulfide–Dibenzoyl peroxide.

Methylthiomethyl (MTM) ethers.[1] These ethers (**6**, 109–110) can be prepared by reaction of alcohols with dimethyl sulfide (8 equiv.) and dibenzoyl peroxide (4 equiv.) in CH_3CN at 0° (75–90% yield). Excess dimethyl sulfide is required since some of the sulfide is oxidized to dimethyl sulfoxide.

[1] J. C. Medina, M. Salomon, and K. S. Kyler, *Tetrahedron Letters*, **29**, 3773 (1988).

Dimethyl sulfoxide–Phenyl dichlorophosphate ($C_6H_5O\overset{O}{\overset{\|}{P}}Cl_2$).

β-Chloroalkyl sulfides.[1] A reagent (**1**), obtained from DMSO and $C_6H_5OPOCl_2$ in CH_2Cl_2 at −20°, forms *trans*-adducts with alkenes in which the methylthio group is attached to the less-substituted carbon atom.

Cyclohexene + 1 $\xrightarrow[80\%]{CH_2Cl_2,\ -20 \rightarrow 20^\circ}$ trans-1-chloro-2-(methylthio)cyclohexane (Cl, SCH_3)

$$CH_3(CH_2)_nCH{=}CH_2 \xrightarrow{80\%} CH_3(CH_2)_nCH(Cl){-}CH_2SCH_3$$

n = 11 or 15

[1] H.-J. Liu and J. M. Nyangulu, *Tetrahedron Letters*, **29**, 5467 (1988).

Dimethylsulfoxonium methylide, $CH_2{=}S(O)(CH_3)_2$ (**1**).

Cyclopropanation of lactams.[1] This reagent adds to the chiral bicyclic lactams **2** (**13**, 18–19) to form *endo*-cyclopropanated adducts (**3**) in >93% de. These adducts are hydrolyzed to optically active cyclopropanes (**4**).

2 $\xrightarrow{1}$ **3** $\xrightarrow[81-88\%]{H_2SO_4,\ CH_3OH,\ \Delta}$ **4**

2		3
R = H	64%	95% ee
R = C_6H_5	58%	93% ee

[1] A. I. Meyers, J. L. Romine, and S. A. Fleming, *Am. Soc.*, **110**, 7245 (1988).

Dimethyl 1,2,4,5-tetrazine-3,6-dicarboxylate, (tetrazine ring bearing two $COOCH_3$ groups) (**1**), m.p. 173–175°.

The tetrazine is prepared in several steps from ethyl diazoacetate.

Diels–Alder reactions; pyrroles.[1] This azadiene undergoes [4+2] cycloaddition with unactivated dienophiles to form dimethyl 1,2-diazine-3,6-dicarboxylates, which are reduced by zinc in acetic acid with ring contraction to dimethyl pyrrole-2,5-dicarboxylates.

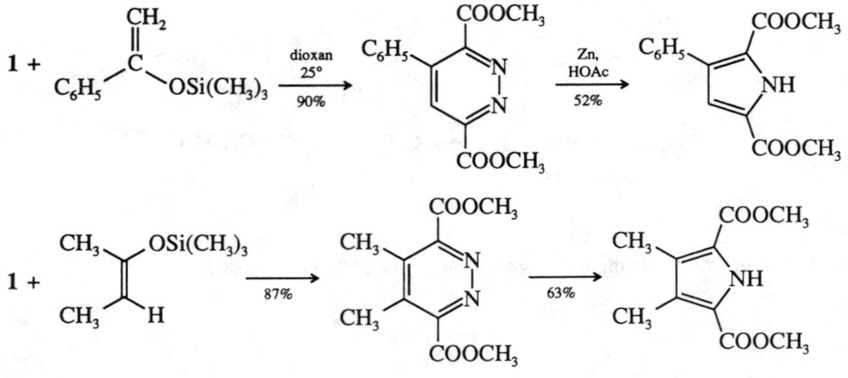

[1] D. L. Boger, J. S. Panek, and M. Patel, *Org. Syn.*, submitted (1988).

Dimethylzinc, $Zn(CH_3)_2$.

Enolate acylation and alkylation.[1] The yield from acylation and alkylation of lithium ketone enolates is markedly improved by addition of dimethylzinc, which suppresses α-proton exchange. The effective reagent may be a lithium alkoxydimethylzincate.

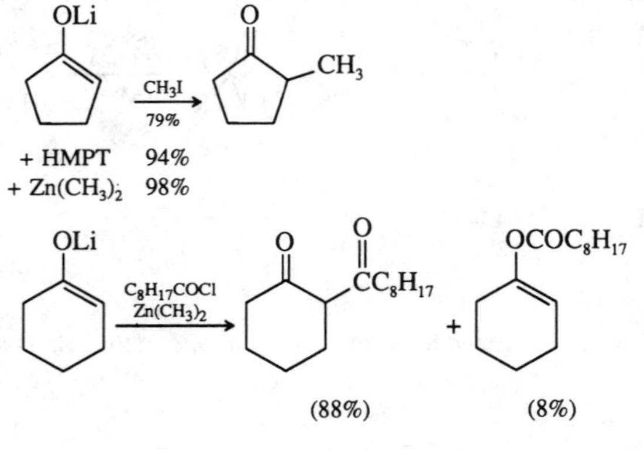

[1] Y. Morita, M. Suzuki, and R. Noyori, *J. Org.*, **54**, 1785 (1989).

3,5-Dinitroperoxybenzoic acid, $(O_2N)_2C_6H_3CO_3H$ (**1**).

Cleavage of isoxazolines to β-hydroxy ketones.[1] This cleavage can be effected with either 3,5-dinitroperbenzoic acid (**1**) or trifluoroperacetic acid using about 5 equiv. of the peracid at 25°. Use of more drastic conditions results in Baeyer–Villigar oxidation of the β-hydroxy ketones to acylated diols.

C_6H_{13} isoxazoline CH_3 $\xrightarrow[56\%]{1(5\text{ equiv.}),\ CH_2Cl_2,\ 25°}$ $C_6H_{13}CH(OH)CH_2COCH_3$

$\xrightarrow[\Delta]{70\%\ 1(3\text{ equiv.}),\ CH_2Cl_2}$ $C_6H_{13}CH(OH)CH_2OAc$ + $C_6H_{13}CH(OAc)CH_2OH$ (7:3)

[1] P. Park and A. P. Kozikowski, *Tetrahedron Letters*, **29**, 6703 (1988).

(Dioctyl)sodium sulfosuccinate,

$CH_3(CH_2)_3CH(C_2H_5)CH_2O_2CCH_2CHCO_2CHCH_2CO_2CH_2CH(C_2H_5)(CH_2)_3CH_3$ (with SO_3Na substituent)
(AOT,**1**).

Selective peptide synthesis.[1] This surfactant can effect synthesis of peptides from certain amino acids in isooctane/water with dicyclohexylcarbodiimide (DCC) or dioctadecylcarbodiimide (DODCI). This coupling is limited to amino acids with hydrophobic groups such as leucine, $(CH_3)_2CHCH_2CH(NH_2)COOH$. Thus Bzl-Trp and $TrpOCH_3$ couple to form a dipeptide when activated with DCC in isooctane/water to the extent of 11%, but when AOT is also present the dipeptide Bzl–Trp–Trp–OCH_3 is formed in 60% yield. In the absence of water, no reaction occurs. This coupling effected with a surfactant suggests that DCC and the hydrophobic amino acid are aligned at the micellar surface in such a way that both polar groups are in close promixity and that formation of the activated ester intermediate is facilitated. Coupling is not promoted when a water-soluble carbodiimide replaces DCC or DODCI. This technique can be used to effect selective coupling depending on the basicity of two different amino acids.

[1] D. Ranganathan, G. P. Singh, and S. Ranganathan, *Am. Soc.*, **111**, 1144 (1989).

1,4-Dioxene, 13, 112–113.

α-Hydroxy and α-keto acids.[1] The products (**1**) of reaction of aldehydes or ketones with lithiated 1,4-dioxene are oxidized by PCC to afford esters of α-hydroxy acids (**2**). Oxidation of the adducts with Fetizon's reagent provides an enone (**4**),

1 $\xrightarrow[58\%]{PCC}$ $CH_3(C_6H_5)C(OH)CO_2(CH_2)_2OCHO$ $\xrightarrow{NaOH}$ $CH_3(C_6H_5)C(OH)COOH$ (**2**)

$$\underset{\mathbf{3}}{(CH_3)_2CH\text{–}CH(OH)\text{–}(\text{dihydrodioxin})} \xrightarrow[90-95\%]{Ag_2CO_3,\ \text{Celite}} \underset{\mathbf{4}}{(CH_3)_2CH\text{–}C(=O)\text{–}(\text{dihydrodioxin})} \xrightarrow[54\%]{PCC}$$

$$(CH_3)_2CHC(=O)C(=O)O(CH_2)_2OCHO \xrightarrow{NaOH} \underset{\mathbf{5}}{(CH_3)_2CHC(=O)COOH}$$

which is oxidized by PCC to an α-keto ester, which is hydrolyzed to an α-keto acid (**5**).

[1] M. Fetizon, P. Goulaouic, and I. Hanna, *Tetrahedron Letters*, **29**, 6261 (1988).

Diphenyliodonium chloride, $(C_6H_5)_2ICl$, **1**, 341; **2**, 178.

Phenyl arenedithiocarboxylates.[1] These esters can be prepared by reaction of diphenyliodonium chloride or bromide with sodium arenedithiocarboxylates.

$$ArC(=S)SNa + (C_6H_5)_2ICl \xrightarrow[68-88\%]{t\text{-BUOH, }70^\circ} ArC(=S)SC_6H_5$$

[1] Z.-C. Chen, Y.-Y. Jin, and R.-Y. Yang, *Synthesis*, 723 (1988).

Diphenylphosphinic chloride (**7**, 138).

β-Lactams from β-amino acids.[1] β-Amino acids cyclize to β-lactams on treatment with $(C_6H_5)_2P(O)Cl$ and triethylamine (1:1) in CH_3CN at 80° usually with preservation of the stereochemistry of the amino acid. However, yields are rather low when the amino group is primary.

$$C_6H_5O(CH_3)C(COOH)\text{–}CH(C_6H_5)NHBzl \xrightarrow[98\%]{(C_6H_5)_2P(O)Cl,\ N(C_2H_5)_3,\ CH_3CN,\ 80^\circ} \text{β-lactam: } C_6H_5O,\ CH_3,\ H,\ C_6H_5,\ N\text{–}Bzl,\ C{=}O$$

[1] S. Kim, P. H. Lee, and T. A. Lee, *J.C.S. Chem. Comm.*, 1242 (1988).

Diphenylselenium hydroxyacetate $(C_6H_5)_2Se(OH)(OCOCH_3)$ (**1**).

The reagent is obtained by reaction of $(C_6H_5)_2SeO$ with acetic anhydride in THF (80% yield). It loses HOAc at 80°.

Hypoiodite reaction.[1] This reaction is usually carried out with iodine and HgO (**5**, 347–348). A new version uses iodine in combination with this diphenylselenurane, and this version gives essentially only cyclic ethers without any iodohydrins or other products encountered with other heavy metal acetates.

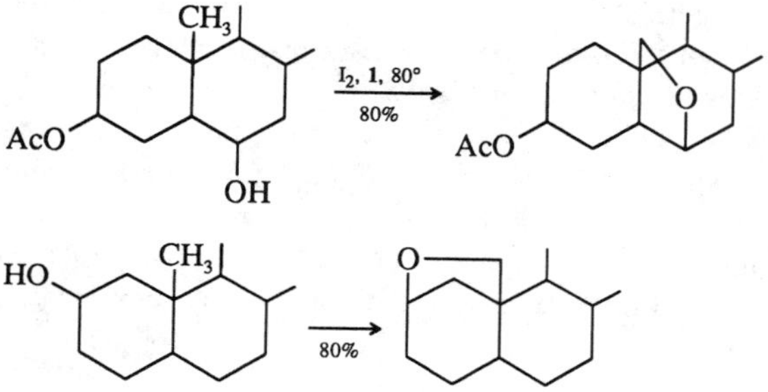

I_2, **1**, 80°
80%

80%

[1] R. L. Dorta, C. G. Francisco, R. Freire, and E. Suárez, *Tetrahedron Letters*, **29**, 5429 (1988).

Diphenylsilane–Palladium(0)–Zinc chloride.

Reduction of allylic acetates. This three-component system is effective for reduction of simple allylic acetates and even of 3-acetoxyglycals, particularly if $Pd[P(C_6H_5)_3]_4$ is replaced by tetrakis(tri-*p*-tolylphosphine)palladium(0).

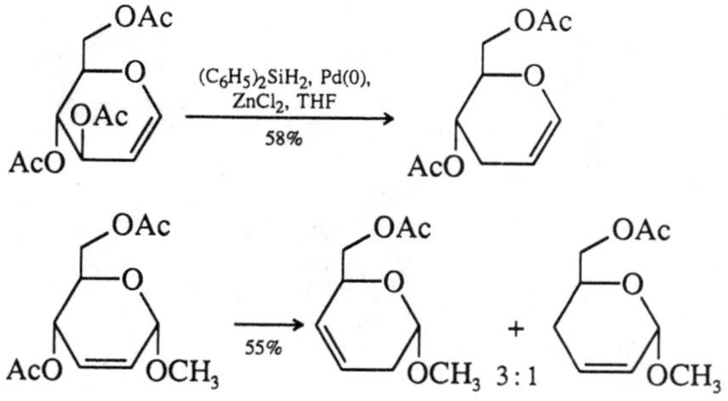

$(C_6H_5)_2SiH_2$, Pd(0), $ZnCl_2$, THF
58%

55%
3 : 1

[1] N. Greenspoon and E. Keinan, *J. Org.* **53**, 3723 (1988).

Diphosphorus tetraiodide, P_2I_4.

Reductive desulfuration of dithioacetals.[1] This reaction of dithioacetals or ketals can be effected with P_2I_4.

[1] Y. Shigemasa, M. Ogawa, H. Sashiwa, and H. Saimoto, *Tetrahedron Letters*, **30**, 1277 (1989).

Di-2-pyridyl carbonate, (1).

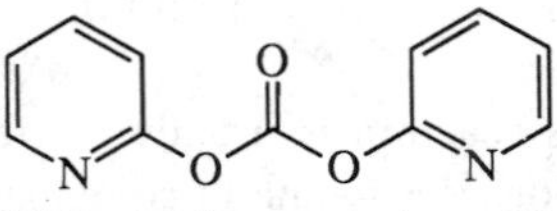

The reagent is obtained (90% yield) by reaction of 2-hydroxypyridine with phosgene and $N(C_2H_5)_3$ in CH_2Cl_2/toluene.[1]

Esterification.[2] This reagent in combination with a catalytic amount of 4-dimethylaminopyridine (DMAP) is very effective for esterification of carboxylic acids with alcohols or thiols at room temperatures. However, reaction of aromatic and hindered acids requires several days at room temperature. French chemists report that only this method is useful for esterification of the protected baccatin III derivative (**2**) with (2R,3S)-N-benzoyl-O-(1-ethoxyethyl)-3-phenylisoserine (**3**) to provide the protected taxol derivative (**4**). A reaction conducted at 73° for 100 hours with 6 equiv. of **1** and 2 equiv. of DMAP produced **4** in 80% yield. Natural taxol, a cancer chemotherapeutic agent, is obtained by removal of the protective groups at C_2' and C_7 of **4**.

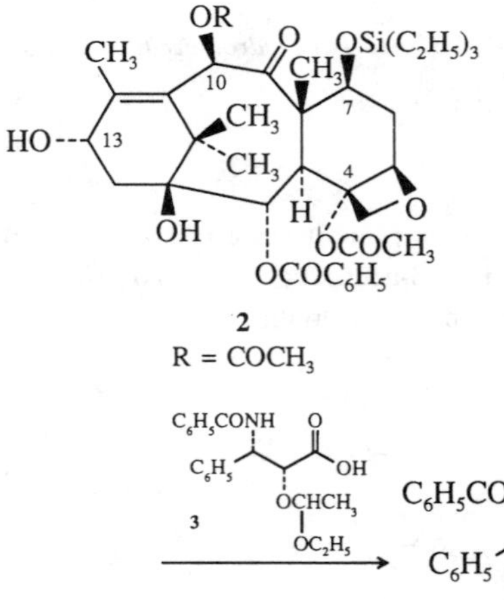

2
R = $COCH_3$

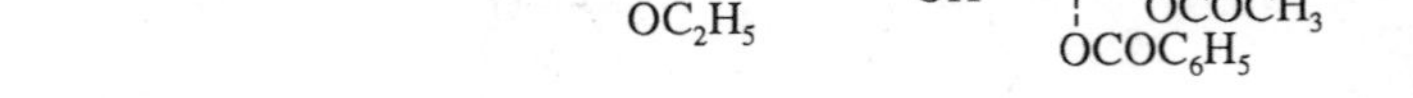

3

4

[1] S. Kim, J. I. Lee, and Y. K. Ko, *Tetrahedron Letters*, **25**, 4943 (1984).
[2] J.-N. Denis, A. E. Green, D. Guénard, F. Guéritte-Voegelein, L. Mangatal, and P. Potier, *Am. Soc.*, **110**, 5917 (1988).

Disodium tetracarbonylferrate.

Aromatic ketones (*cf.*, **4**, 461, 463).[1] The acyliron complexes (**2**), prepared from $Na_2Fe(CO)_4$ as shown, have rather low reactivity with alkyl halides. However, in the presence of Pd(0) and $ZnCl_2$ they react readily with aryl iodides, particularly in the polar solvent N-methyl-2-pyrrolidone (NMP).

$$\text{(I)}\ \ RX + Na_2Fe(CO)_4 \xrightarrow{P(C_6H_5)_3,\ NMP} RCOFe(CO)_3[P(C_6H_5)_3]$$

1 → 2

$$RCOFe(CO)_3[P(C_6H_5)_3]\ (\mathbf{2}) \xrightarrow[55-85\%]{ArX,\ Pd[P(C_6H_5)_3],\ ZnCl_2} RCOAr$$

Example:

$$BuBr \longrightarrow \xrightarrow[71\%]{C_6H_5I} BuCOC_6H_5$$

[1] T. Koga, S. Makinouchi, and N. Okukado, *Chem. Letters*, 1141 (1988).

N-Dodecylpyridoxal (DPL), (1). [structure: pyridinium ring with CHO, HOH_2C, O^-, CH_3, and N^+–$C_{12}H_{25}$]

Transamination. Transamination with pyridoxal or analogs requires a metal catalyst such as Cu(II). However, transamination can be effected with DPL in combination with hexadecyltrimethylammonium chloride. By using these two reagents phenylglycine undergoes transamination with 2-oxoglutaric acid (equation I).

$$\text{(I)}\ \ C_6H_5CH(NH_3^+)COO^- + HOOCCH_2CH_2C(=O)COOH \underset{}{\overset{DPL,\ RN(CH_3)_3Cl}{\underset{30°,\ pH\ 8-9}{\rightleftharpoons}}} C_6H_5COCOOH + HOOCCH_2CH_2CH(NH_3^+)COO^-$$

[1] H. Kondo, H. Tanamachi, and J. Sunamoto, *Chem. Letters*, 2013 (1988).

E

Ephedrine.

Imidazolidinones. The reaction of urea with (−)-ephedrine hydrochloride provides a chiral imidazolidinone (**1**),[1] which can be used as a chiral auxiliary for

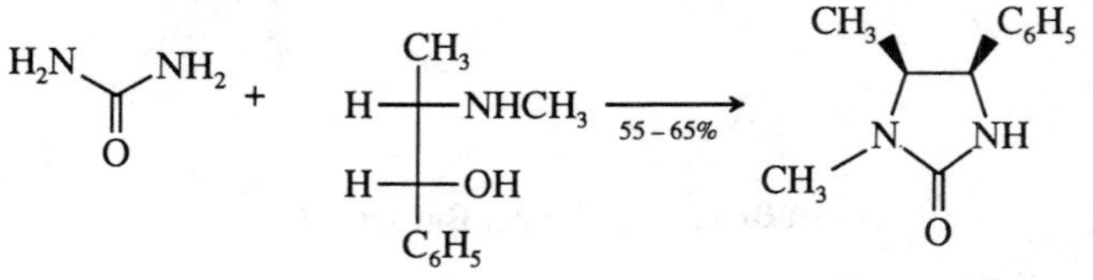

H_2N NH_2 + CH$_3$ H—NHCH$_3$ H—OH C$_6$H$_5$ (1R, 2S) —55–65%→ CH$_3$ C$_6$H$_5$ N NH CH$_3$ O

(1R, 2S)

1, α_D −44.5°

effecting stereoselective synthesis of citronellic acid (**5**). Thus reaction of **1** with the acid chloride **2** gives the unsaturated amide **3**. Addition of a methyl cuprate to **3** proceeds with high asymmetric induction (**12**, 251) to furnish an adduct (**4**) that is hydrolyzed to (−)-citronellic acid (**5**).[2]

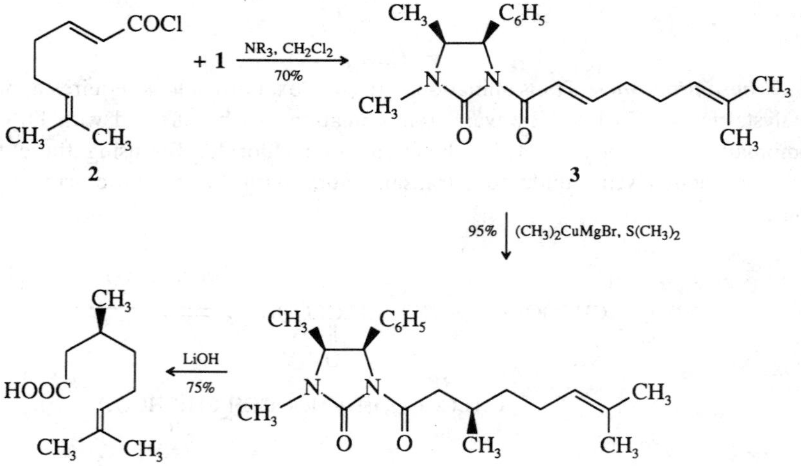

COCl, CH$_3$, CH$_3$ **2** + **1** —NR_3, CH_2Cl_2, 70%→ **3**

95% ↓ $(CH_3)_2CuMgBr$, $S(CH_3)_2$

4 —LiOH, 75%→ HOOC ... (−)-**5**

Enantioselective protonation.[3] (R)- and (S)-α-Damascone (**4**) have been prepared by a Grignard reaction followed by enantioselective protonation with **1**[1] and **2**,[3] both available from (−)- or (+)-ephedrine. Thus protonation of the ketone enolate **3** with (+)-**1** or (−)-**2** furnishes (S)- or (R)-α-damascone (**4**), respectively.

(+)–**1** (–)–**2**

(±) 1) BuLi 2) CH_2=$CHCH_2MgCl$ → **3** (E/Z = 9 : 1)

(+)–**1**, Al_2O_3 60% ; (–)–**2**, Al_2O_3 60%

(S)–**4** (58% ee) (R)–**4** (70% ee)

[1] H. Roder, G. Helmchen, E. M. Peters, K. Peters, and H. G. v. Schnering, *Angew. Chem. Int. Ed.*, **23**, 898 (1984).
[2] E. Stephan, G. Pourcelot, and P. Cresson, *Chem. Ind.*, 562 (1988).
[3] C. Fehr and J. Galindo, *Am. Soc.*, **110**, 6909 (1988).

(Ethoxycarbonyl)nitrene.

Ethoxycarbonylamino acids.[1] (Ethoxycarbonyl)nitrene (**2**), generated by thermolysis of N(ethoxycarbonyl)-N,O-bis(trimethylsilyl)hydroxylamine (**1**),[2] reacts with ketene silyl acetals to form N-protected α-amino esters (**3**) in 25–70% yield.

$$C_2H_5OC(=O)N(Si(CH_3)_3)OSi(CH_3)_3 \ (\mathbf{1}) \xrightarrow[-[(CH_3)_3Si]_2O]{90^\circ} [C_2H_5OC(=O)N] \ (\mathbf{2}) \xrightarrow{CH_3CH=C(OSi(CH_3)_3)(OC_2H_5)}$$

$$[\text{aziridine: } H, CH_3; OSi(CH_3)_3, OC_2H_5; N\text{-}COOC_2H_5] \xrightarrow[70\%]{H_3O^+} CH_3CH(NHCOOC_2H_5)COOC_2H_5 \ (\mathbf{3})$$

[1] M. Antonietta, L. Pellacani, and P. A. Tardella, *J. Chem. Res.* (*S*), 304 (1988).
[2] Y. H. Chang, F.-T. Chiu, and G. Zon, *J. Org.*, **46**, 342 (1981).

Ethylene glycol.

Diels–Alder reactions.[1] The laboratories of Breslow and of Grieco have reported that water can enhance the rate of Diels–Alder reactions of dienes that possess carboxylic acid or similar hydrophilic groups (**12**, 314). Liotta *et al.*[1] have examined solvent effects on cycloaddition reactions of benzoquinones with dienes substituted with a relatively hydrophobic group, and report significant rate enhancement in ethylene glycol relative to benzene (26:1) or even to reactions in the absence of a solvent. They attribute the solvent effect to aggregation of the diene and the quinone.

CH_3 OCO_2CH_3 + CH_3 O CH_3 O $\xrightarrow{95°}$ CH_3O_2CO O CH_3 CH_3 H CH_3 O

C_6H_6 (93 hr.) 25%
$HOCH_2CH_2OH$ (22 hr.) 71%

[1] T. Dunams, W. Hoekstra, M. Pentaleri, and D. Liotta, *Tetrahedron Letters*, **29**, 3745 (1988).

Ethyl hydrogen malonate, $HOOCCH_2COOC_2H_5$. Preparation.[1]

2,4-Pentadienoates.[2] Half esters of malonic acid react with α,β-enals in pyridine containing dimethylaminopyridine to form these diunsaturated esters, with generally high (2E)-selectivity.

CH_3 CHO + $HOOCCH_2COOCH_3$ $\xrightarrow[95\%]{\text{DMAP, Py, 60°}}$ CH_3 $COOCH_3$

(2E/2Z = 95:5)

[1] R. E. Strube, *Org. Syn. Coll. Vol. IV.*, 417 (1963).
[2] J. Rodriguez and B. Waegell, *Synthesis*, 534 (1988).

Ethyl nitrite, C_2H_5ONO (**1**). The nitrite is available as a 15% solution in C_2H_5OH (Aldrich).

Ketoximes from styrenes.[1] In the presence of $Co(DH)_2Cl(Py)^2$ [DH = monoanion of dimethylglyoxime] and $(C_2H_5)_4NBH_4$, styrenes react with ethyl nitrite to form ketoximes in nearly quantitative yield.

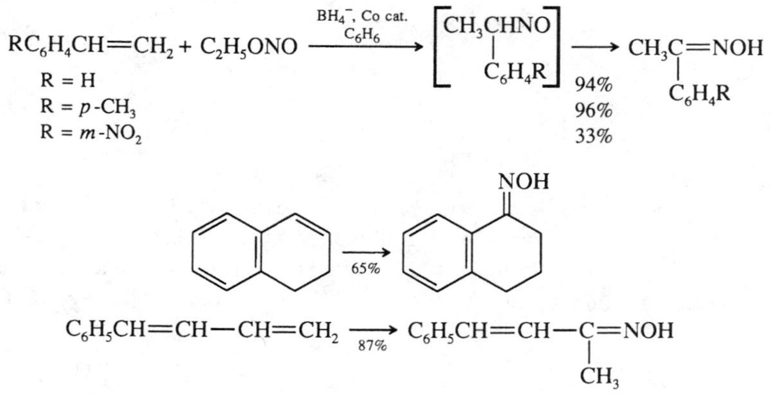

$$RC_6H_4CH{=}CH_2 + C_2H_5ONO \xrightarrow[C_6H_6]{BH_4^-,\ Co\ cat.} \left[\begin{matrix}CH_3CHNO\\ |\\ C_6H_4R\end{matrix}\right] \longrightarrow \begin{matrix}CH_3C{=}NOH\\ |\\ C_6H_4R\end{matrix}$$

R = H, 94%
R = p-CH_3, 96%
R = m-NO_2, 33%

NOH

$\xrightarrow{65\%}$

$$C_6H_5CH{=}CH{-}CH{=}CH_2 \xrightarrow[87\%]{} C_6H_5CH{=}CH{-}\underset{\displaystyle CH_3}{\underset{|}{C}}{=}NOH$$

[1] T. Okamoto, K. Kobayashi, S. Oka, and S. Tanimoto, *J. Org.*, **53**, 5089 (1988).
[2] G. N. Schrauzer, *Inorg. Synth.*, **11**, 61 (1968).

Ethyl triphenylphosphonoacetate, $(C_6H_5)_3P{=}CHCOOC_2H_5$ (**1**).

Reaction with N-protected 2-hydroxypyrrolidines or -piperidines. The Wittig reaction with these substrates provides α-alkylated pyrrolidines or piperidines after base-catalyzed intramolecular Michael addition.[1]

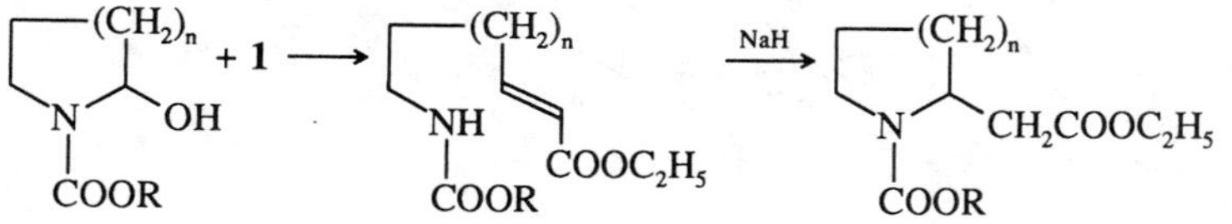

$(CH_2)_n$ ... N ... OH, COOR + **1** → $(CH_2)_n$ ... NH ... $COOC_2H_5$, COOR $\xrightarrow{NaH}$ $(CH_2)_n$ ... N ... $CH_2COOC_2H_5$, COOR

2, n = 1	80%	**3**	78%	**4**
n = 2	71%		87%	

[1] T. Nagasaka, H. Yamamoto, H. Hayashi, M. Watanabe, and F. Hamaguchi, *Heterocycles*, **29**, 155 (1989).

F

Ferric chloride.

Dimerization of ketone enolates. Frazier and Harlow[1] have noted that ketone enolates (LDA or KH) couple to 1,4-diketones when oxidized by dry $FeCl_3$ in DMF. Yields range from 25 to 70%. Paquette *et al.*[2] have extended this reaction to intramolecular coupling of a bisenolate to a cyclopropane. Thus addition of a

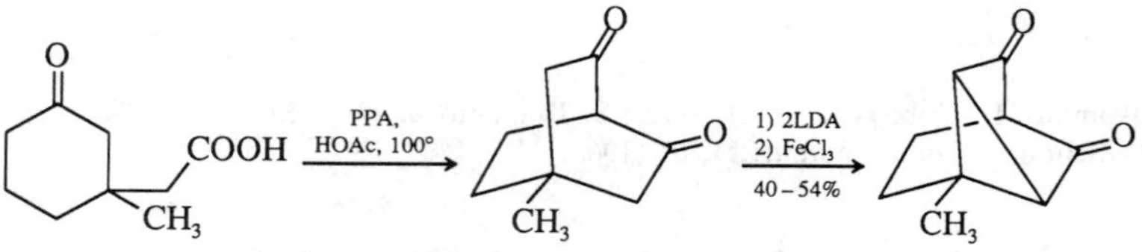

PPA, HOAc, 100° → **1**; 1) 2LDA 2) $FeCl_3$ 40–54% → **2**

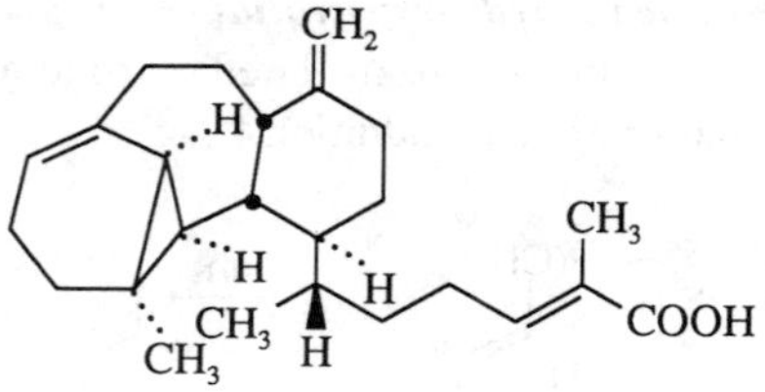

3

THF solution of the dienolate of **1** to a solution of dry $FeCl_3$[3] in DMF provides the cyclopropane **2** in about 50% yield. The product was prepared as a precursor in a projected synthesis of ceroubenic acid-III (**3**), a metabolite of certain wasps.

Cyclic amides.[4] Reaction of N-acetyloxyamides with $FeCl_3$ (2 equiv.) and HOAc (1 equiv., essential for solubilization of $FeCl_3$ in CH_2Cl_2) affords a species that effects amidation (inter- or intramolecular) of an aryl group.

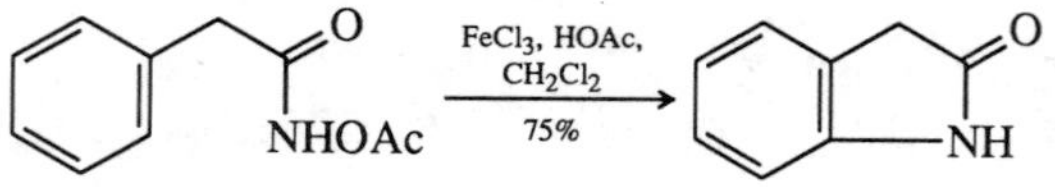

NHOAc → $FeCl_3$, HOAc, CH_2Cl_2, 75% → NH

[1] R. H. Frazier, Jr. and R. L. Harlow, *J. Org.*, **45**, 5408 (1980).
[2] M.-A. Poupart, G. Laosalle, and L. A. Paquette, *Tetrahedron Letters*, **29**, 269 (1988); *Org. Syn.*, submitted (1988).

[3] Commercial $FeCl_3$ is dried by treatment with thionyl chloride in refluxing DMF and then heated under high vacuum at 40°.
[4] M. Cherest and X. Lusinchi, *Tetrahedron Letters*, **30**, 715 (1989).

Ferric chloride–Acetic anhydride, 6,260.

t-Butyl ethers.[1] The *t*-butyl group is the most useful means for protection of ω-hydroxyalkyl lithium or Grignard reagents because the bulk hinders chelation of the metal with the ethereal oxygen. The *t*-butyl ethers can be prepared in quantitative yield by reaction in hexane with isobutene catalyzed by the solid acid catalyst Amberlyst H-15 (Aldrich, Fluka).

Deprotection can be effected at 25° in 90–95% yield by cleavage with the $FeCl_3/Ac_2O$ system in ether as the solvent. Under these conditions no isomerization of diene or enyne units is observed. After cleavage, addition of Na_2HPO_4 converts the $FeCl_3$ to insoluble $FePO_4$.

[1] A. Alexakis, M. Gardette, and S. Colin, *Tetrahedron Letters*, **29**, 2951 (1988).

Ferric chloride–Alumina. This supported reagent is prepared[1] by stirring $FeCl_3$ with basic aluminum oxide in CH_2Cl_2 for one hour followed by removal of the solvent.

Ene catalyst.[2] This catalyst affords high yields as well as high stereoselectivity in ene reactions. Thus the alkylidenemalonate **1**, prepared by Knoevenagel reaction of dimethyl malonate with (R)-citronellal, undergoes an ene cyclization in the presence of $FeCl_3/Al_2O_3$ to give essentially a single *trans*-disubstituted cyclohexane (**2**). This product was converted in several steps to the optically pure cadinane

$(CH_3OOC)_2CH_2$ + (R) $\xrightarrow{82\%}$ **1**

$\xrightarrow[92\%]{FeCl_3/Al_2O_3}$ **2**

2 $\xrightarrow{\text{several steps}}$ **3**

veticadinol (**3**). The ene cyclization is used to provide the desired *trans*-fusion of the decalin ring system.

[1] L. F. Tietze and U. Beifuss, *Synthesis*, 359 (1988).
[2] L. F. Tietze, U. Beifuss, J. Antel, and G. M. Sheldrick, *Angew. Chem. Int. Ed.*, **27**, 703 (1988).

Fluorine.

Oxidation of pyridinecarboxylic acids.[1] Reaction of a pyridinecarboxylic acid or the ester with 10% F_2 in N_2 results in oxidation to the corresponding 2-pyridone in 50–75% yield. An N-fluoropyridinium salt is probably an intermediate.

F_2, CH_3CN, H_2O, 0°; 56%

[1] M. Van Der Puy, D. Nalewajek, and G. E. Wicks, *Tetrahedron Letters*, **29**, 4389 (1988).

N-Fluoropyridinium trifluoromethanesulfonate, $^-$OTf (**1**), m.p. 185–187°, sensitive to moisture.

The salt is prepared by reaction of pyridine with sodium triflate and fluorine diluted in nitrogen or helium (caution: F_2 is very toxic and corrosive).

α-Fluorination of ketones.[1] This salt can fluorinate moderately reactive substances such as alkyl or silyl enol ethers and vinyl acetates. Of particular interest,

$OSi(CH_3)_3$ —1, CH_2Cl_2; 87%→

70%

(α/β = 1:2)

it reacts with the trienol silyl ether of a 3,11,17-triketosteroid to give selectively the 9α-fluoro-3,11,17-ketosteroid (50% yield).

The fluorinating power of this triflate can be markedly enhanced by introduction of chloro or alkoxycarbonyl groups into the ring system. Thus the triflates **2** and **3** are particularly potent fluorinating agents.

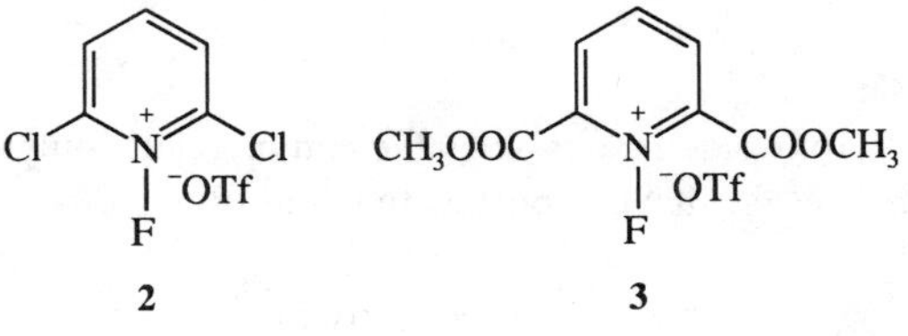

Cl Cl N⁺ $^{-}$OTf F **2** CH_3OOC $COOCH_3$ N⁺ $^{-}$OTf F **3**

[1] T. Umemoto, K. Tomita, and K. Kawada, *Org. Syn.*, submitted (1988); *Tetrahedron Letters*, **27**, 3271, 4465 (1986).

Formaldehyde–Sodium iodide.

Mannich cyclization.[1] Reaction of the alkynylamine **1** with formaldehyde, sodium iodide (3 equiv.), and camphorsulfonic acid (CSA, 1 equiv.) results directly in the piperidine **2** in 81–90% yield. In the absence of NaI, no cyclization is observed even after a reaction at 100° for 3 hours. The vinylic iodide group is useful for elaboration to exocyclic tetrasubstituted double bonds.

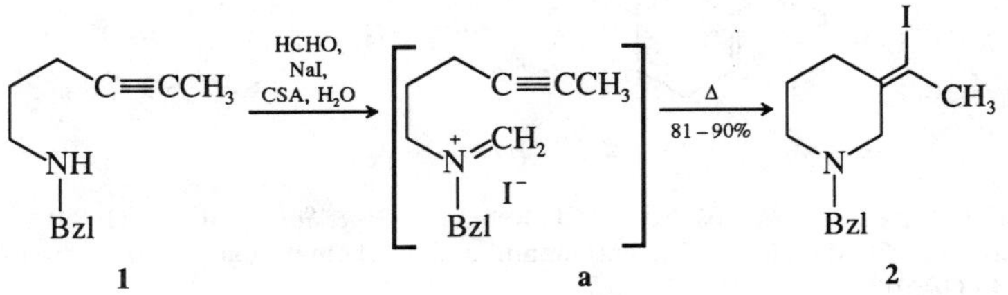

$C{\equiv}CCH_3$ NH Bzl **1** $\xrightarrow{HCHO,\ NaI,\ CSA,\ H_2O}$ [$C{\equiv}CCH_3$ $\overset{+}{N}{=}CH_2$ I^- Bzl] **a** $\xrightarrow[81-90\%]{\Delta}$ I CH_3 N Bzl **2**

[1] H. Arnold, L. E. Overman, M. J. Sharp, and M. C. Witschel, *Org. Syn.*, submitted (1989).

G

Glycolaldehyde (1).

This simplest hydroxy aldehyde is available commercially only as the dimer (**2**), m.p. 80–90°. Rebek *et al.*[1] have reported that the dimer dissociates to **1** in the presence of an acid. In the presence of LDA (4 equiv.) **2** reacts with the enolate of **3** and $ZnCl_2$ to form the lactone **4** in useful yield. A similar reaction of **2** with the ketone enolate of **4** forms a keto epoxide (**5**).[2]

[1] J. Wolfe, D. Nemeth, A. Costero, and J. Rebek, Jr. *Am. Soc.*, **110**, 983 (1988).
[2] P. Dinprasert, C. Mahidol, C. Thebtaranonth, and Y. Thebtaranonth, *Tetrahedron Letters*, **30**, 1149 (1989).

H

B-Halodiisopinocampheylboranes. Ipc_2BCl and Ipc_2BBr are prepared by reaction of Ipc_2BH with HCl and HBr, respectively. Ipc_2BI is obtained from reaction with I_2.

Cleavage of epoxides. As with simpler haloboranes, these reagents cleave epoxides of cycloalkenes to give, after nonoxidative workup, halohydrins in 65–90% yield. When carried out at −78 to −100°, the cleavage can show high enantioselectivity. Thus the halodiisopinocampheylboranes derived from (+)-pinene react with the oxide of cyclohexene or of cyclopentene to furnish (1R,2R) halo-

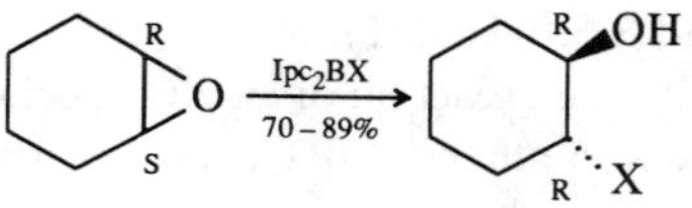

Ipc_2BX, 70–89%

X = Cl 22% ee
X = I 91% ee

hydrins in 22–91% ee. This reaction is a further example of the use of *meso*-compounds for asymmetric synthesis.[1]

[1] N. N. Joshi, M. Srebnik, and H. C. Brown, *Am. Soc.*, **110**, 6246 (1988).

3-Heptadecylmonoperphthalic acid, $CH_3(CH_2)_{16}$— (**1**).

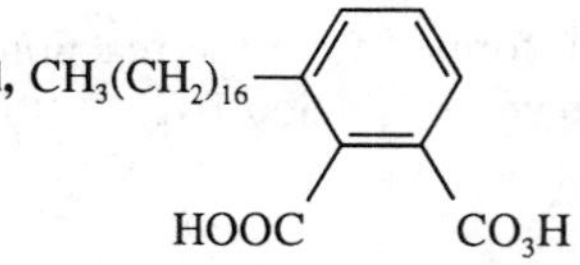

HOOC CO₃H

This surfactant type of a peracid is obtained in several steps from the maleic anhydride adduct **2** of heneicosadiene.

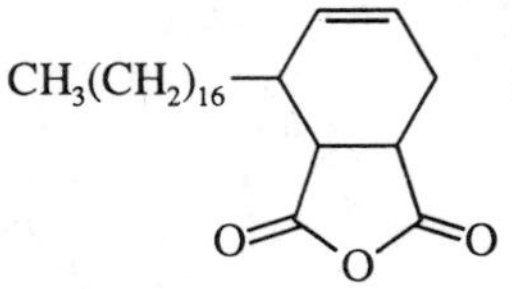

$CH_3(CH_2)_{16}$— **2**

Baeyer–Villiger oxidation.[1] This peracid converts β-ionone (**3**) selectively into the product of Baeyer–Villiger oxidation (**5**). Oxidation of **3** with monoperphthalic acid or *m*-chloroperbenzoic acid yields the epoxide (**4**) of **3** or of the

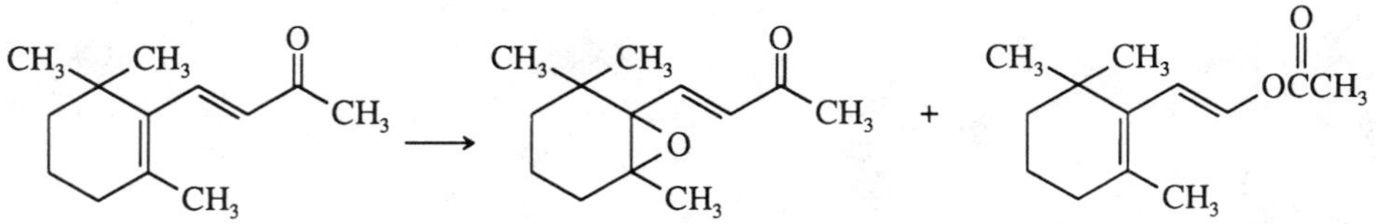

3 → 4 + 5

	4 : 5
$ClC_6H_4CO_3H$	90 : 10
$ClC_6H_4CO_3H$ + SDS	100 : 0
1 + $NaHCO_3$	11 : 83

Baeyer–Villiger product. Addition of sodium dodecyl sulfate (SDS) to the oxidation with *m*-chloroperbenzoic acid results in complete regiospecific epoxidation.

[1] Y. Fujise, K. Fujiwara, and Y. Ito, *Chem. Letters*, 1475 (1988).

Hexamethyldisilane, $(CH_3)_3SiSi(CH_3)_3$ (**1**).

Allyltrimethylsilanes.[1] The reaction of an allyl alcohol (1 equiv.) with CH_3Li (1.5 equiv.) and then with **1** in ether/HMPT (1:4) at 80° furnishes allyltrimethylsilanes. They are also formed from the reactions shown in equations I and II via the intermediates shown in equation III.

$$\text{(I)}\quad CH_2{=}CHCH{=}O + BuLi \xrightarrow[48\%]{\mathbf{1},\ \text{ether}} BuCH{=}CHCH_2Si(CH_3)_3\ (E)$$

$$\text{(II)}\quad n\text{-}C_5H_{11}CHO + CH_2{=}CHLi \xrightarrow[40\%]{\mathbf{1},\ \text{ether}} n\text{-}C_5H_{11}CH{=}CHCH_2Si(CH_3)_3\ (E)$$

$$\text{(III)}\quad \text{cyclohexanone} \xrightarrow{CH_2=CHLi} \left[\text{1-vinylcyclohexyl-OLi} \xrightarrow{\mathbf{1}} \text{1-vinylcyclohexyl-OSi}(CH_3)_3 + LiSi(CH_3)_3 \right] \xrightarrow{75\%} \text{cyclohexylidene}{=}CHCH_2Si(CH_3)_3$$

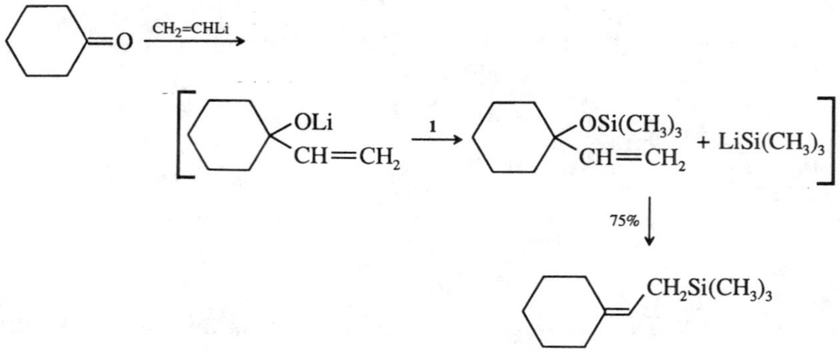

[1] J. R. Hwu, L. C. Lin, and B. R. Liaw, *Am. Soc.*, **110**, 7252 (1988).

Hexamethyldisilathiane, $S[Si(CH_3)_3]_2$ (**1**), b.p. 91–95°/100 mm, toxic.

The reagent is prepared[1] most conveniently in 80–88% yield by reduction of sulfur with sodium naphthalenide in THF (ultrasonic irradiation) to form Na_2S *in situ*, which is then allowed to react with chlorotrimethylsilane.

$$S + 2Na^+(C_{10}H_8)^{\dot{-}} \longrightarrow [Na_2S] \xrightarrow[80-88\%]{ClSi(CH_3)_3} \mathbf{1}$$

Cyclic sulfides and thiolactones.[2] On treatment with an alkyllithium, **1** is converted into a silane with liberation of "Li_2S" (or an equivalent), which reacts with $Br(CH_2)_nBr$ to form cyclic sulfides. Reaction of "Li_2S" with an ω-bromoacyl chloride provides thiolactones.

A similar reduction of selenium with a metal naphthalenide provides an *in situ* synthesis of alkali metal selenides and diselenides.[3]

[1] J.-H. So and P. Boudjouk, *Org. Syn.*, submitted (1988).
[2] K. Steliou, P. Salama, and J. Corriveau, *J. Org.*, **50**, 4969 (1985).
[3] D. P. Thompson and P. Boudjouk, *ibid.*, **53**, 2109 (1988).

Hexamethylphosphoric triamide $[(CH_3)_2N]_3P{=}O$, (HMPT).

[2,3]Wittig rearrangement.[1] Reaction of the chiral sulfone, **1**, from L-arabinose, with allyllithium (3 equiv.) in THF followed by addition of HMPT results in reaction of the aldehyde formed *in situ* by loss of phenylsulfinate with remaining allyllithium to give the alcohol **2** as a mixture of epimers. The *anti*-isomer (**2a**) undergoes completely regioselective epoxidation under the conditions of Mihelich (**11**,97–98), but with moderate diastereoselectivity to give **3**.

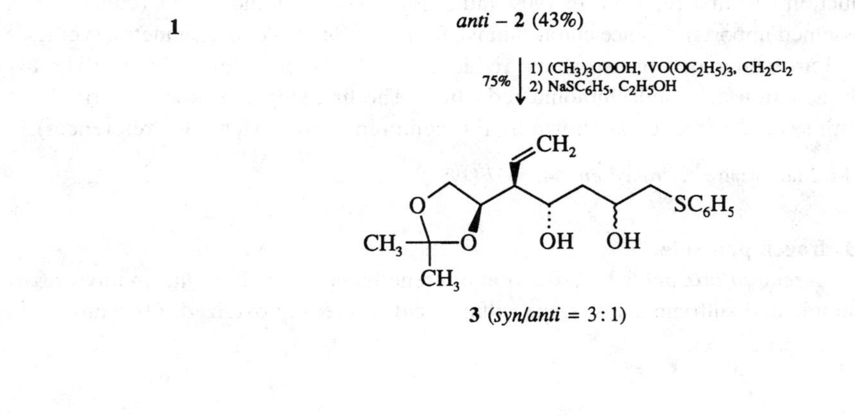

1) 3 C_3H_5Li
2) HMPT

1

+ *syn* – **2** (31%)

anti – **2** (43%)

75%
1) $(CH_3)_3COOH$, $VO(OC_2H_5)_3$, CH_2Cl_2
2) $NaSC_6H_5$, C_2H_5OH

3 (*syn*/*anti* = 3:1)

Sequential dialkylation.[2] The chiral γ-butyrolactone **1** can undergo sequential dialkylation, particularly when lithiation is conducted in THF in the presence of

$$(C_6H_5)_3CO\text{-lactone } \mathbf{1} \xrightarrow[59\%]{\text{1) LDA, HMPT, } CH_3I;\ \text{2) LDA, HMPT, } C_2H_5I} \mathbf{2}\ (99:1)$$

HMPT at −78°. Either diastereomer of **2** can be obtained in high optical purity by inverting the order of introduction of R^1X and R^2X.

[1] R. Brückner, *Tetrahedron Letters*, **29**, 5747 (1988).
[2] K. Tomioka, Y.-S. Cho, F. Sato, and K. Koga, *J. Org.*, **53**, 4094 (1988).

Hydridotris(triphenylphosphine)copper hexamer (1), 14,175.

Preparation:

$$NaOC(CH_3)_3 + CuCl \xrightarrow{C_6H_5CH_3} [CuOC(CH_3)_3]_4 \xrightarrow[H_2]{P(C_6H_5)_3} \underset{\mathbf{1}\ (50-65\%)}{[(C_6H_5)_3PCuH]_6} + NaCl + (CH_3)_3COH$$

The hydride hexamer as a solid is fairly stable to air, but decomposes readily in solution.[1]

[1] D. M. Brestensky, D. E. Huseland, C. McGettigan, and J. M. Stryker, *Tetrahedron Letters*, **29**, 3749 (1988).

Hydriodic acid.

Reduction of sulfoxides to thioethers.[1] Use of hydrogen halides for this reduction was first reported in 1909 and is still a viable method. This reduction has assumed importance since chiral sulfinyl groups are valuable in asymmetric syntheses and are eliminated in two steps: reduction to the ether followed by catalytic hydrogenation or metal/ammonia reduction. The first step can now be carried out with several reagents, as shown by this comprehensive review (349 references).

[1] M. Madesclaire, *Tetrahedron*, **44**, 6537 (1988).

Hydrogen peroxide.

Arenesulfinic acids.[1] Oxidation of arenethiols generally results in mixtures of sulfinic and sulfonic acids, since sulfinic acids are easily oxidized. Oxidation with

H_2O_2 (30%) in a basic medium provides sodium sulfinates, which are stable to H_2O_2.

$$C_6H_5SH + H_2O_2 \xrightarrow{NaOH, H_2O,\ C_2H_5OH} [C_6H_5SO_2Na] \xrightarrow[86\%]{HCl,\ H_2O} C_6H_5SO_2H$$

[1] T. Kamiyama, S. Enomoto, and M. Inoue, *Chem. Pharm. Bull.*, **36**, 2652 (1988).

Hydrogen peroxide–Benzeneseleninic acid, $C_6H_5SeO_2H$.

ArCHO → ArOH.[1] Aromatic aldehydes undergo Baeyer–Villiger reaction when treated with H_2O_2 (30%) in the presence of several selenium compounds as catalysts, of which areneseleninic acids are most effective. The resulting aryl formates are readily hydrolyzed to phenols.

[1] L. Syper, *Synthesis*, 167 (1989).

Hydrogen peroxide–Potassium carbonate.

Oxidation of 2-nitrocycloalkanones.[1] These substrates are oxidized by H_2O_2 (30%) and K_2CO_3 in CH_3OH to dicarboxylic acids or keto acids, depending on whether the NO_2 group is secondary or tertiary.

$$\text{2-nitro-4-}C(CH_3)_3\text{-cyclohexanone} \xrightarrow[88\%]{H_2O_2,\ K_2CO_3,\ CH_3OH} HOOC(CH_2)_2CH(C(CH_3)_3)CH_2COOH$$

$$\text{2-nitro-2-}CH_3\text{-cyclohexanone} \longrightarrow HOOC(CH_2)_4C(=O)CH_3$$

[1] R. Ballini, E. Marcantoni, M. Petrini, and G. Rosini, *Synthesis*, 915 (1988).

Hydrogen peroxide–Potassium fluoride.

α-Alkylbenzyl alcohols.[1] An interesting synthesis of optically active alcohols of this type from acylsilanes involves enantioselective reduction as the first step. Thus the acylsilane **1** is reduced by diborane and a chiral α-amino alcohol [(S)-

$$CH_3\overset{O}{\overset{\|}{C}}Si(C_6H_5)_3 \ (\mathbf{1}) \xrightarrow[71\%]{B_2H_6,\ ROH^*} \text{(H)(OH)}C(CH_3)Si(C_6H_5)_3 \ (\mathbf{2},\ 94\%\ ee)$$

$$\mathbf{2} \xrightarrow[\Delta\ (100\%)]{Ac_2O,\ DMAP\ (95\%)} \text{(H)}(C_6H_5)C(CH_3)Si(C_6H_5)_2OAc \ (\mathbf{3}) \xrightarrow[50\%]{H_2O_2,\ KF,\ KHCO_3,\ CH_3OH} \text{(H)}(C_6H_5)C(CH_3)OH \ (\mathbf{4})$$

(−)-2-amino-3-methyl-1,1-diphenylbutanol-1][2] to **2** in 94% ee, probably because of the bulky nature of the silyl group. Brook rearrangement of the acetate of **2** provides **3**, which is then subjected to oxidation with retention of configuration (**12**, 243–244).

[1] J. D. Buynak, J. B. Strickland, T. Hurd, and A. Phan, *J.C.S. Chem. Comm.*, 89 (1989).
[2] S. Itsuno, M. Nakano, K. Miyazaki, H. Masuda, K. Ito, A. Hirao, and S. Nakahama, *J.C.S. Perkin I*, 2039 (1985).

Hydrogen peroxide–Trichloroacetic acid.

Cleavage of acid-labile protective groups.[1] The reaction of H_2O_2 (70%, FMC) and Cl_3CCOOH in CH_2Cl_2/*t*-butyl alcohol converts a dimethyl acetal (**1**) into a hydroperoxy methyl acetal (**a**), which can be isolated but which for convenience (and safety) is reduced to the aldehyde **2** in 80% overall yield. The same conditions can effect oxidative cleavage of tetrahydropyranyl and trityl ethers.

$$RCH(OCH_3)_2 \ (\mathbf{1}) \xrightarrow{CH_2Cl_2,\ (CH_3)_3COH,\ H_2O_2,\ Cl_3CCOOH} \left[RCH\begin{matrix}OCH_3\\OOH\end{matrix}\right] (\mathbf{a}) \xrightarrow[80\%]{S(CH_3)_2} RCHO \ (\mathbf{2})$$

The paper discusses the safety precautions necessary in all manipulations with hydrogen peroxide.

[1] A. G. Myers, M. A. M. Fundy, and P. A. Lindstrom, Jr., *Tetrahedron Letters*, **29**, 5609 (1988).

Hydrosilanes–Tetrabutylammonium fluoride.

Reduction of aldehydes or ketones.[1] This reduction can be effected with hydrosilanes, particularly sterically hindered silanes such as $(C_6H_5)_2SiH_2$ and

$C_6H_5(CH_3)_2SiH$, in the presence of Bu_4NF or tris(diethylamino)sulfonium difluorotrimethylsilicate (TASF, **13**, 336) in aprotic polar solvents, particularly HMPT. The actual reagent may well be a hexavalent fluorosilicate involving HMPT. CsF and KF are far less effective than Bu_4NF. With proper control of all the parameters, near quantitative yields can be achieved, and high stereoselectivity is also possible, as in the case with methylcyclohexanone.

2-Methylcyclohexanone → (Bu_4NF, HMPT) → cis- and trans-2-methylcyclohexanol

$C_6H_5(CH_3)_2SiH$ 99% 76:24
$(C_6H_5)_2CH_3SiH$ 81% 94:6

This reduction is *anti*-selective in the reduction of α-oxy and α-amino ketones. This contrasts with *syn*-selectivity for metal hydride reagents and for hydrosilanes in trifluoroacetic acid.

$C_6H_5COCH(OAc)CH_3$ → ($C_6H_5(CH_3)_2SiH$, Bu_4NF; 95%) → $C_6H_5CH(OH)CH(CH_3)OAc$ (anti) + (syn) 95:5

$C_6H_5COCH(CH_3)CON(C_2H_5)_2$ → (98%) → $C_6H_5CH(OH)CH(CH_3)CON(C_2H_5)_2$ (anti) + (syn) >99:1

[1] M. Fujita and T. Hiyama, *J. Org.*, **53**, 5405, 5415 (1988).

β-Hydroxyethyltriphenylarsonium bromide, $[(C_6H_5)_3\overset{+}{A}sCH_2CH_2OH]Br^-$ (**1**). The salt is prepared in 69% yield by reaction of $(C_6H_5)_3As$ and $BrCH_2CH_2OH$ at 110°.

2,3-Epoxy-3-arylpropanols.[1] The salt reacts with aryl aldehydes in THF containing a trace of H_2O and solid KOH to form these epoxides in 60–85% yield via a solid–liquid phase-transfer process.

ArCHO + **1** → (60–85%) → ArCH—CHCH$_2$OH (epoxide, O bridging)

(E/Z = 74–91:26–9)

[1] L. Shi, W. Wang, and Y.-Z. Huang, *Tetrahedron Letters*, **29**, 5295 (1988).

Hydroxylamine hydrochloride.

Pyridine synthesis.[1] Dihydropyrans (**1**), formed by cycloaddition of aryl enones with vinyl ethers (**12**,561), can serve as protected 1,5-dicarbonyl compounds and react with hydroxylamine (2 equiv.) to form 2,4-disubstituted pyridines (**2**).

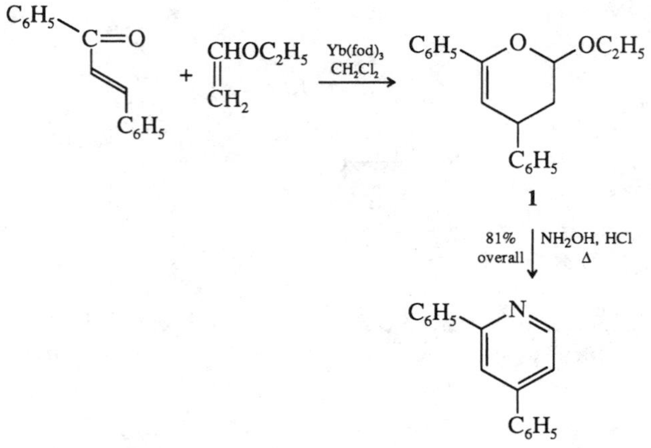

C_6H_5 $C=O$ C_6H_5 + $CHOC_2H_5$ CH_2 $\xrightarrow{Yb(fod)_3,\ CH_2Cl_2}$ C_6H_5 O OC_2H_5 C_6H_5 **1**

81% overall; NH_2OH, HCl, Δ

C_6H_5 N C_6H_5 **2**

At least one aryl group in the enone is essential. The reaction provides the skeleton of streptonigrinoids (equation I).

(I) N O CH_3O CH_3O OCH_3 $\xrightarrow[25-30\%]{2\ steps}$ N N CH_3O CH_3O OCH_3

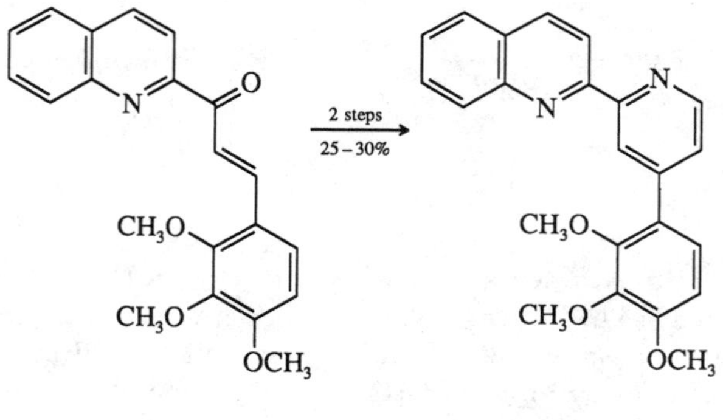

[1] M. A. Ciufolini and N. E. Byrne, *J.C.S. Chem. Comm.*, 1230 (1988).

Hydroxylamine-O-sulfonic acid, H_2NOSO_3H.

Isothiazoles.[1] These heterocycles can be prepared by reaction of α-acetylenic aldehydes or ketones with hydroxylamine-O-sulfonic acid and then with sodium hydrosulfide in a buffered aqueous solution.

$$HC(=O)C{\equiv}CH \xrightarrow{H_2NOSO_3H} \left[HC(=NOSO_3H){-}C{\equiv}CH \right] \xrightarrow[42\%]{NaSH,\ NaHCO_3} \text{isothiazole} + Na_2SO_4$$

[1] F. Lucchesini, N. Picci, M. Pocci, A. De Munno, and V. Bertini, *Heterocycles*, **29**, 97 (1989).

(R)-3-Hydroxybutyric acid (1), **14**,178.

Amination.[1] The reaction of the dioxanone **2**, derived from (R)-3-hydroxybutanoic acid, with di-*t*-butyl azodicarboxylate (**14**,115–116) provides the derivative **3** in high diastereomeric excess. The same amination of an alkyl (R)-3-hydroxy-

$$\mathbf{2} \xrightarrow[95\%]{BocN{=}NBoc} \mathbf{3}\ (99\%\ de)$$

$$\mathbf{4} \xrightarrow[58\%]{BocN{=}NBoc} \mathbf{5}\ (64\%\ de) \xrightarrow[50\%]{} \mathbf{6}$$

butanoate (**4**) provides **5** in 58% chemical yield and in 64% optical yield. Deprotection and hydrogenation of **5** provides D-allothreonine (**6**) in 50% yield.

[1] J. P. Genet, S. Juge, and S. Mallart, *Tetrahedron Letters*, **29**, 6765 (1988).

I

Iodine.

5-Alkylidene-4,5-dihydrofurans.[1] Reaction of 2-alkenyl 1,3-dicarbonyl compounds (**1**) with I_2 effects cyclization to iodoalkyldihydrofurans (**2**). Dehydroiodination (DBU) of **2** results in 5-alkylidene-4,5-dihydrofurans (**3**), which undergo acid-catalyzed isomerization to furans (**4**).

OC_2H_5 O= O= =CHCH$_3$ CH_3 **1**, *trans/cis* = 85:15 —I_2, Na_2CO_3, CH_2Cl_2, 98%→ **2** (85:15) [O=C(OC$_2$H$_5$), CH$_3$, O, $\overset{*}{C}$HCH$_3$, I] —DBU, 99%→ **3** (E, 98%, from *erythro*-2) [O=C(OC$_2$H$_5$), CH$_3$, O, =C(CH$_3$)H] —70% H_3O^+→ **4** [O=C(OC$_2$H$_5$), CH$_3$, O, CH_2CH_3]

ArCH(CH₃)COOH. These acids (or the esters) have been obtained by a 1,2-aryl migration of propiophenones (*cf.* **12**, 452). This rearrangement is effected most easily with iodine or ICl_3 in trimethyl orthoformate.[2,3]

$$p\text{-RC}_6\text{H}_4\overset{\text{O}}{\overset{\|}{\text{C}}}\text{CH}_2\text{CH}_3 \xrightarrow{I_2,\ CH(OCH_3)_3} \left[p\text{-RC}_6\text{H}_4\text{C}(\text{OCH}_3)_2\text{CH(I)CH}_3\right] \xrightarrow{I_2} p\text{-RC}_6\text{H}_4\overset{\text{CH}_3}{\overset{|}{\text{C}}}\text{HCOOCH}_3 + \text{CH}_3\text{I}$$

R = H 66%
R = *i*-Bu 98%

Reactions with I₂ on alumina. A mixture of an arene and iodine when shaken with alumina previously dehydrated at 400° can undergo iodination at temperatures above 110°. The method is useful for reactive arenes. Thus azulene can be converted to the 1,3-diiodo derivative in quantitative yield.

The reaction of I_2/Al_2O_3 with alkenes even in the presence of NaCl does not result in iodo chlorides but in alkyl iodides. Evidently HI is generated by reaction of I_2 with surface hydroxyls. This reaction can even take place with nonactivated alumina, but yields are higher (30–85%) with activated Al_2O_3. This reaction does not occur on silica gel.

The reaction of I_2/Al_2O_3 with alkynes furnishes a practical route to (E)-*vic*-diiodoalkenes in 82–92% yield. Yields are lower (25–50%) if the alkyne is substituted by a carboxyl group.[4]

1-Iodo-1-alkynes.[5] These compounds are usually prepared by iodination of lithium acetylides, but they can also be obtained in 70–90% yield by iodination of 1-alkynes in DMF catalyzed by CuI under phase-transfer conditions (Bu_4NCl, Na_2CO_3, or K_2CO_3 as base).

[1] R. Antonioletti, F. Bonadies, and A. Scettri, *Tetrahedron Letters*, **29**, 4987 (1988).
[2] T. Yamauchi, K. Hattori, K. Nakao, and K. Tamaki, *J. Org.*, **53**, 4858 (1988).
[3] Similar results have been reported in a patent [*C. A.*, **105**, 6321p (1986)].
[4] R. M. Pagni, G. W. Kabalka, R. Boothe, K. Gaetano, L. J. Stewart, R. Conaway, C. Dial, D. Gray, S. Larson, and T. Luidhardt, *J. Org.*, **53**, 4477 (1988).
[5] T. Jeffery, *J.C.S. Chem. Comm.*, 909 (1988).

Iodine–Copper(II) acetate.

cis-Diols. The reaction of an alkene with I_2 and $Cu(OAc)_2$ in HOAc provides a mixture of products which is hydrolyzed by KOH in aqueous CH_3OH to a *cis*-diol. In the case of Δ^2-cholestene, this reaction results in the more hindered *cis*-diol. Yields are drastically lower if propionic acid replaces acetic acid.

1) I_2, $Cu(OAc)_2$, HOAc
2) KOH, CH_3OH/H_2O
80%

91%

(2β, 3β-diol)

[1] C. A. Horiuchi and J. Y. Satoh, *Chem. Letters*, 1209 (1988).

Iodine–Hydrogen peroxide.

Phenols → quinones. Oxidation of phenols to quinones can be effected with 1 eq. of H_2O_2 (60%) and excess iodine in methanol at room temperature. Yields are generally 70–98%. The actual oxidant is probably iodine and the function of

H_2O_2 is to reoxidize the HI formed to I_2. The oxidation of 2,6-disubstituted phenols proceeds in high yield if bromine replaces iodine in this oxidation. This oxidation probably involves a 4,4-dibromocyclohexadienone intermediate, which can be isolated from oxidation of 2,6-di-*t*-butylphenol.

[1] F. Minisci, A. Citterio, E. Vismara, F. Fontana, S. DeBernardinis, and M. Correales, *J. Org.*, **54**, 728 (1989).

Iodine–Mercury(II) oxide.

Ring expansion of cycloalkanones.[1] The hypoiodite reaction can also be used for a four-atom ring expansion of cyclic ketones via the corresponding lactols (cf., **13**, 150).

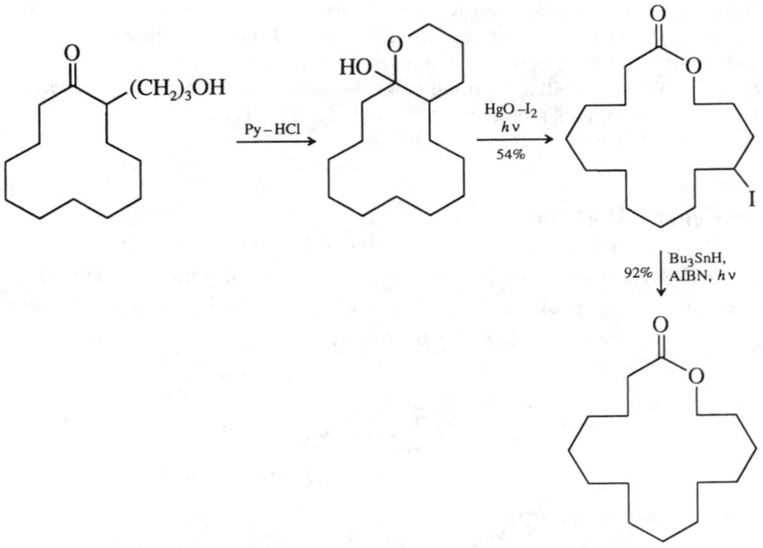

[1] H. Suginome and S. Yamada, *Chem. Letters*, 245 (1988).

Iodobenzene dichloride [Phenyliodine(III) dichloride], $C_6H_5ICl_2$ (**1**).

***Remote chlorination* (*6*, 298–300).** Breslow has extended his remote functionalization of steroids to a double functionalization at C_9 and C_{17} of cholestanol. Thus the ester (**2**) of cholestanol, prepared from a *p*-iodophenylnicotinic acid, when treated with **1** (1.2 equiv.) is chlorinated selectively at C_9; chlorination of the same ester with 3 equiv. of $C_6H_5ICl_2$ results in chlorination at C_9 and at C_{17} in quantitative yield. The *para*-iodo group of the ester plays an important role in this remote chlorination.

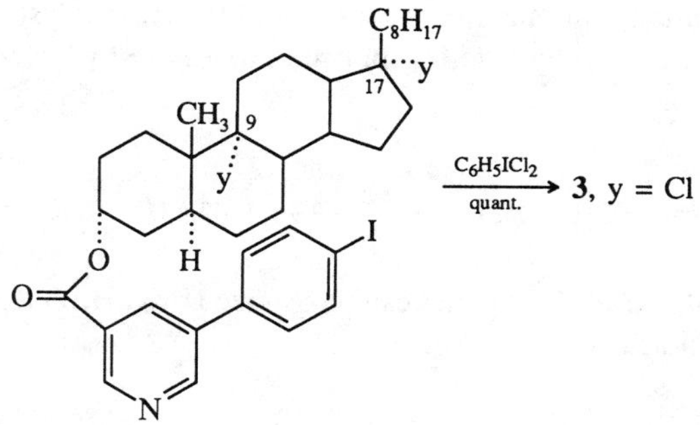

$$\mathbf{2},\ y = H \xrightarrow[\text{quant.}]{C_6H_5ICl_2} \mathbf{3},\ y = Cl$$

2, y = H

[1] R. Batra and R. Breslow, *Tetrahedron Letters*, **30**, 535 (1989).

Iodosobenzene tetrafluoroborate, $(C_6H_5\overset{+}{I})_2O\ 2BF_4^-$ (**1**). This yellow salt, m.p. 160° dec., is obtained by reaction of $C_6H_5I(OAc)_2$ with HBF_4 (75% yield.)

It converts enol silyl ethers (**2**) into 1,4-diketones (**3**).[1]

$$\underset{\mathbf{2}}{C_6H_5C(OSi(CH_3)_3){=}CH_2} \xrightarrow{\mathbf{1}} \underset{\mathbf{3}}{C_6H_5COCH_2CH_2COC_6H_5}$$

$$\mathbf{2} \xrightarrow{C_6H_5IO\cdot HBF_4} \underset{\mathbf{4}}{C_6H_5\overset{+}{I}CH_2COC_6H_5BF_4^-}$$

In contrast, a reagent obtained by reaction of C_6H_5IO with HBF_4 in CH_2Cl_2 at −20°, and formulated as $C_6H_5IO{\cdot}HBF_4$, converts **2** into an unstable iodonium salt **4**.[2]

[1] V. V. Zhdankin, R. Tykwinski, R. Caple, B. Berglund, A. S. Koz'min, and N. S. Zefirov, *Tetrahedron Letters*, **29**, 3717 (1988).
[2] *Idem, ibid.*, **29**, 3703 (1988).

Iodosylbenzene, C_6H_5IO.

Amines → Imines.[1] This reagent can effect dehydrogenation of secondary amines activated by an aryl group or an α,β-double bond to imines. The rate and the yield are increased by addition of $Cl_2Ru[P(C_6H_5)_3]_3$ and molecular sieves. In this respect iodosylbenzene differs from *t*-butyl hydroperoxide, which also effects this dehydrogenation, but only in the presence of the Ru catalyst (**13**, 54). Primary amines are also oxidized by C_6H_5IO, but the imines are hydrolyzed *in situ* to the carbonyl precursor.

$$C_6H_5CH{=}CHCH_2NHC_6H_5 \xrightarrow[86\%]{C_6H_5IO,\ Ru(II),\ CH_2Cl_2,\ 25°} C_6H_5CH{=}CHCH{=}NC_6H_5$$

Oxidation of amines.[2] Amines can be oxidized by C_6H_5IO to nitriles, ketones, and lactams as indicated.

$$n\text{-}C_5H_{11}CH_2NH_2 \xrightarrow[57\%]{2C_6H_5IO,\ CH_2Cl_2} n\text{-}C_5H_{11}C{\equiv}N$$

Cyclohexylamine $\xrightarrow[49\%]{C_6H_5IO}$ cyclohexanone

Pyrrolidine $\xrightarrow[49\%]{C_6H_5IO}$ 2-pyrrolidinone

It can effect oxidative decarboxylation of cyclic amino acids, possibly by dehydrogenation to an imine.

Pyroglutamic acid (COOH) $\xrightarrow[27\%]{2C_6H_5IO}$ succinimide

Vinyliodonium salts.[3] Reaction of vinylsilanes (**2**) with iodosylbenzene catalyzed by BF_3 etherate results in vinyliodonium tetrafluoroborates (**3**) in generally

$(CH_3)_3C$–cyclohexenyl–$Si(CH_3)_3$ (**2**) $\xrightarrow[80\%]{C_6H_5IO,\ BF_3\cdot O(C_2H_5)_2,\ CH_2Cl_2,\ 0°}$ $(CH_3)_3C$–cyclohexenyl–$I^+(C_6H_5)\ BF_4^-$ (**3**) $\xrightarrow[55\%]{NaNO_2}$ $(CH_3)_3C$–cyclohexenyl–NO_2

3 $\xrightarrow[73-90\%]{R_2CuLi}$ $(CH_3)_3C$–cyclohexenyl–R

high yield. These salts, in contrast to vinylsilanes, react very readily with nucleophiles with the iodine (III) substituent as the leaving group.

Alkenyliodonium tetrafluoroborates (**14**, 10–11). On treatment with base these salts can undergo elimination of iodobenzene to generate a carbenoid that undergoes 1,5-C—H insertion (equation I). This elimination can compete with a 1,2-

(I) CH_3, $SnBu_3$ — 1) PhIO, $BF_3 \cdot O(C_2H_5)_2$; 2) $NaBF_4$; 80% → CH_3, BF_4^- $I^+C_6H_5$ — $KOC(CH_3)_3$; 61% → H, O, CH_3, H

rearrangement of a β-alkyl group to produce an alkyne as in the case of the salt **2**. This α-elimination of alkenyliodonium salts is believed to involve a free carbene

CH_3, $C_6H_5\overset{+}{I}$, BF_4^-, C_6H_{13}, O (**2**) — $N(C_2H_5)_3$; 67% → CH_3, C_6H_{13} + $CH_3C{\equiv}C(CH_2)_3\overset{O}{\overset{\|}{C}}C_6H_{13}$

67:33

rather than a carbenoid, since the (E)- and (Z)-isomers of **3** on treatment with base produce essentially the same products.[4]

(II) C_8H_{17}, Bu, $I^+C_6H_5$, BF_4^- — $KOC(CH_3)_3$, 0° → C_8H_{17}, CH_3 + Bu, C_5H_{11}

(E)-**3**	63%	49:51
(Z)-**3**	65%	51:49

α-Oxo triflates.[5] Reaction of trimethylsilyl enol ethers with C_6H_5IO and trimethylsilyl trifluoromethanesulfonate (TMSOTf) in CH_2Cl_2 at −78° affords α-oxo triflates in 55–75% yield.

$$\mathrm{ArC(OTMS){=}CHCH_3 + C_6H_5IO + TMSOTf \xrightarrow[77\%]{} ArC(=O)CH(CH_3)OTf}$$

[1] P. Müller and D. M. Gilabert, *Tetrahedron*, **44**, 7171 (1988).
[2] R. M. Moriarty, R. K. Vaid, M. P. Duncan, M. Ochiai, M. Inenaga, and Y. Nagao, *Tetrahedron Letters*, **29**, 6913, 6917 (1988).
[3] M. Ochiai, K. Sumi, Y. Takaoka, M. Kunishima, Y. Nagao, M. Shiro, and E. Fujita, *Tetrahedron*, **44**, 4095 (1988).
[4] M. Ochiai, Y. Takaoka, and Y. Nagao, *Am. Soc.*, **110**, 6565 (1988).
[5] R. M. Moriarty, W. R. Epa, R. Penmasta, and A. K. Awasthi, *Tetrahedron Letters*, **30**, 667 (1989).

N-Iodosuccinimide (NIS).

Iodoacetalization. Reaction of a vinyl ether with NIS and an alcohol provides an iodoacetal, which undergoes elimination of HI when treated with base to provide a ketene acetal. This sequence is particularly useful for synthesis of unsymmetrical ketene acetals.

$$\mathrm{C_2H_5OCH{=}CH_2 \xrightarrow[80\%]{NIS,\ (CH_3)_3COH} C_2H_5OCH(CH_2I)OC(CH_3)_3 \xrightarrow[57\%]{KOC(CH_3)_3,\ THF,\ 25^\circ} C_2H_5OC(=CH_2)OC(CH_3)_3}$$

[1] D. S. Middleton and N. S. Simpkins, *Syn. Comm.*, **19**, 21 (1989).

Ion-exchange resins.

Cleavage of hydrazones, oximes, and semicarbazones. Amberlyst 15, an ion-exchange resin with sulfonic acid groups, is a superior reagent for exchange of nitrogeneous derivatives of carbonyl compounds with acetone. Yields are 85–95% with reactions that proceed at room temperature.[1]

Esterification.[2] Acids are esterified in high yield when dissolved in methanol containing Amberlyst 15. The reaction is effected at 25° in 4–14 hours with yields >80%. Chiral acids are esterified without epimerization.

[1] R. Ballini and M. Petrini, *J.C.S. Perkin I*, 2563 (1988).
[2] M. Petrini, R. Ballini, E. Marcantoni, and G. Rosini, *Syn. Comm.*, **18**, 847 (1988).

Iron^{3+}-Montmorillonite (Fe-Mont) or Aluminum^{3+}-Montmorillonite (Al-Mont).

Preparation.[1]

N-Silyldihydropyridines.[2] Pyridines substituted at the *meta*-position by an electron-withdrawing group react with silyl ketene acetals in the presence of Fe- or Al-Mont (**1**) to give a 4-substituted N-silyl-1,4-dihydropyridine in high yield.

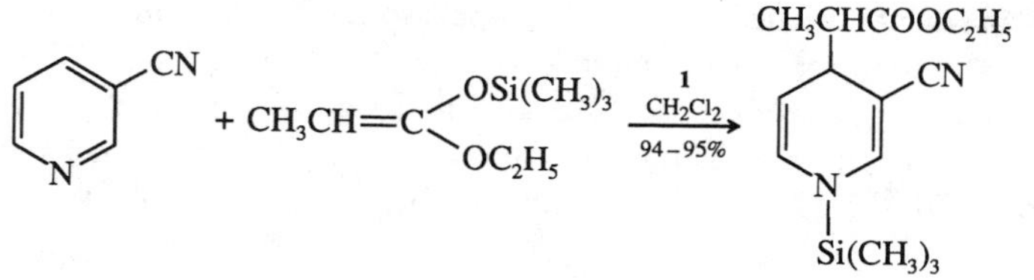

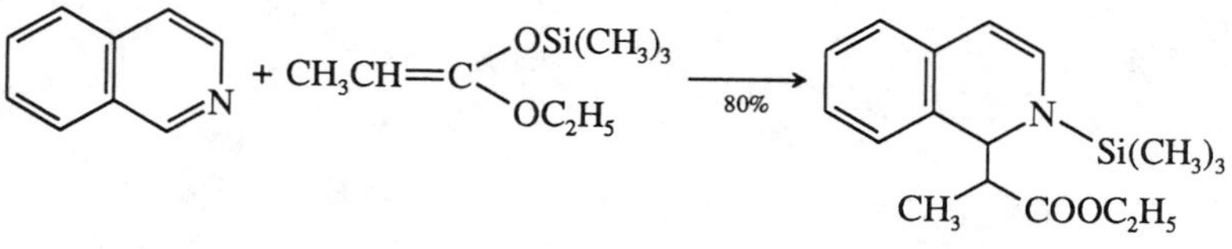

CN; + $CH_3CH{=}C(OSi(CH_3)_3)(OC_2H_5)$; 1, CH_2Cl_2, 94–95%; $CH_3CHCOOC_2H_5$; CN; N; $Si(CH_3)_3$

N; + $CH_3CH{=}C(OSi(CH_3)_3)(OC_2H_5)$; 80%; N; $Si(CH_3)_3$; CH_3; $COOC_2H_5$

[1] M. Kawai, M. Onaka, and Y. Izumi, *Bull. Chem. Soc. Japan*, **61**, 1237 (1988).
[2] M. Onaka, R. Ohno, and Y. Izumi, *Tetrahedron Letters*, **30**, 747 (1989).

2,3-Isopropylidene-2,3-dihydroxy-1,4-bis(diphenylphosphine)butane (DIOP), **4**, 273; **5**, 360–361; **9**, 259–260; **13**, 153.

Asymmetric hydrogenation of itaconic acids.[1] Japanese chemists have prepared a new bisphosphine ligand (**2**), which is more efficient than DIOP for asymmetric hydrogenation of itaconic acids when complexed with rhodium. It is available in four steps from 4-bromo-2,6-dimethylphenol.

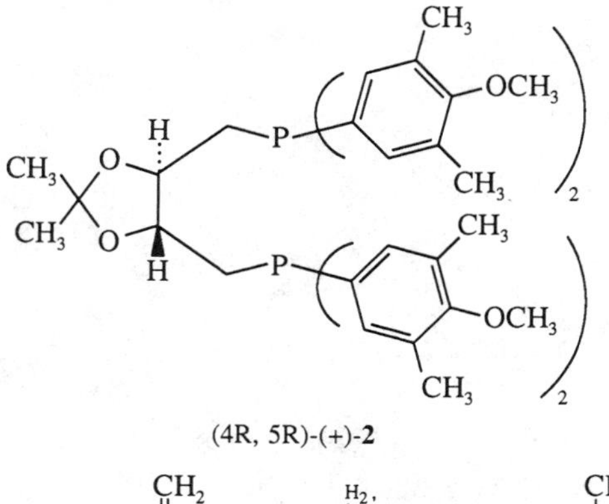

CH_3; OCH_3; CH_3; 2; H; P; CH_3; O; CH_3; O; H; P; CH_3; OCH_3; CH_3; 2

(4R, 5R)-(+)-**2**

HOOC, CH_2, COOH; H_2, Rh^+-(+)-**2**, 100%; HOOC, CH_3, *, COOH

S, 86–91% ee

HOOC, C_6H_5, COOH; 79%; HOOC, C_6H_5, *, COOH

S, 96% ee

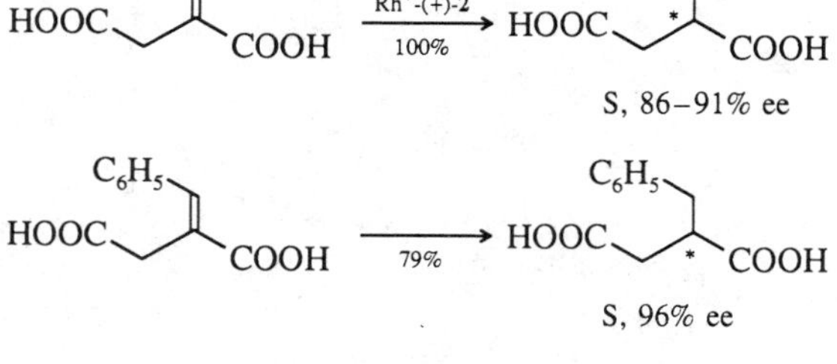

The enhanced enantioselectivity of **2** in respect to DIOP is considered to derive from the *para*-methoxy group and methyl groups.

DIOP analogs.[2] Japanese chemists have prepared a symmetrical DIOP (**3**) in which the phosphine is substituted by a *para*-dimethylamino group and two unsymmetrical DIOPS (**4** and **5**), and have examined the effect on asymmetric

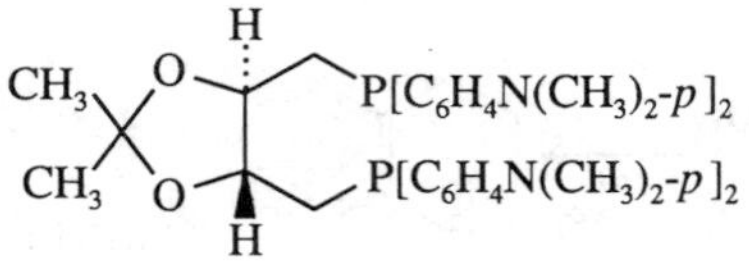

3, $\alpha_D + 23°$

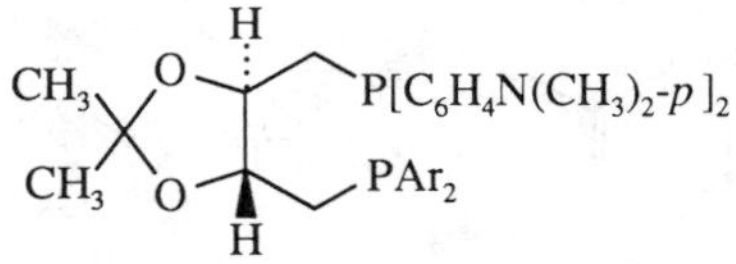

4, Ar = C_6H_5, α_D -1.70°
5, Ar = $C_6H_4CH_3$-*m*, $\alpha_D = 8.4°$

hydrogenation of rhodium complexes formed from $[Rh(COD)Cl]_2$ or from $Rh^+(COD) \cdot BF_4^-$. Catalysts prepared from **3**, **4**, and **5** show greater chemical reactivity than the catalyst prepared from DIOP and somewhat higher enantioselectivity in hydrogenation of itaconic acid and dimethyl itaconate. Therefore, an electron-rich phosphine has some effect on the rate and enantioselectivity.

[1] T. Morimoto, M. Chiba, and K. Achiwa, *Tetrahedron Letters*, **30**, 735 (1989).
[2] T. Morimoto, M. Chiba, and K. Achiwa, *ibid.*, **29**, 4755 (1988).

L

(S)-Lactic acid, HOOC(H)C(CH$_3$)OH **(S)-1.**

Asymmetric Diels–Alder reaction.[1] The (2S)-2-*t*-butyl-5-methylene-1,3-dioxolan-4-one (**3**), prepared from pivaldehyde and (S)-(+)-lactic acid (**1**), reacts with cyclopentadiene at 25° to form only two of the four possible Diels–Alder adducts,

$(CH_3)_3CCHO$ + (S)-**1** $\xrightarrow{\text{TsOH}}$ (2S, 5S)-**2** (62% de) $\xrightarrow[\text{2) } N(C_2H_5)_3 \text{ (60\%)}]{\text{1) NBS (98\%)}}$ (2S)-**3**

(2S)-**3** + cyclopentadiene $\xrightarrow[86\%]{25°}$ (1R, 2R, 4R)-**4** + (1S, 2R, 4S)-**4** (96:4)

both formed by 96% face selectivity for the methylene group and with 96% *exo*-selectivity. A reaction catalyzed by $Cl_2Ti(O\text{-}i\text{-}Pr)_2$ at −20° gives (1R,2R,4R)-**4**, (1S,2R,4S)-**4**, and (1S,2S,4S)-**4** in the ratio 59:1:40 in 62% chemical yield. Thus a Lewis-acid catalyst lowers the overall yield, increases *exo*-selectivity, but decreases π-face selectivity. Similar results obtain in reactions catalyzed by $C_2H_5AlCl_2$.

[1] J. Mattay, J. Mertes, and G. Maas, *Ber.*, **122**, 327 (1989).

(S)-Lactic aldehyde, HCOCH(OH)CH$_3$ **(1)**. Preparation.[1]

Optically active β-lactams.[2] The reaction of the enolate of *t*-butyl butanoate with the optically active silylimine **2**, prepared from silylated **1**, provides the optically active β-lactam **3**. The optical purity and configuration of **3** was established by conversion to the antibiotic carbapenem (+)-PS-5 (**4**).

$$1 \xrightarrow{LiN[SiCH_3)_3]_2} \mathbf{2} \xrightarrow[\text{2) HF, } CH_3CN \text{ (97\%)}]{\text{1) } C_2H_5CH{=}C(O\text{-}t\text{-}Bu)(OLi) \text{ (61\%)}} \mathbf{3}$$

(+)–PS–5 (**4**)

[1] M. Hirama, I. Nishizaki, T. Shigemoto, and S. I. Ito, *J.C.S. Chem. Comm.*, 393 (1986).
[2] G. Cainelli, M. Panunzio, D. Giacomini, G. Martello, and G. Spunta, *Am. Soc.*, **110**, 6879 (1988).

Lead(II) bromide/Aluminum, $PbBr_2/Al$ (**1**).

Addition of CX_4 to CHO.[1] This bimetallic system effects addition of CX_4 to aldehydes in DMF at 25° in >75% yield. $PbBr_2$ can be replaced by $PbCl_2$, Pb, or $SnCl_2$ with only slightly lower yields. The system also effects a 1,2-elimination from

$$RCHO + CX_4 \xrightarrow[75-98\%]{\mathbf{1}\text{, DMF, 25°}} RCH(OH)CX_3 \quad (X=Cl, Br)$$

$$RCH(OH)CX_3 \xrightarrow{\mathbf{1}\text{, HCl, } H_2O\text{, 50–60°}} RCH{=}CX_2 \text{ (70–90\%) or } RCH(OH)CX_2H \text{ (10–50\%)}$$

the products to afford 1,1-dihaloalkenes and/or a product of reduction of one halo group.

[1] H. Tanaka, S. Yamashita, M. Yamanoue, and S. Torii, *J. Org.*, **54**, 444 (1989).

L-*t*-Leucine, *t*-butyl ester, $(CH_3)_3CCH(NH_2)COOC(CH_3)_3$ (**1**). L-*t*-Leucine is available from Aldrich (expensive).

Asymmetric double alkylation of α,β-enaldimines. A synthesis of (+)-ivalin, an antileukemic sesquiterpene (**4**), is based on this stereoselective reaction. The

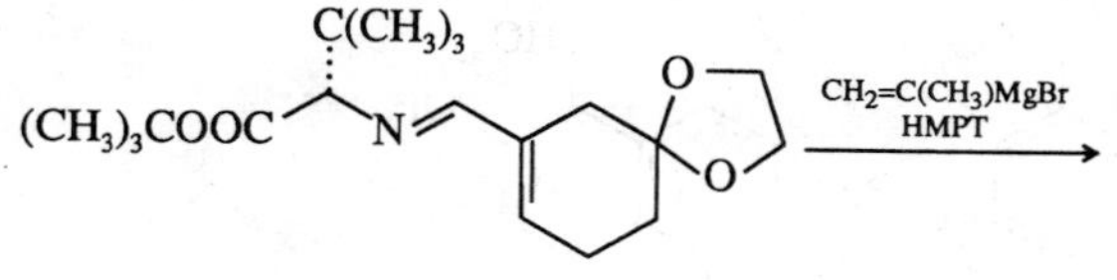

$C(CH_3)_3$
$(CH_3)_3COOC$ N O O
$CH_2{=}C(CH_3)MgBr$
HMPT
2

$C(CH_3)_3$
$(CH_3)_3COC$ N O O
O MgBr
C CH_3 CH_2
CH_3I
HMPT
35%
OHC CH_3 1 5 O O 2
C CH_3 CH_2 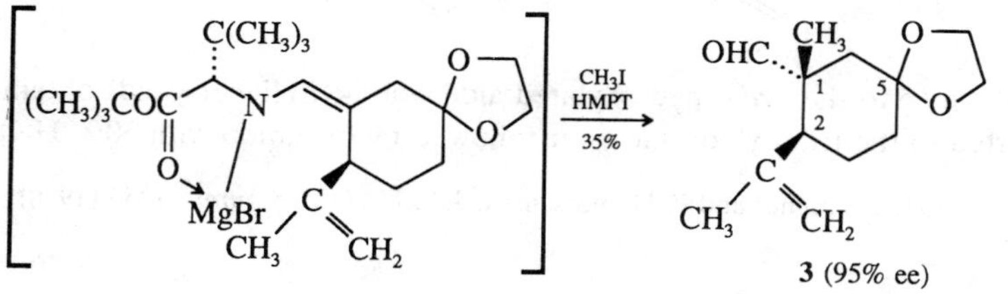
3 (95% ee)

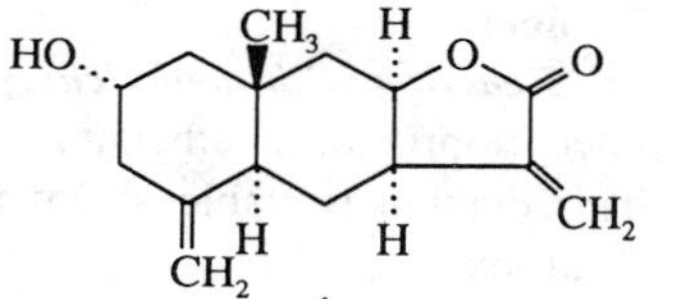

HO CH_3 H O O
CH_2 H H CH_2
4

aldimine **2**, prepared from **1**, reacts with a Grignard reagent and then with methyl iodide to form the aldehyde **3** and an epimer at C_1 as the major products. The product (**3**) was used to furnish the middle ring of (+)-ivalin (**4**).

[1] K. Tomioka, F. Masumi, T. Yamashita, and K. Koga, *Tetrahedron*, **45**, 643 (1989).

N-Lithio-N,N′,N′-trimethylethylenediamine, $(CH_3)_2N(CH_2)_2N(Li)CH_3$ (**1**)

This reagent is prepared by sonication of a mixture of N,N,N′-trimethylethylenediamine and lithium in THF.

ortho-*Hydroxylation of ArCHO.*[1] The reaction of an aldehyde (**2**) with **1** and then with BuLi (or BuCl, Li) provides the adduct **3**, which is oxidized by O_2 or

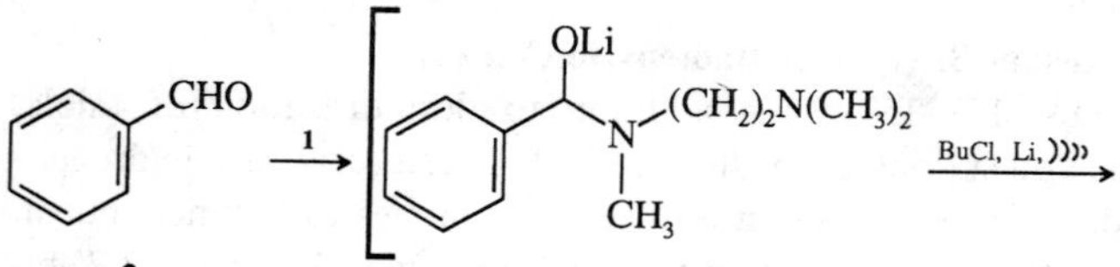

CHO
2
1
OLi
N $(CH_2)_2N(CH_3)_2$
CH_3
BuCl, Li,)))

OLi
N $(CH_2)_2N(CH_3)_2$
Li CH_3
3
oxid.
70%
CHO
OH
4

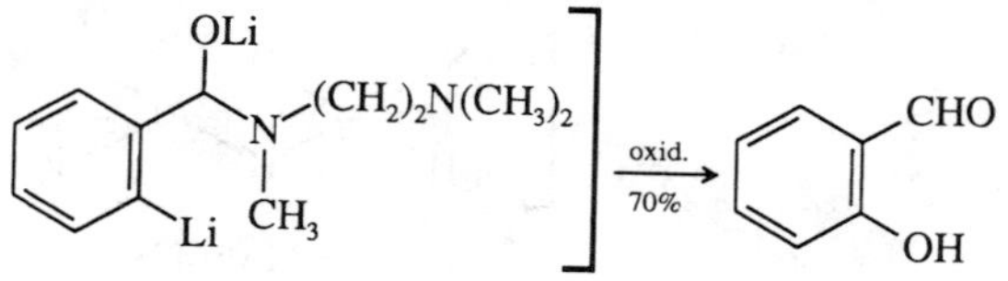

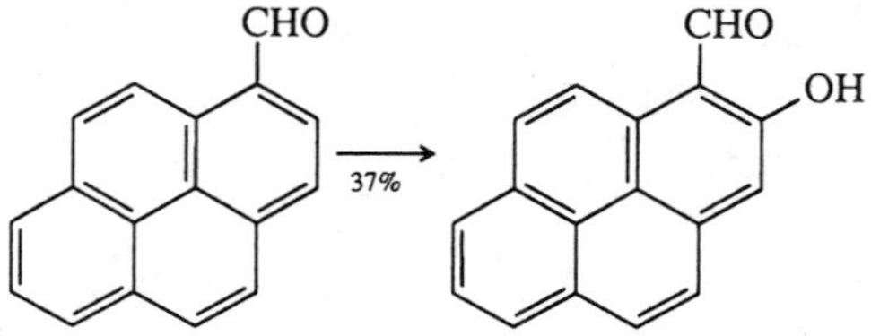

CHO CHO OH 37%

MoOPh at 0° to the *ortho*-hydroxylated aldehyde (**4**). Higher yields obtain if **3** is converted to the tributylboronic ester followed by oxidation with 30% H_2O_2.

[1] J. Einhorn, J.-L. Luche, and P. Demerseman, *J.C.S. Chem. Comm.*, 1350 (1988).

Lithium.

Reaction with methylenecyclopropane.[1] Lithium powder reacts with methylenecyclopropane in ether at 25° to form 2,4-dilithio-1-butene (**2**) as the only product. This dianion is stable at 25° for 27 days; it is a precursor to a methylene-1,3-dicarboxylic acid.

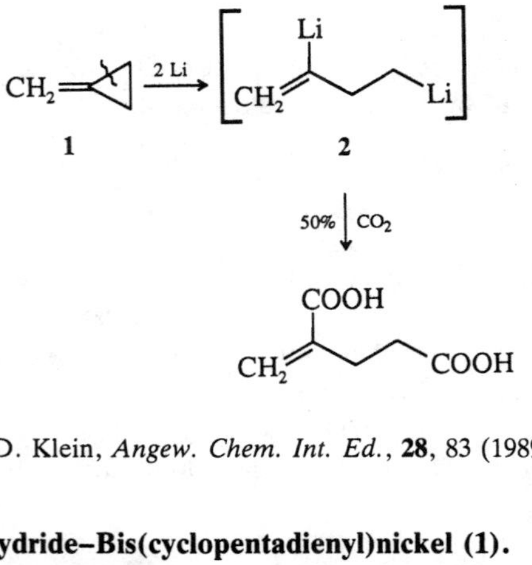

CH_2 2 Li Li CH_2 Li **1** **2** 50% CO_2 COOH CH_2 COOH

[1] A. Maercker and K.-D. Klein, *Angew. Chem. Int. Ed.*, **28**, 83 (1989).

Lithium aluminum hydride–Bis(cyclopentadienyl)nickel (1).

Desulfurization (**13**, 158–159).[1] In addition to desulfurization of thiols, thioethers, sulfoxides, and sulfones, Cp_2Ni-$LiAlH_4$ (1:1) can serve as a hydrogenation catalyst for reduction of alkenes; it also reduces enones to ketones, but in low yield. In general, it is similar in reactivity to Raney nickel and to (2,2′-bi-

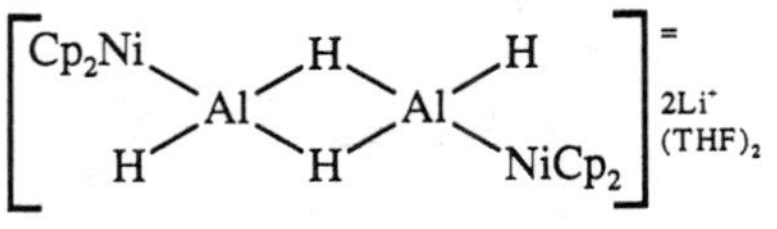

Cp_2Ni H H Al Al H H $NiCp_2$ = $2Li^+$ $(THF)_2$

$Cp_2NiAlH_2Li \cdot THF$

pyridyl)(1,5-cyclooctadiene)nickel, $LiAlH_2(THF)_n \cdot C_{10}H_8N_2 \cdot Ni$.[2] The spectroscopic properties of **1** suggest that the active species has a bridged structure.

[1] M.-C. Chan, K.-M. Cheng, K. M. Ho, C. T. Ng, T. M. Yam, B. S. L. Wang, and T.-Y. Luh, *J. Org.*, **53**, 4466 (1988).
[2] J. J. Eisch, L. E. Hallenbeck, and K. I. Han, *Am. Soc.*, **108**, 7763 (1986).

Lithium aluminum hydride–Lithium iodide.

syn-1,3-Diols.[1] $LiAlH_4$ alone shows no stereoselectivity in reduction of β-hydroxy ketones, but addition of LiI provides a *syn*-selective reagent for reduction of β-alkoxy ketones and β-alkoxy-β′-hydroxy ketones.

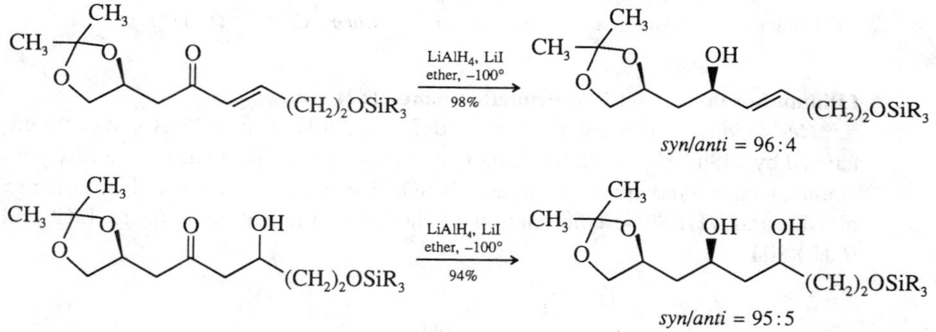

LiAlH₄, LiI ether, −100° 98%

syn/anti = 96:4

LiAlH₄, LiI ether, −100° 94%

syn/anti = 95:5

[1] Y. Mori, M. Kuhara, A. Takeuchi, and M. Suzuki, *Tetrahedron Letters*, **29**, 5419 (1988); Y. Mori, A. Takeuchi, H. Kageyama, and M. Suzuki, *ibid.*, **29**, 5423 (1988).

Lithium N-benzyltrimethylsilyl amide (LSA), $(CH_3)_3SiNLiBzl$.

Cyclization of 2,6-octadienoic esters.[1] LSA adds 1,4 to crotonoates. Extension of this reaction to these dienoic esters (**1** and **2**) results in cyclization to cyclopentanes (**3**) and cyclohexanes (**4**), respectively.

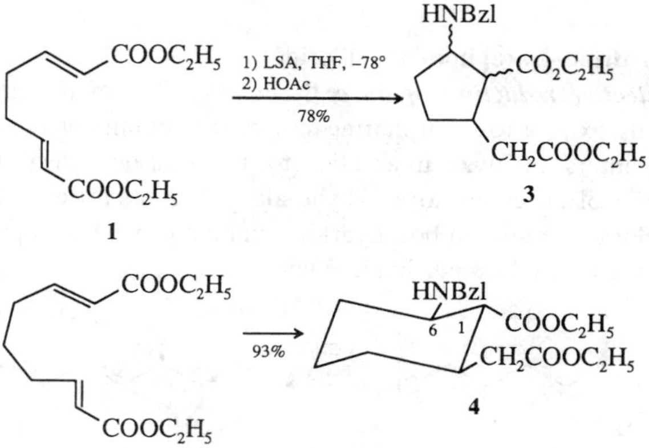

1) LSA, THF, −78° 2) HOAc 78%

1 **3**

93%

2 **4**

This tandem conjugate addition followed by β-elimination was used to effect synthesis of dihydronepatalactone (**7**) from (**5**) via **6**.

$CO_2C_2H_5$ $\xrightarrow{\text{1) LSA; 2) } CH_3I, K_2CO_3}$ $COOC_2H_5$, $COOC_2H_5$, H, CH_3 $\longrightarrow$ CH_3, H, O, O, H, CH_3

CH_3 $COOC_2H_5$

5 **6** (74:26) **7**

[1] T. Uyehara, N. Shida, and Y. Yamamoto, *J.C.S. Chem. Comm.*, 113 (1989).

Lithium borohydride–Chlorotrimethylsilane (1:2)

Reductions.[1] The reactivity of $LiBH_4$ (or $NaBH_4$) in THF is markedly enhanced by $ClSi(CH_3)_3$. The resulting reagent reduces amino acids to optically pure amino alcohols, and amides to amines. The active agent is believed to be a complex of BH_3 and THF. The combination of these two reagents also forms LiCl and $(CH_3)_3SiH$.

$$(CH_3)_2CHCH(NH_2)CO_2H \xrightarrow[91\%]{NaBH_4,\ ClSi(CH_3)_3} (CH_3)_2CHCH(NH_2)CH_2OH$$

$$C_6H_5CONH_2 \xrightarrow[91\%]{} C_6H_5CH_2NH_2$$

$$ArCH{=}CHNO_2 \xrightarrow[91\%]{} ArCH_2CH_2NH_2$$

[1] A. Giannis and K. Sandhoff, *Angew. Chem. Int. Ed.*, **28**, 218 (1989).

Lithium borohydride–Europium(III) chloride.

Enantioselective reduction of an α,β-enone.[1] One of the final steps in a synthesis of palytoxin, a toxin of marine soft corals containing 115 carbon atoms and 60 chiral centers, involves, in addition to the usual deprotections, enantioselective reduction of an enone to an allylic alcohol. A mixture (1:1) is obtained with borohydrides, but lithium borohydride combined with $EuCl_3$ provides an 8:1 mixture, with the desired isomer being favored.

CH_3, CH_3, OH, O, CH_3 $\xrightarrow{LiBH_4,\ EuCl_3}$ CH_3, CH_3, OH, OH, CH_3

(8:1)

[1] Y. Kishi, *Chemica Scripta*, **27**, 573 (1987).

Lithium *t*-butylhydroperoxide, $(CH_3)_3COOLi$ (**1**). The reagent is generated *in situ* by reaction of BuLi or CH_3Li with *t*-butyl hydroperoxide.

Epoxidation of enones, enals, α,β-unsaturated esters.[1] Epoxidation of these substrates is generally effected with alkaline hydrogen peroxide, but can also be effected with this new reagent with high stereoselectivity.

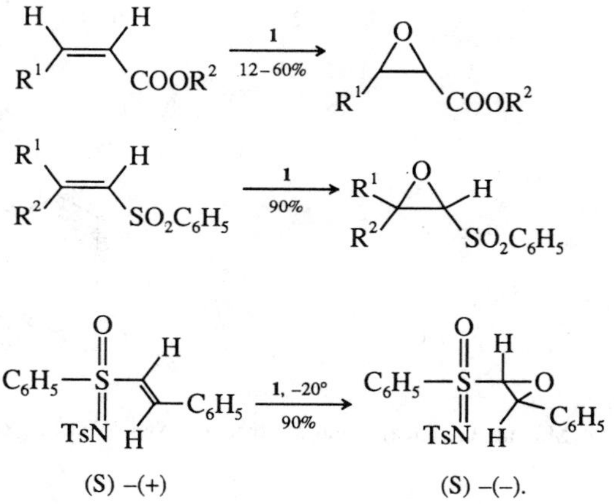

(S) –(+) → (S) –(–).

[1] O. Meth-Cohn, C. Moore, and H. C. Taljaard, *J.C.S. Perkin I*, 2663 (1988).

Lithium cyclohexylisopropylamide (LCI).

Thioxanthen-9-ones. This base is somewhat superior to LDA for lithiation of thiosalicylamides (**1**) and for genertion of benzynes from halobenzenes. Thus treatment of **1** and a benzyne precursor (2 equiv.) and 3 equiv. of the base (LCI) results in thioxanthen-9-one **2** in yields as high as 98%. Various methoxy- and dimethoxy-substituted derivatives of **2** can be obtained by this condensation. This synthesis can be applied to synthesis of acridones and xanthen-9-selonones.

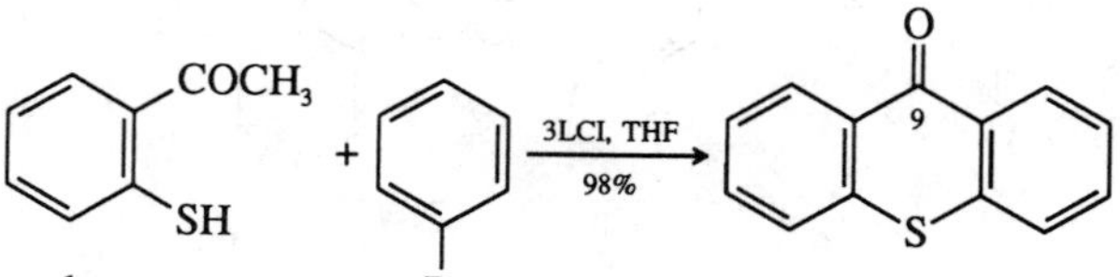

[1] M. Watanabe, M. Date, M. Tsukazaki, and S. Furukawa, *Chem. Pharm. Bull.*, **37**, 36 (1989).

Lithium diethylamide.

Dehydrohalogenation. The key step in a synthesis of the 16-membered lactone norpyrenophorin (**4**) is the selective monodehydrohalogenation of the dibromo acid **2**, obtained from the tetrahydropyranylacetic acid **1**, with LDA (2 equiv.).

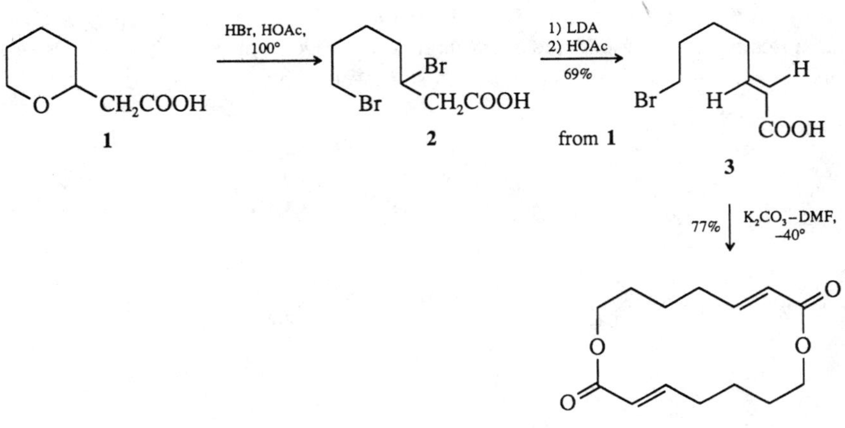

CH_2COOH **1** → (HBr, HOAc, 100°) → Br, Br, CH_2COOH **2** → (1) LDA 2) HOAc, 69% from **1**) → Br, H, H, COOH **3** → (K_2CO_3–DMF, 40°, 77%) → **4**

[1] L. Maciejewski, M. Martin, G. Ricart, and J. Brocard, *Syn. Comm.*, **18**, 1757 (1988).

Lithium diisopropylamide.

9-Phenanthrols via directed lithiation.[1] Biphenyl diethyl amides substituted by a 2′-methyl group (**1**), prepared by cross coupling of arylboronic acids and aryl bromides,[2] undergo directed lithiation to provide 9-phenanthrols.

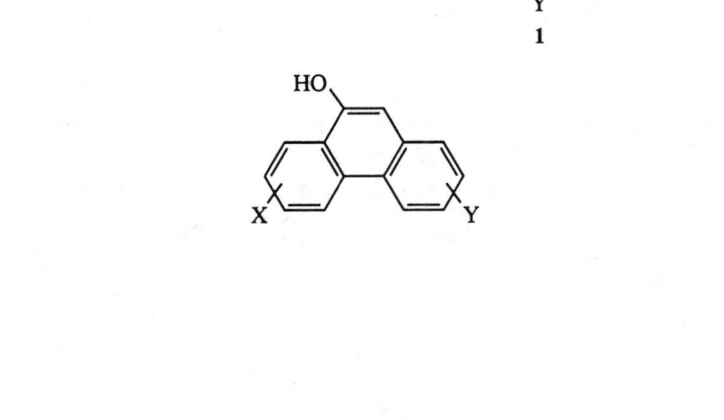

X, $(HO)_2B$, NR_2, O + CH_3, Br, Y → (Pd(O), Na_2CO_3, DME, 67–98%) → X, O, CNR_2, CH_3, Y **1** → (LDA, THF, 0→25°, 78–98%) → HO, X, Y

The coupling of the arylboronic acid **1** with *o*-bromobenzaldehydes (**2**) leads directly to phenanthridines (**3**).[3]

$$\mathbf{1} + \mathbf{2} \xrightarrow[45-89\%]{Pd(O),\ Na_2CO_3} \mathbf{3}$$

Dianions of aldols. The reaction of 2 equiv. of LDA with aldols results in the distal dianion, which can be trapped with $ClSi(CH_3)_3$.

$$HOCH_2CH(CH_3)COCH_3 \xrightarrow[80\%]{1)\ 2LDA\ \ 2)\ ClSi(CH_3)_3} (CH_3)_3SiOCH_2CH(CH_3)C(OSi(CH_3)_3)=CH_2$$

$$\xrightarrow{74\%}$$

These dianions react with ketones to give β,β′-dihydroxy ketones in 55–95% yield.[4]

$$CH_3CH(CH_2OH)COCH_3 + C_6H_5CHO \xrightarrow[93\%]{2LDA} HOCH_2CH(CH_3)COCH_2CH(OH)C_6H_5 \quad (1:1)$$

[1] J. Fu, M. J. Sharp, and V. Snieckus, *Tetrahedron Letters*, **29**, 5459 (1988).
[2] M. J. Sharp, W. Cheng, and V. Snieckus, *ibid.*, **28**, 5093, 5097 (1987).
[3] M. A. Siddiqui and V. Snieckus, *ibid.*, **29**, 5463 (1988).
[4] V. A. Martin and K. F. Albizate, *J. Org.*, **53**, 5986 (1988).

Lithium diselenide, Li_2Se_2 (**1**). This reagent is prepared conveniently by reaction of Se(0) with Li in THF at 20° with diphenylacetylene as electron carrier.

Diselenides. These compounds can be prepared in satisfactory yield by reaction of **1** with alkyl halides, epoxides, or lactones in THF at 20°.

[1] L. Syper and J. Mtochowski, *Tetrahedron*, **44**, 6119 (1988).

Lithium N-lithiomethyldithiocarbamates (1).

The reagent is prepared from a secondary methylamine:

$$RNHCH_3 \xrightarrow[2)\ CS_2]{1)\ BuLi} \underset{\underset{S=CSLi}{|}}{RNCH_3} \xrightarrow[-78 \rightarrow -25°]{sec\text{-}BuLi} \underset{\underset{S=CSLi}{|}}{RNCH_2Li}\ (1)$$

N-Butylaminomethylation. The reagent **1**, R=Bu, reacts with electrophiles at −25° to form an intermediate (**a**) that is hydrolyzed to a N-alkylbutylamine.

$$\underset{\underset{S=CSLi}{|}}{BuNCH_2Li} \xrightarrow[-25°]{BuI} \left[\underset{\underset{S=CSLi}{|}}{BuNCH_2Bu}\right]_{\mathbf{a}} \xrightarrow[85\%]{H_3O^+} BuNHCH_2Bu$$

$$\downarrow 87\%\ \ 1)\ C_6H_5Si(CH_3)_2Cl;\ 2)\ H_3O^+$$

$$BuNHCH_2Si(CH_3)_2C_6H_5$$

[1] H. Ahlbrecht and D. Kornetzky, *Synthesis*, 775 (1988).

Lithium naphthalenide (1).

Disilenes.[1] Disilenes (**3**) can be prepared from 1,2-dichloro-1,1,2,2-tetrasubstituted disilanes (**2**) by reaction with lithium naphthalenide. Most disilenes are unstable to air, but **3** is fairly stable. Surprisingly, it is oxygenated to give **4**, rather than the expected 1,3-cyclodisiloxane (**5**).

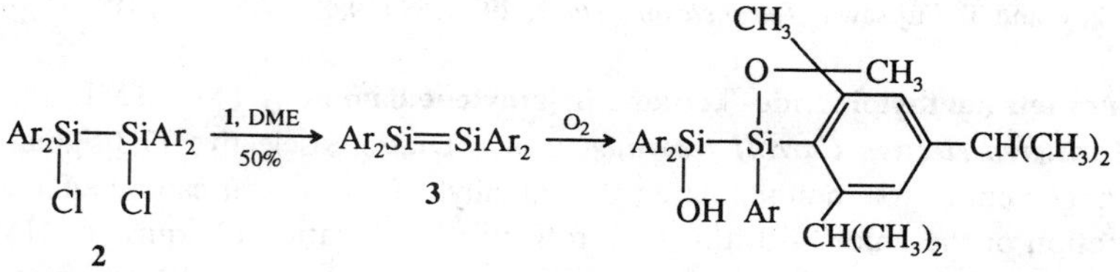

$$Ar_2Si(Cl)-Si(Cl)Ar_2 \ (\mathbf{2}) \xrightarrow[50\%]{1,\ DME} Ar_2Si{=}SiAr_2 \ (\mathbf{3}) \xrightarrow{O_2} Ar_2Si(OH)-Si(Ar)-C_6H(CH(CH_3)_2)_2C(CH_3)_2O \ (\mathbf{4})$$

$Ar = C_6H_2CH(CH_3)_2$-2,4,6

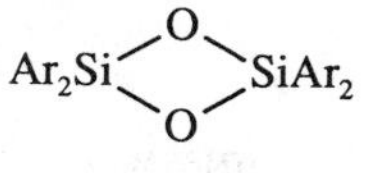

$$Ar_2Si(\mu\text{-}O)_2SiAr_2 \ (\mathbf{5})$$

[1] H. Watanabe, K. Takeuchi, K. Nakajima, Y. Nagai, and M. Goto, *Chem. Letters*, 1343 (1988).

Lithium tetraalkylcerates, R_4CeLi. These reagents are prepared *in situ* by reaction of an alkyllithium (4 equiv.) with $CeCl_3$ at $-78° \rightarrow -20°$.

Alkylolefins. These cerates convert epoxides into alkylated olefins. DME is

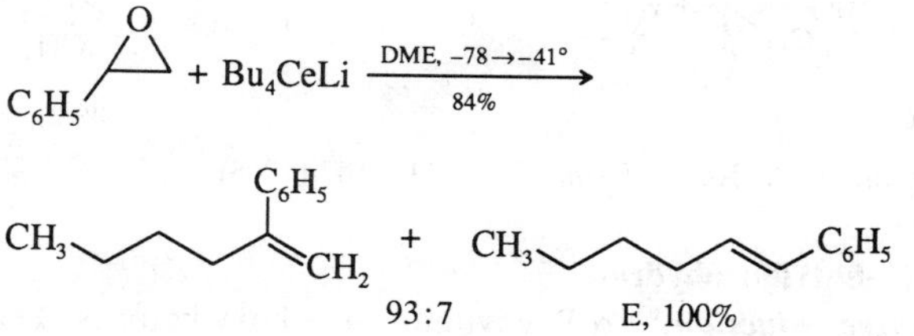

$$C_6H_5\text{-oxirane} + Bu_4CeLi \xrightarrow[84\%]{DME,\ -78\rightarrow-41°} CH_3(CH_2)_4C(C_6H_5){=}CH_2 + CH_3(CH_2)_4CH{=}CHC_6H_5$$

93:7 (E, 100%)

the solvent of choice, and yields and regioselectivity are improved by added TMEDA. This reaction was used for synthesis of a sesquiterpene (equation I).

$$\text{(I)} \quad p\text{-}CH_3C_6H_4\text{-oxirane} + [(CH_3)_2C{=}CHCH_2CH_2]_4CeLi \xrightarrow[43\%]{} p\text{-}CH_3C_6H_4C({=}CH_2)CH_2CH_2CH{=}C(CH_3)_2$$

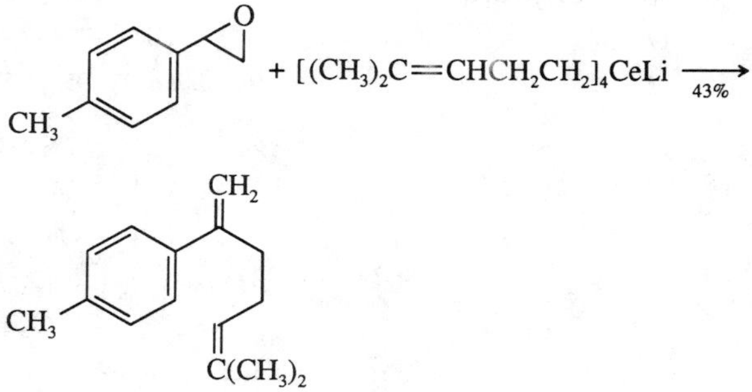

[1] Y. Ukaji and T. Fujisawa, *Tetrahedron Letters*, **29**, 5165 (1988).

Lithium tetramethylpiperide–Tetramethylethylenediamine (LTMP–TMEDA).

Diastereoselective Carroll rearrangement. Diastereoselective [3,3]sigmatropic rearrangement of the monoenolate (**1**) of an allylic β-keto ester can be effected by generation of the anion with this base followed by silylation (2 equiv. of HMPT) to form the silyl ketene acetal **2**. This derivative rearranges at 40° to give, after hydrolysis, a β-keto acid, isolated as the methyl ester **3**, bearing a quatarnary carbon center.

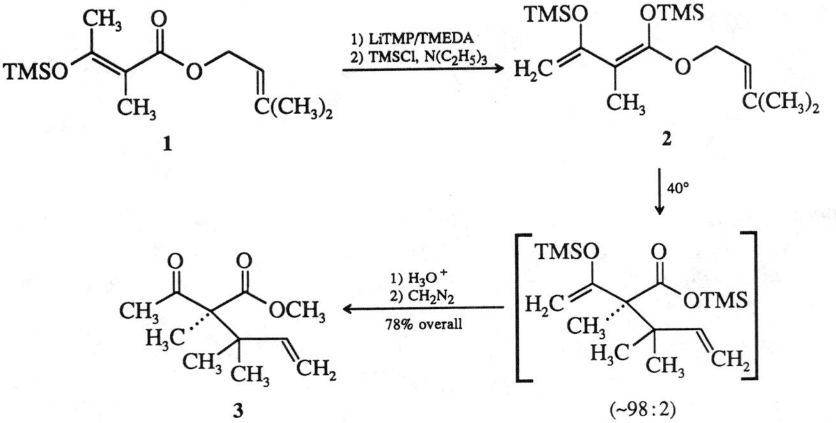

1) LiTMP/TMEDA
2) TMSCl, $N(C_2H_5)_3$

40°

1) H_3O^+
2) CH_2N_2
78% overall

(~98:2)

[1] J. C. Gilbert and T. A. Kelly, *Tetrahedron*, **44**, 7587 (1988).

Lithium tri-*sec*-butylborohydride.

syn-Selective reduction.[1] α-Phenylthio-β-methoxy ketones (**1**) are reduced to *syn*-α-phenylthio-β-methoxy alcohols (**2**) by either lithium tri-*sec*-butylborohydride

H^-, THF
59%

2 (*syn*/*anti* = 92–95:8–5)

1) 2-fluoro-1-ethylpyridinium BF_4^-, $N(C_2H_5)_3$
2) LiI

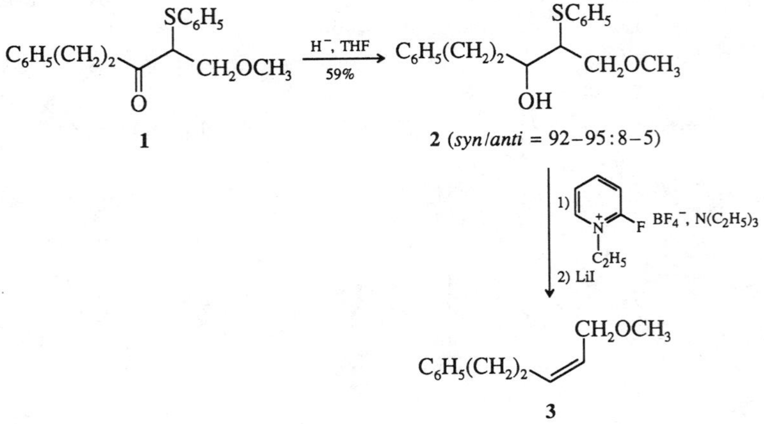

or lithium triethylborohydride. The products are precursors to (Z)-allyl alcohols (**3**).[2]

Diastereoselective reduction of acyclic α,β-dialkoxy ketones.[3] Reduction of the ketone **2** with this hydride results in the alcohol **3** with high *syn*-selectivity (97:3). The same *syn*-selectivity is observed with lithium trisiamylborohydride, but the yield is lower. The product was used for a synthesis of (+)-**4**, a pheromone of a species of ants.

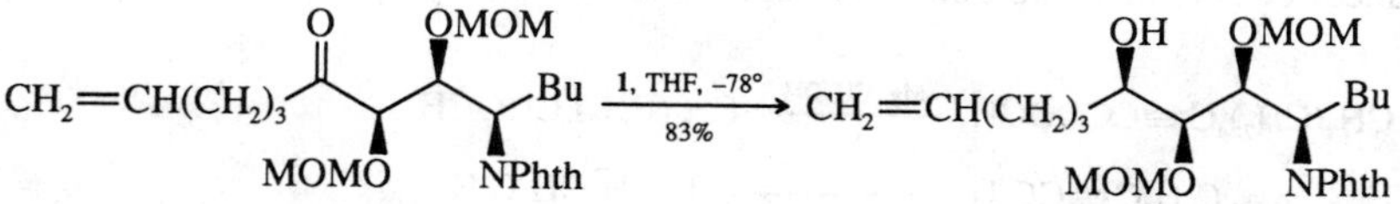

$CH_2{=}CH(CH_2)_3$ O OMOM Bu MOMO NPhth — 1, THF, −78°, 83% → $CH_2{=}CH(CH_2)_3$ OH OMOM Bu MOMO NPhth

2, α_D-16.4 **3** (97:3)

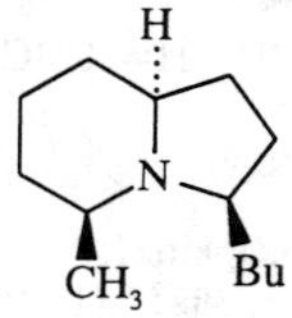

H N CH_3 Bu

(+)-**4**

[1] M. Shimagaki, A. Suzuki, T. Nakata, and T. Oishi, *Chem. Pharm. Bull.*, **36**, 3138 (1988).
[2] T. Mukaiyama and M. Imaoka, *Chem. Letters*, 413 (1978).
[3] N. Yamazaki and C. Kibayashi, *Tetrahedron Letters*, **29**, 5767 (1988).

M

Magnesium–Methanol.

Reduction of α,β-acetylenic esters.[1] Mg/CH_3OH reduces these esters (but not the corresponding acids) to saturated esters in 68–72% yield. This method also reduces double or triple bonds conjugated to two phenyl groups.

$$CH_3(CH_2)_4C{\equiv}CCOOCH_3 \xrightarrow[76\%]{Mg,\ CH_3OH} CH_3(CH_2)_6COOCH_3$$

$$C_6H_5C{\equiv}CC_6H_5 \xrightarrow[85\%]{} C_6H_5CH_2CH_2C_6H_5$$

$$C_6H_5CH{=}CHCH{=}CHC_6H_5 \xrightarrow[96\%]{} C_6H_5CH_2CH{=}CHCH_2C_6H_5 + C_6H_5CH{=}CHCH_2CH_2C_6H_5 \quad (5.5:1)$$

[1] R. O. Hutchins, Suchismita, R. E. Zipkin, I. M. Taffer, R. Sivakumar, A. Monaghan, and E. M. Elisseou, *Tetrahedron Letters*, **30**, 55 (1989).

Magnesium anthracene, (1)

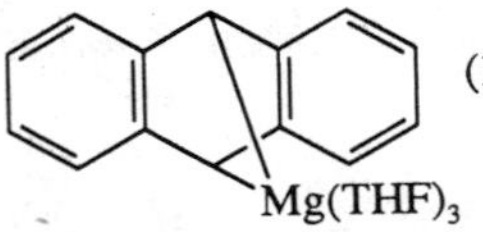

Mg(THF)$_3$

This complex is obtained by reaction of Mg with anthracene in THF at 20–60°. The THF can be displaced by DME or dioxane.

On protonolysis, this complex is converted quantitatively into 9,10-dihydroanthracene. In reactions with CO_2 or THF it behaves like a diorganomagnesium compound (equation I).[1]

(I) H, COOH / H, COOH $\xleftarrow{2CO_2}$ 1 $\xrightarrow{2\ \text{THF}}$ H, $(CH_2)_4OH$ / H, $(CH_2)_4OH$

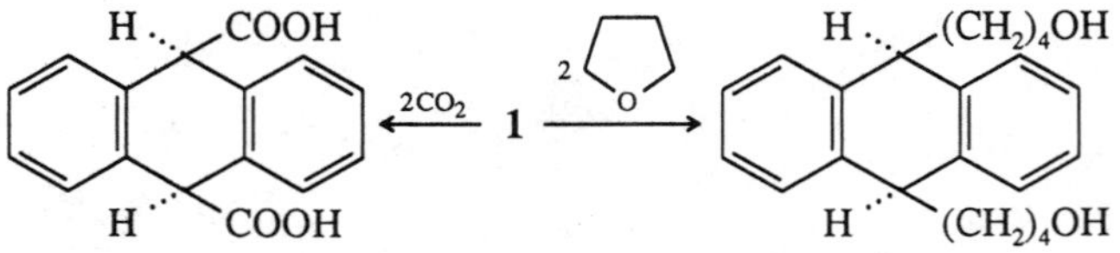

[1] B. Bogdanovic, *Acc. Chem. Res.*, **21**, 261 (1988).

Magnesium bromide.

Alkyne/ester coupling.[1] A key step in a synthesis of the antibiotic monensin (**4**) involves coupling of an alkyne (**2**), derived from a degradation product of the natural product, with an activated anhydride (**1**), also an oxidative degradation product of monensin. This coupling was achieved via the magnesium acetylide of **2**, and proceeds in 87% yield. The resulting ynone (**3**) was converted into **4** by

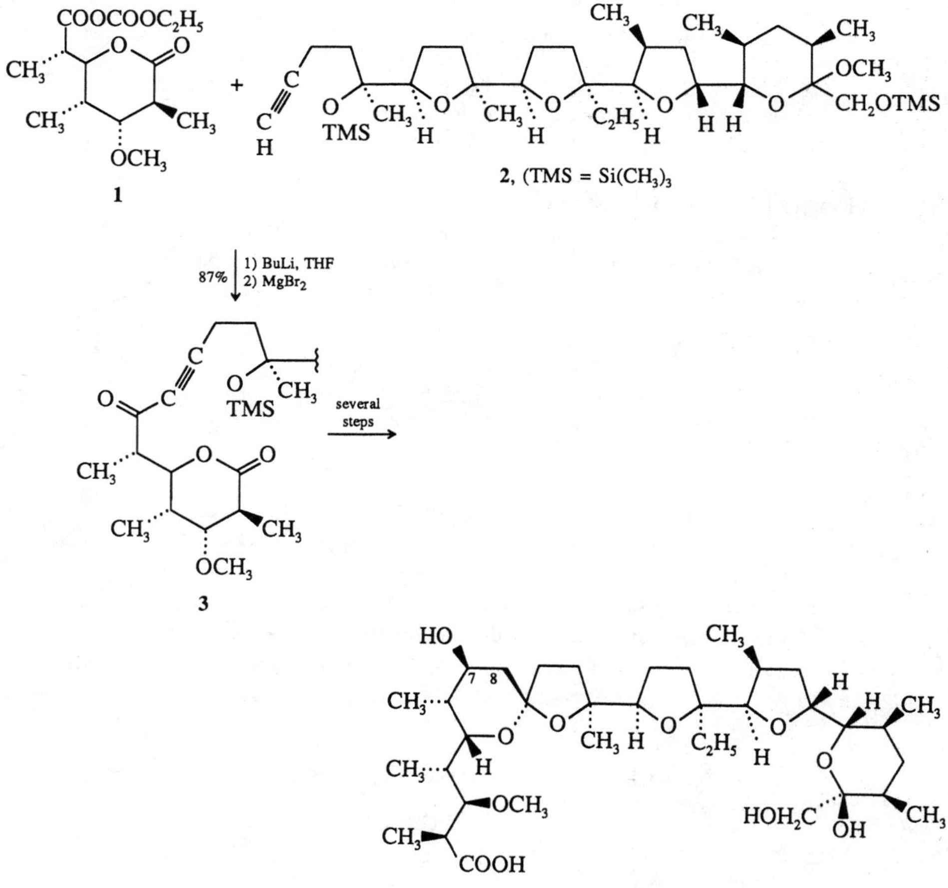

1

2, (TMS = $Si(CH_3)_3$

87% 1) BuLi, THF 2) $MgBr_2$

3

several steps

4

hydrolysis of the lactone ring, spirocyclization, and stereoselective reduction of the oxo group at C_7 to the axial alcohol (potassium tri-*sec*-butylborohydride).

Cleavage of an azetidinone epoxide.[2] $MgBr_2$, prepared from Mg with $BrCH_2CH_2Br$, cleaves the epoxide **1** regiospecifically to the diastereomeric bromohydrins **2**. The N-silyl derivative of **1** is stable under the same conditions. The ketone corresponding to **2** is a useful precursor to a carbapenem.

$$\underset{\textbf{1}}{\text{(}R_3SiO\text{)}CH_3CH\text{–azetidinone(NH, C=O)–}CH_2CH\text{–}CH_2\text{ (epoxide, –O–)}} \xrightarrow[88\%]{MgBr_2,\ THF,\ 0\text{–}5^\circ} \underset{\textbf{2}}{\text{(}R_3SiO\text{)}CH_3CH\text{–azetidinone(NH, C=O)–}CH_2CHCH_2Br\text{ (OH)}}$$

[1] D. Cai and W. C. Still, *J. Org.*, **53**, 4641 (1988).
[2] Y. Ueda and S. C. Maynard, *Tetrahedron Letters*, **29**, 5197 (1988).

Magnesium chloride–Sodium iodide, 13, 66.

β-Keto amides. The carboxylation of ketones catalyzed by MgI_2 to form β-keto acids has been extended to a preparation of β-keto amides from ketones and aliphatic isocyanates. $AlCl_3$, $SnCl_4$, $FeCl_3$, and $ZnCl_2$ are not useful catalysts.

$$C_6H_5\overset{O}{\overset{\|}{C}}C_2H_5 + BuNCO \xrightarrow[26\%]{MgI_2,\ N(C_2H_5)_3,\ CH_3CN} C_6H_5\overset{O}{\overset{\|}{C}}\underset{CH_3}{\underset{|}{C}H}\overset{O}{\overset{\|}{C}}NHBu$$

[1] L. C. Lasley and B. B. Wright, *Syn. Comm.*, **19**, 59 (1989).

Manganese(III) acetate–Diphenyl disulfide (1).

Hydroxysulphenylation. The reaction of these two reagents in CH_2Cl_2/trifluoroacetic acid with allylic esters gives *vic*-trifluoroacetoxy sulfides, which are hydrolyzed to *vic*-hydroxy sulfides. The reaction with allyl acetate (equation I)

$$\text{(I)}\quad CH_2{=}CHCH_2OAc + Mn(OAc)_3 + C_6H_5SSC_6H_5 \xrightarrow{1)\ CH_2Cl_2/CF_3COOH} \underset{(71\%)}{C_6H_5SCH_2\underset{OH}{\underset{|}{C}H}CH_2OAc} + \underset{(11\%)}{C_6H_5SCH_2\underset{OAc}{\underset{|}{C}H}CH_2OH}$$

$$\text{(II)}\quad CH_2{=}CHCH_2OCOCF_3 + Mn(OAc)_3 + C_6H_5SSC_6H_5 \xrightarrow[98\%]{1;\ OH^-} HOCH_2\underset{SC_6H_5}{\underset{|}{C}H}CH_2OH$$

affords two monoacetates, both presumably formed via an episulfonium intermediate. In contrast, reaction with allyl trifluoroacetate (equation II) affords a single *anti*-Markovnikov adduct. A similar regiocontrol obtains in sulfenylation of N-allylacetamide and N-allyltrifluoroacetamide with $Mn(OAc)_3$ and $C_6H_5SSC_6H_5$.

The addition to allyl nitrile also shows some preference for the *anti*-Markovnikov adduct (equation III). However, both products undergo a Ritter-type hydrolysis of the nitrile group to provide the same lactone.

$$\text{(III)}\ CH_2{=}CHCH_2CN + Mn(OAc)_3 + (C_6H_5S)_2 \longrightarrow$$

$$\underset{(66\%)}{HOCH_2\underset{\displaystyle SC_6H_5}{\underset{|}{C}}HCH_2CN} + \underset{(20\%)}{C_6H_5SCH_2\underset{\displaystyle OH}{\underset{|}{C}}HCH_2CN}$$

$$\xrightarrow[93\%,\ 70\%]{CoCl_2,\ H_2O,\ 170^\circ} \text{4-(phenylthio)dihydrofuran-2(3H)-one } (C_6H_5S\text{-substituted }\gamma\text{-lactone})$$

[1] Z. K. M. Abd El Samii, M. I. Al Ashmawy, and J. M. Mellor, *J.C.S. Perkin I*, 2509, 2517, 2523 (1988).

Manganese dioxide.

RC≡N → RCONH₂.[1] Hydration of nitriles to amides can be effected in good to high yield with MnO_2 on SiO_2, prepared by stirring silica gel with $MnSO_4$ and $KMnO_4$ dissolved in water. Benzylic oxidation can be a competing reaction.

$$t\text{-BuC}{\equiv}N \xrightarrow[93\%]{MnO_2/SiO_2,\ C_8H_{18},\ \Delta} t\text{-BuCONH}_2$$

$$C_6H_5CH_2C{\equiv}N \longrightarrow \underset{(35\%)}{C_6H_5CH_2CONH_2} + C_6H_5\overset{O}{\overset{\|}{C}}-\overset{O}{\overset{\|}{C}}NH_2$$

Ultrasound activation.[2] Commercial MnO_2 is a weak oxidant, but can be markedly activated by ultrasonic irradiation and agitation in the case of oxidation of allylic or benzylic alcohols.

$$C_6H_5CH{=}CHCH_2OH \xrightarrow[0\%]{MnO_2,\ CH_3CN} C_6H_5CH{=}CHCHO$$

+)))), 17° 100%

$$CH_3(CH_2)_6CH_2OH \xrightarrow[0\%]{} CH_3(CH_2)_6CHO$$

+)))), 50° 5%

[1] K.-T. Liu, M.-H. Shih, H.-W. Huang, and C.-J. Hu, *Synthesis*, 715 (1988).
[2] T. Kimura, M. Fujita, and T. Ando, *Chem. Letters*, 1387 (1988).

Mercury.

Dehydrodimerization. On excitation with a mercury vapor lamp, mercury is converted to an excited state, Hg*, which can convert a C—H bond into a carbon radical and a hydrogen atom. This process can result in dehydrodimerization, which has been known for some time, but which has not been synthetically useful because of low yields when carried out in solution. Brown and Crabtree[1] have shown that this reaction can be synthetically useful when carried out in the vapor phase, in which the reaction is much faster than in a liquid phase, and in which very high selectivities are attainable. Secondary C—H bonds are cleaved more readily than primary ones, and tertiary C—H bonds are cleaved the most readily. Isobutane is dimerized exclusively to 2,2,3,3-tetramethylbutane. This dehydrodimerization is also applicable to alcohols, ethers, and silanes. Cross-dehydrodimerization is also possible, and is a useful synthetic reaction.

Cyclohexane + CH_3OH $\xrightarrow{Hg^*}$ bicyclohexyl + cyclohexyl-CH_2OH + $HOCH_2CH_2OH$

Cyclohexane + 1,3,5-trioxane → 2-cyclohexyl-1,3,5-trioxane $\xrightarrow{H_3O^+}$ cyclohexyl-CHO

[1] S. H. Brown and R. H. Crabtree, *Am. Soc.*, **111**, 2935, 2946 (1989).

Mercury(II) acetate.

Hydroxymercuration. Reaction of the iridoid glucoside aucubin (**1**) with $Hg(OAc)_2$ results in exclusive attack of the enol ether double bond to give **2**. Demercuration with $NaBH_4$ in the usual way affords isoeucommiol (**3**). However

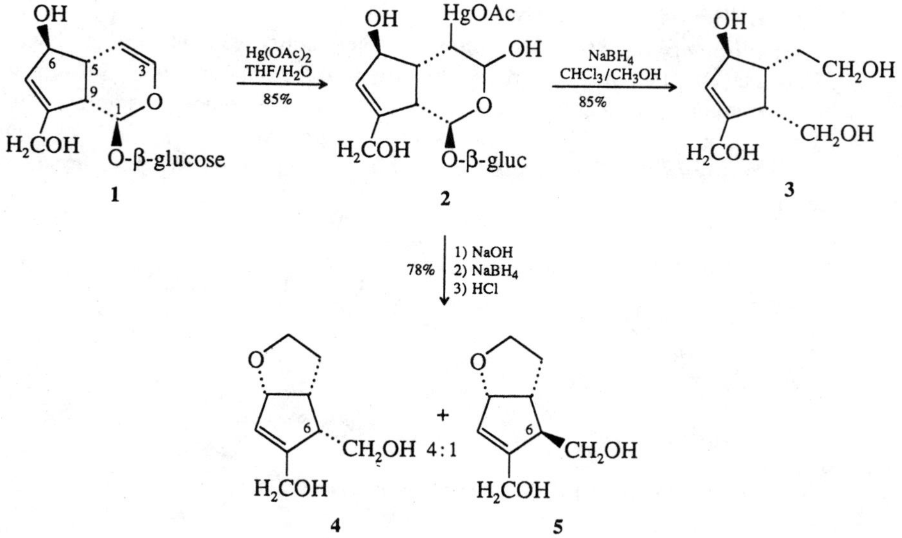

OH
6 5 3
9 1 O
H_2COH O-β-glucose
1

$Hg(OAc)_2$
THF/H_2O
85%

OH HgOAc
OH
O
H_2COH O-β-gluc
2

$NaBH_4$
$CHCl_3/CH_3OH$
85%

OH
CH_2OH
CH_2OH
H_2COH
3

78%
1) NaOH
2) $NaBH_4$
3) HCl

O
6
CH_2OH 4 : 1
H_2COH
4

+

O
6
CH_2OH
H_2COH
5

slow reduction of **2** in a basic medium is attended by epimerization at C_6 to give a mixture of two products that undergo acid-catalyzed cyclization to **4** and **5**, precursors to modified prostaglandins.[1]

Oxidation of cyclic* t*-amines.[2] $Hg(OAc)_2$ in combination with ethylenediaminetetraacetic acid (EDTA) is the classical reagent for dehydrogenation of *t*-amines (**1**, 648). This reaction can be used to convert cyclic *t*-amines to lactams.

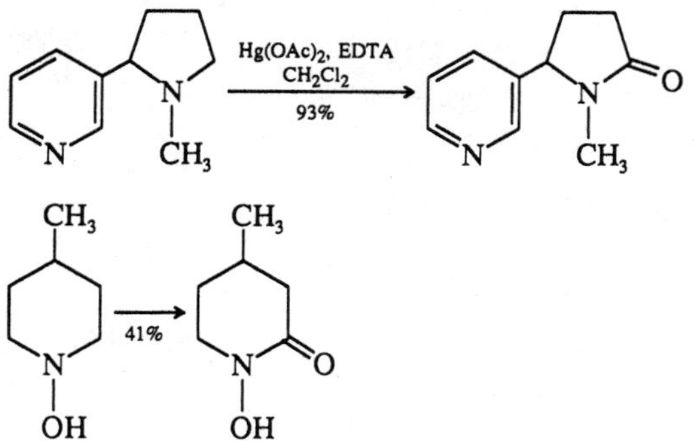

N N
CH_3

$Hg(OAc)_2$, EDTA
CH_2Cl_2
93%

N N O
CH_3

CH_3
N
OH

41%

CH_3
N O
OH

[1] G. Carnevale, E. Davini, C. Iavarone, and C. Trogolo, *J. Org.*, **53**, 5343 (1988).
[2] E. Wenkert and E. C. Angell, *Syn. Comm.*, **19**, 1331 (1988).

Mercury(II) chloride.

Cyclization of δ,ε-unsaturated amines. These amines (**1**), when treated with $HgCl_2$ or $Hg(OAc)_2$, can cyclize predominantly to either *cis-* or *trans*-2,5-disubstituted pyrrolidines (**2**), depending on the solvent. The *cis*-isomer predominates in reactions when THF or $CHCl_3$ is the solvent, whereas use of THF-H_2O (1:1) provides mostly *trans*-**2**. The presence of water also improves the yield in some reactions.

1) $HgCl_2$
2) $NaBH_4$

1		*trans*-2	*cis*-2
THF	36%		10:90
THF/H_2O	64%		87:13

[1] M. Tokuda, Y. Yamada, and H. Suginome, *Chem. Letters*, 1289 (1988).

Mercury(II) sulfate.

Hydration or elimination of glycals. Glycals can undergo hydration or elimination when treated with $HgSO_4$ in dilute sulfuric acid or aqueous acetic acid.[1] However, D-glucal **1** is converted only into the (1,2-dihydroxyethyl)furan (**2**).[2] This product has been used for a short synthesis of L-hexoses. Thus **2** can be converted to the monobenzoate **3** with inversion (Mitsunobu). This product is converted into

$HgSO_4$, H_2SO_4, 97%; C_6H_5COOH, $P(C_6H_5)_3$, DEAD, 92%; Br_2 in CH_3OH

1 → **2**, α_D +38° → **3** → **4**

the L-hexenulose **4**, a known precursor to several L-hexoses including L-glucose and L-galactose.

[1] F. Gonzalez, S. Lesage, and A. S. Perlin, *Carbohydrate Res.*, **42**, 267 (1975).
[2] F. M. Hauser, S. R. Ellenberger, and W. P. Ellenberger, *Tetrahedron Letters*, **29**, 4939 (1988).

Methanesulfonic anhydride.

Friedel–Crafts methanesulfonylation.[1] This reaction is commonly conducted with methanesulfonyl chloride, but this reagent gives poor yields in the case of halo- or nitroarenes. Reasonably good yields can be obtained by use of methanesulfonyl anhydride and a catalytic amount of acid.

$$C_6H_5Cl + (CH_3SO_2)_2O \xrightarrow[85\%]{CH_3SO_3H,\ 100^\circ} o\text{-}ClC_6H_4SO_2CH_3 + p\text{-}ClC_6H_4SO_2CH_3 \quad (1:2)$$

[1] M. Ono, Y. Nakamura, S. Sato, and I. Itoh, *Chem. Letters*, 395 (1988).

α-Methoxyallene.

Cyclopentannelation (**12**, 310). This reaction was mistakenly ascribed to α-methoxyallene, which fails to undergo cyclopentannelation under the cited conditions. Actually, the reagent is (methoxymethoxy)allene. A few other allenic ethers can be used, but all must have a departing group capable of forming a stable cation.[1]

[1] M. A. Tius, J.-B. Ousset, D. P. Astrab, A. H. Fauq, and S. Trehan, *Tetrahedron Letters*, **30**, 923 (1989).

B-Methoxydiisopinocampheylborane (1). Preparation.[1]

syn-β-*Methoxyhomoallyl alcohols.*[2] The reagent **2**, prepared by reaction at −78° of lithiated allyl methyl ether with **1**, derived from (+)-α-pinene, reacts with aldehydes to provide *syn*-β-methoxyhomoallyl alcohols (**3**) in 88% ee.

$$(Ipc)_2BOCH_3\ (\mathbf{1}) + CH_2{=}CHCH_2OCH_3 \xrightarrow[-78^\circ]{sec\text{-BuLi}} Li^+\ (Ipc)_2B^-(OCH_3)CH_2CH{=}CHOCH_3\ (\mathbf{2})$$

$$\mathbf{2} \xrightarrow[57-72\%]{RCHO} RCH(OH)CH(OCH_3)CH{=}CH_2\ \ \mathbf{3}\ (88-90\%\ ee)$$

[1] H. C. Brown and B. Singaram, *J. Org.*, **49**, 945 (1984).
[2] H. C. Brown, P. K. Jadhav, and K. S. Bhat, *Am. Soc.*, **110**, 1535 (1988).

[(Methoxydimethylsilyl)(trimethylsilylmethyl]lithium, $(CH_3)_2Si(OCH_3)—CHLiSi(CH_3)_3$ (1)

The reagent is prepared from the corresponding silane with *t*-butyllithium.

Vinylsilanes.[1] The reaction of **1** with aldehydes or ketones provides vinylsilanes via a Peterson-type reaction.

$$R^1R^2C=O + \mathbf{1} \xrightarrow[60-85\%]{-78^\circ} R^1R^2C=CHSi(CH_3)_3$$

(E/Z = 2–3:1)

[1] T. F. Bates and R. D. Thomas, *J. Org.*, **54**, 1784 (1989).

(E,Z)-1-Methoxy-2-methyl-3-(trimethylsilyloxy)-1,3-pentadiene, 13, 180–181.
Improved preparation[1]:

$$C_2H_5COC_2H_5 \xrightarrow[55-60\%]{HCO_2C_2H_5,\ NaH,\ C_6H_6} C_2H_5COC(CH_3)=CHOH$$

$$\xrightarrow[70-75\%]{TsOH,\ CH_3OH,\ C_6H_6} C_2H_5COC(CH_3)=CHOCH_3$$

$$\xrightarrow[76-83\%]{(CH_3)_3SiOTf,\ N(C_2H_5)_3} CH_3CH=C(OSi(CH_3)_3)C(CH_3)=CHOCH_3$$

[1] D. C. Myles and M. H. Bigham, *Org. Syn.*, submitted (1988).

Methoxy(phenylthio)methane.

1,4-Dicarbonyl compounds. The adducts (**2**) of the anion (**1**) of methoxy(phenylthio)methane and aldehydes (**12**, 316) are converted into α-sulfenyl aldehydes (**3**) on treatment with methanesulfonyl chloride and triethylamine (1:1). Reaction of **3** with a Wittig or Wittig–Horner reagent followed by isomerization

$$\text{RCHO} \xrightarrow[\ ]{\text{CH}_3\text{O}\text{CH(Li)}\text{SC}_6\text{H}_5\ (1)} \text{RCH(OH)CH(SC}_6\text{H}_5\text{)OCH}_3\ (\mathbf{2}) \xrightarrow[70-90\%]{\text{CH}_3\text{SO}_2\text{Cl},\ \text{N(C}_2\text{H}_5)_3} \text{RCH(SC}_6\text{H}_5\text{)CHO}\ (\mathbf{3})$$

$$\mathbf{3} + (C_6H_5)_3P{=}CHCOOC_2H_5 \xrightarrow[50-85\%]{1)\ \text{THF, }0^\circ;\ 2)\ t\text{-BuOK}} (C_6H_5S)(R)C{=}CHCHCOOC_2H_5\ (\mathbf{4})$$

$$\mathbf{4} \xrightarrow[70-90\%]{CF_3COOH,\ H_2O} RC(=O)CH_2CH_2COOC_2H_5\ (\mathbf{5})$$

of the double bond (t-BuOK) provides a γ-phenylthio-β,γ-unsaturated carbonyl compound (**4**), which is converted into a γ-oxo ester (**5**) on treatment with trifluoroacetic acid. 1,4,7-Tricarbonyl compounds can be prepared by reduction of the ester group of **4** to an aldehyde group and repetition of the sequence.

1,3- or 1,4-Diketones.[2] The acetates of adducts (**1**) of methoxy-(phenylthio)methyllithium to aldehydes (**12**, 316) react with enol silyl ethers catalyzed by a Lewis acid to form β-methoxy-γ-phenylthio ketones (**2**) with a novel migration of the phenylthio group. These products can be converted into either 1,3- or 1,4-diketones as shown in the formulation.

$$C_6H_5CH(OH)CH(SC_6H_5)OCH_3\ (\mathbf{1}) \xrightarrow[64\%]{1)\ \text{BuLi};\ 2)\ \text{AcCl};\ 3)\ CH_2{=}C(C_6H_5)OSi(CH_3)_3,\ SnCl_4} C_6H_5CH(SC_6H_5)CH(OCH_3)CH_2C(=O)C_6H_5\ (\mathbf{2})$$

$$\mathbf{2} \xrightarrow[51\%]{1)\ ClC_6H_4CO_3H;\ 2)\ NaHCO_3,\ \Delta} C_6H_5CH_2C(=O)CH_2C(=O)C_6H_5$$

$$\mathbf{2} \xrightarrow[96\%]{(CH_3)_3COK,\ -78^\circ} C_6H_5C(SC_6H_5){=}CHCH_2C(=O)C_6H_5 \xrightarrow[71\%]{H_3O^+} C_6H_5C(=O)CH_2CH_2C(=O)C_6H_5$$

(E/Z = 1:9)

[1] T. Sato, H. Okazaki, J. Otera, and H. Nozaki, *Am. Soc.*, **110**, 5209 (1988).
[2] T. Sato, M. Inoue, S. Kobara, J. Otera, and H. Nozaki, *Tetrahedron Letters*, **30**, 91 (1989).

4-Methoxy-2,2,6,6-tetramethyl-1-oxopiperidinium chloride, (1), **13**, 183.

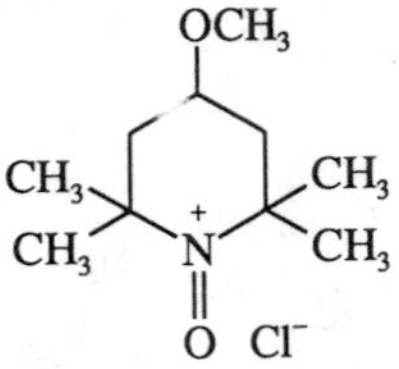

Selective oxidation of hydroxy groups of esters.[1] The reagent fails to oxidize a primary hydroxyl group located in the α- or β-position to an ester group but does oxidize an hydroxyl group in a more remote position.

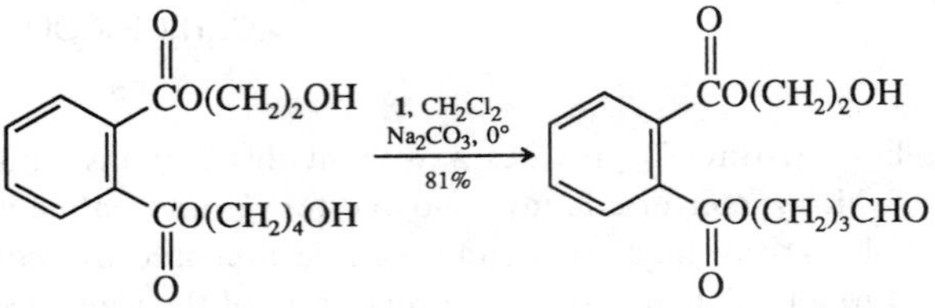

$$\xrightarrow[81\%]{1,\ CH_2Cl_2,\ Na_2CO_3,\ 0°}$$

[1] M. Yamaguchi, T. Takata, and T. Endo, *Tetrahedron Letters*, **29**, 5671 (1988).

Methylaluminum bis(2,6-di-*t*-butyl-4-methylphenoxide) (MAD, **1**), **13**, 203.

Conjugate reduction of enones.[1] This reduction can be effected with 2 equiv. of Li(Bu)(*i*-Bu)$_2$AlH (**2**) and of MAD. The choice of solvents can affect the selectivity. Sterically hindered ketones undergo 1,4-reduction very readily.

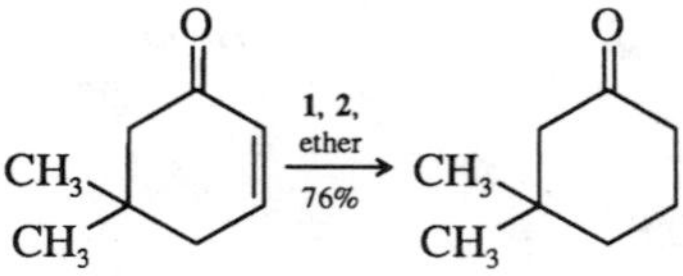

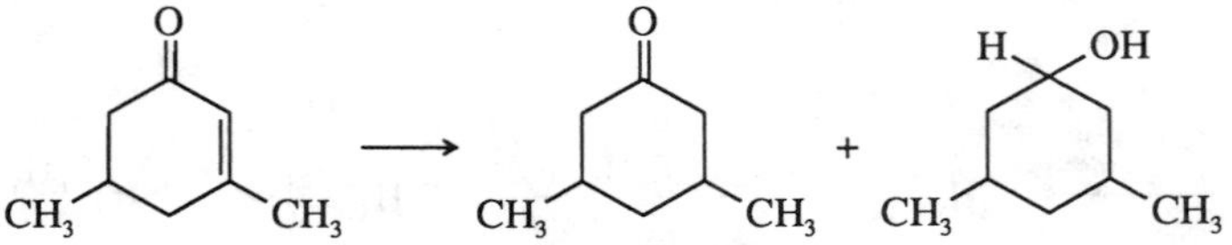

$$\xrightarrow[76\%]{1,\ 2,\ ether}$$

(45%) (49%)

$$(E)\text{-}C_6H_5CH{=}CHC(O){-}C_6H_2(CH_3)_3 \xrightarrow[84\%]{} C_6H_5CH_2CH_2C(O){-}C_6H_2(CH_3)_3$$

Conjugate addition to quinone monoacetals.[2] These substrates undergo only 1,2-addition with $(CH_3)_2CuLi$ and most Grignard reagents, but 1,4-addition can be achieved by complexation of the acetal with 2 equiv. of **1**.

CH_3O OCH_3 + RLi $\xrightarrow{1,\ C_6H_5CH_3,\ -78°}$ CH_3O OCH_3 R

R = C_6H_5 78%
= Bu 42%

[1] K. Nonoshita, K. Maruoka, and H. Yamamoto, *Bull. Chem. Soc. Japan*, **61**, 2241 (1988).
[2] A. J. Stern and J. S. Swenton, *J.C.S. Chem. Comm.*, 1255 (1988).

Methylaluminum bis(2,6-diphenylphenoxide) (1); Methylaluminum bis(4-bromo-2,6-di-*t*-butylphenoxide) (2).

$CH_3Al[—O—C_6H_3(C_6H_5)\text{-}2,6]_2$ **1** $CH_3Al[—O—C_6H_2\text{-}4\text{-Br-}2,6\text{-di-}t\text{-Bu}]_2$ **2**

Claisen rearrangement of allyl vinyl ethers.[1] Thermal rearrangement of vinyl ethers (**3**) of *sec*-allylic alcohols in the presence of **1** or **2** leads to the aldehyde **4**. Rearrangement of the optically active **5** is accompanied by transfer of chirality, particularly in the case of **1**.

CH_2 CH_2 *i*-Bu O → *i*-Bu CHO

3 → **4**
1, −20° 85% E/Z = 97:3
2, −78° 64% = 7:93

CH_3 CH_2 R *i*-Bu O → CH_3 R CHO *i*-Bu (E) + CH_3 S CHO Bu-*i* (Z)

5
1 98:2
2 16:84

[1] K. Maruoka, K. Nonoshita, H. Banno, and H. Yamamoto, *Am. Soc.*, **110**, 7922 (1988).

Methyl cyanoformate, 12, 321. Supplier: Aldrich.

Reductive acylation of enones.[1] The lithium enolate generated by reduction of enones with Li/NH_3 is converted to a β-keto ester by reaction with this reagent in ether (THF promotes O-acylation).

Li, NH_3, t-BuOH → [LiO– enolate] → $NCCOOCH_3$, ether, 74% → H, H, $COOCH_3$

The same reaction carried out on R-(−)-carvone (**1**) affords a 3:1 mixture of two esters (equation I).

(I) **1** → 1) Li/NH_3 2) $NCCOOCH_3$, 62% → CH_2, CH_3, CH_3, CO_2CH_3 + 3:1 CH_2, CH_3, CH_3, CO_2CH_3

[1] S. R. Crabtree, L. N. Mander, and S. P. Sethi, *Org. Syn.*, submitted (1988).

N-Methyl-2-dimethylaminoacetohydroxamic acid, $(CH_3)_2NCH_2CON(CH_3)OH$ (**1**).

Transacylation. The reagent effects transacetylation under neutral conditions.

OAc, $COOC_6H_5$ + **1** → OH, $COOC_6H_5$ (91–98%) + $(CH_3)_2NCH_2CON(CH_3)OCOCH_3$ → ^-OH → **1** + CH_3COOH

[1] M. Ono and I. Itoh, *Tetrahedron Letters*, **30**, 207 (1989).

Methyl hydrogen malonate, $HOOCCH_2COOCH_3$ (**1**). Methyl potassium malonate is available from Fluka.

Methyl 2,4-pentadienoates. These diunsaturated esters are obtained in >85% yield by reaction of α,β-enals with **1** in pyridine catalyzed by dimethylaminopyridine.

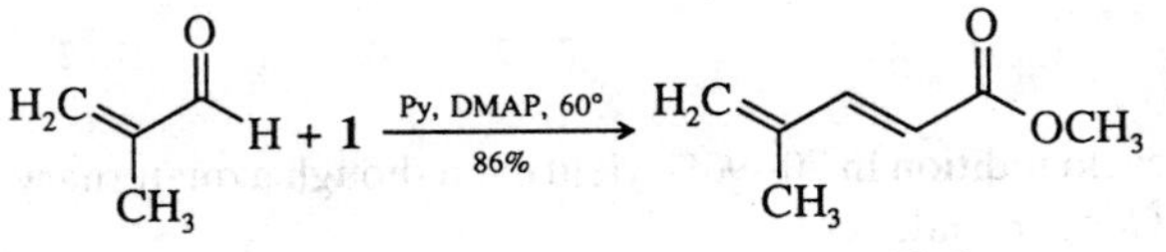
O
H₂C
H + 1
CH₃
Py, DMAP, 60°
86%
O
H₂C
OCH₃
CH₃

(E)-**2**

[1] J. Rodriguez and B. Waegell, *Synthesis*, 534 (1988).

Methyl (R)-3-hydroxybutyrate, (1).

Chiral β-lactams.[1] The reaction of **1** with $(C_2H_5)_2Zn$ in THF provides **a**, which on treatment with lithium hexamethyldisilazane provides **b**, which reacts with the imine **2** to provide a single *cis*-azetidinone (**3**), an intermediate to thienamycin.

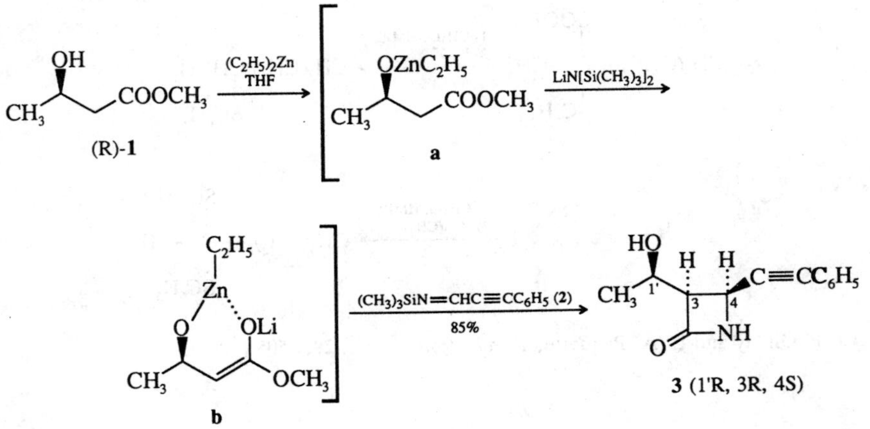
OH
CH₃
COOCH₃
(R)-1
(C₂H₅)₂Zn
THF
OZnC₂H₅
CH₃
COOCH₃
a
LiN[Si(CH₃)₃]₂
C₂H₅
Zn
O
OLi
CH₃
OCH₃
b
(CH₃)₃SiN=CHC≡CC₆H₅ (2)
85%
HO
H H
CH₃
1'
3
4
C≡CC₆H₅
NH
O
3 (1'R, 3R, 4S)

[1] N. Oguni and Y. Ohkawa, *J.C.S. Chem. Comm.*, 1376 (1988).

Methylketeneimines, $CH_3CH{=}C{=}NAr$ (**1**). Synthesis.[1]

[2+2]Cycloaddition with aldehydes; 2-iminooxetanes.[2] Lewis acids are not useful for this cycloaddition, but traces of lanthanide shift reagents, $Yt(fod)_3$, $Yt(hfc)_3$, or $Eu(hfc)_3$ are very efficient catalysts for cycloaddition to a mixture of (*cis*)- and (*trans*)-2-iminooxetanes (**2**) in the case of both aliphatic and aromatic aldehydes.

The Z/E ratio is largely independent of the lanthanide catalyst and of the aldehyde, except in the case of methylketene N-benzylimine. Dimethylketeneimines also

$$1 + RCH{=}O \xrightarrow[65-85\%]{cat.} (Z)\text{-}2 + (E)\text{-}2 \quad (\sim 1:1)$$

(Z)-2 (E)-2

undergo this cycloaddition in 70–90% yield even though a quaternary carbon center is formed at C_3 of the adduct.

[1] Typical preparation: A. Battaglia, A. Dondoni, and P. Giorgianni, *J. Org.*, **47**, 3998 (1982).
[2] G. Barbaro, A. Battaglia, and P. Giorgianni, *ibid*, **53**, 5501 (1988).

Methyllithium.

Decarboxylation of carboxylic acids.[1] Reaction of methyllithium with non-enolizable α-phenyl- or α-phenythio carboxylic acids such as **1** in HMPT results in a Haller–Bauer type fragmentation with loss of CO_2. The intermediate carbanion can be trapped by various electrophiles.

$$CH_3(CH_2)_5\text{—}C(COOH)(SC_6H_5)\text{—}CH_3 \ (\mathbf{1}) \xrightarrow[83\%]{1)\ CH_3Li,\ HMPT;\ 2)\ H_3O^+} CH_3(CH_2)_5CH(SC_6H_5)CH_3$$

$$\mathbf{1} \xrightarrow{1)\ CH_3Li,\ HMPT;\ 2)\ ClSi(CH_3)_3} CH_3(CH_2)_5\text{—}C(Si(CH_3)_3)(SC_6H_5)\text{—}CH_3$$

[1] J. P. Gilday and L. A. Paquette, *Tetrahedron Letters*, **29**, 4505 (1988).

Methyl (E)-4-methoxy-2-oxo-3-butenoate, $CH_3O_2C\overset{O}{\overset{\|}{C}}\text{—}CH{=}CHOCH_3$ (**1**). Preparation.[1]

Diels–Alder reactions; carbohydrates.[2] This β,γ-unsaturated α-keto ester undergoes Diels–Alder reactions with electron-rich dienophiles to provide protected sugars. The thermal and pressure-promoted reaction of **1** with ethyl vinyl ketone provides the *endo*-adduct **2**, which can be converted in two steps into the ethyl β-mannopyranoside **3**. The same reaction of **1** with (Z)-1-acetoxy-2-benzyloxyethylene is even more *endo*-selective, and gives the *endo*-adduct **4**, which can be converted into a derivative of benzyl DL-mannopyranoside.

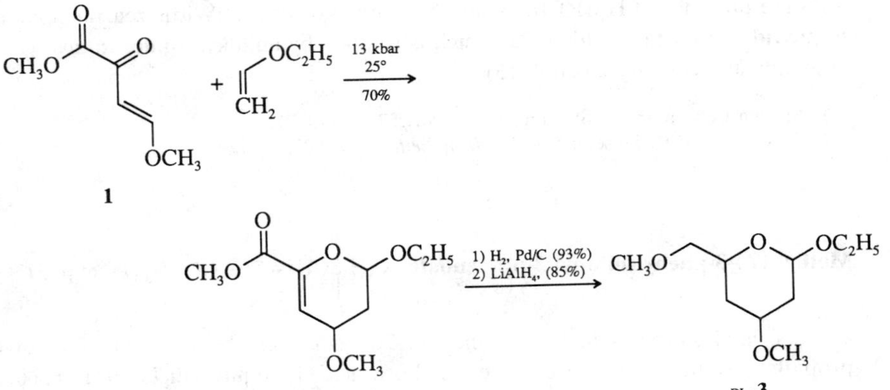

CH_3O ... OC_2H_5 ... CH_2 — 13 kbar, 25°, 70% →

OCH_3

1

CH_3O ... OC_2H_5 — 1) H_2, Pd/C (93%) 2) $LiAlH_4$, (85%) → CH_3O ... OC_2H_5

OCH_3 ... OCH_3

endo-2 (6:1) DL-3

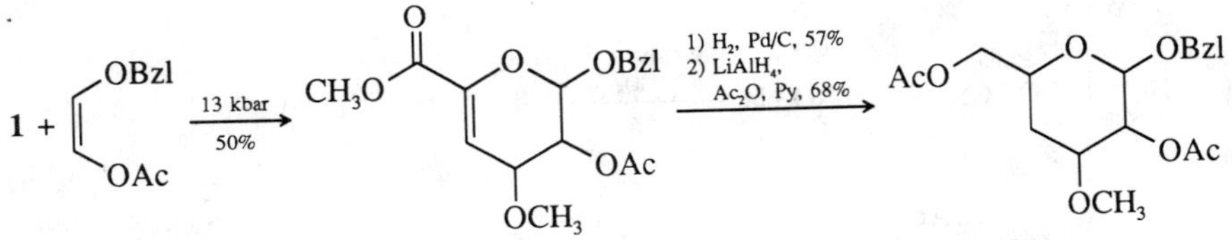

1 + OBzl / OAc — 13 kbar, 50% → CH_3O ... OBzl, OAc, OCH_3 — 1) H_2, Pd/C, 57% 2) $LiAlH_4$, Ac_2O, Py, 68% → AcO ... OBzl, OAc, OCH_3

endo-4 (>45:1) DL-5

[1] W. Trowitzsch, *Z. Naturforsch. B.*, **32B**, 1068 (1977).
[2] D. L. Boger and K. D. Robarge, *J. Org.*, **53**, 5793 (1988).

p-Methylphenylsulfonyldiazomethane (tosyldiazomethane), *p*-$CH_3C_6H_4SO_2CHN_2$ (**1**). Preparation.[1]

sec-Homoallylic alcohols.[2] Reaction of a *sec*-allylic alcohol (**2**) with **1** provides the sulfone **3**, the anion of which can be trapped by an alkyl halide to give

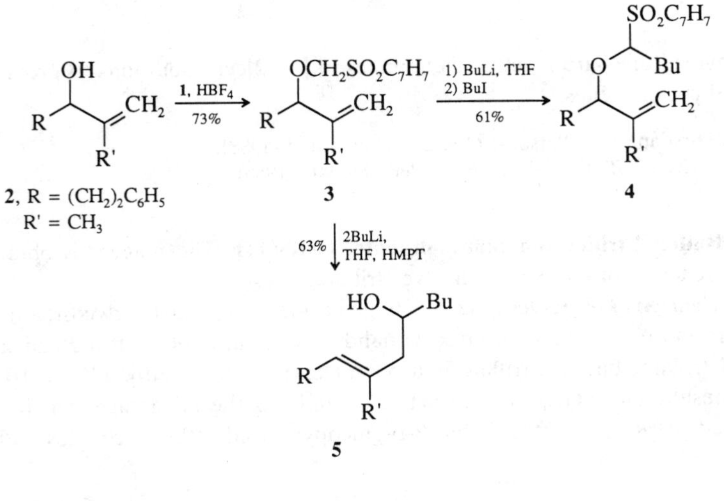

OH, R, CH_2, R' — **1**, HBF_4, 73% → $OCH_2SO_2C_7H_7$, R, CH_2, R' — 1) BuLi, THF 2) BuI, 61% → $SO_2C_7H_7$, O, Bu, R, CH_2, R'

2, R = $(CH_2)_2C_6H_5$, R' = CH_3 **3** **4**

63% ↓ 2BuLi, THF, HMPT

HO, Bu, R, R'

5

4. In the presence of HMPT the anion of **3** undergoes a [2,3]Wittig rearrangement to provide an unstable aldehyde, which is trapped by an alkyllithium to give (E)-homoallylic secondary alcohols (**5**).

[1] A. M. van Leusen and J. Strating, *Org. Syn.*, **57**, 95 (1977).
[2] R. Brückner and B. Peiseler, *Tetrahedron Letters*, **29**, 5233 (1988).

Methyl (Z)-3-phenylsulfonyl-2-propenoate, $C_6H_5S(=O)_2CH{=}CHCOOCH_3$ (**1**), m.p. 51°.

This vinyl sulfone can be obtained in 70–85% yield by reaction of methyl propiolate, sodium benzenesulfinate, and boric acid (1.5 equiv.) in THF/H_2O (1:1) catalyzed by Bu_4NHSO_4.[1]

$$HC{\equiv}CCOOCH_3 + C_6H_5SO_2Na \xrightarrow[70-85\%]{B(OH)_3,\ Bu_4NHSO_4} \mathbf{1}$$

OAc, CH_3, CH_2 + **1** → (79%) OAc, COOCH$_3$, CH_3, $SO_2C_6H_5$ + (1:1) OAc, $COOCH_3$, CH_3, $SO_2C_6H_5$ → ($NaOC_6H_5$) $COOCH_3$, CH_3

Vinyl sulfones are reactive dienophiles in [4+2]cycloadditions to provide substituted cyclohexenes.[2]

[1] G. C. Hirst and P. J. Parsons, *Org. Syn.*, submitted (1988).
[2] P. L. Fuchs and T. F. Braish, *Chem. Rev.*, **86**, 903 (1986).

Methylsulfenyl trifluoromethanesulfonate, CH_3SOTf. The reagent is obtained *in situ* by reaction of CH_3SBr with silver triflate.

Activation of thioglucosides. Thioglucosides are useful for glycosidation since they are more reactive than glycosyl halides. They are known to be activated by methyl triflate, but this triflate is a potent carcinogen. Methylsulfenyl trifluoromethanesulfonate (**1**) is also an activator and has the advantage that it can be prepared *in situ*. Thus the 1-thio-β-D-glucopyranoside (β−**2**) couples with α-D-

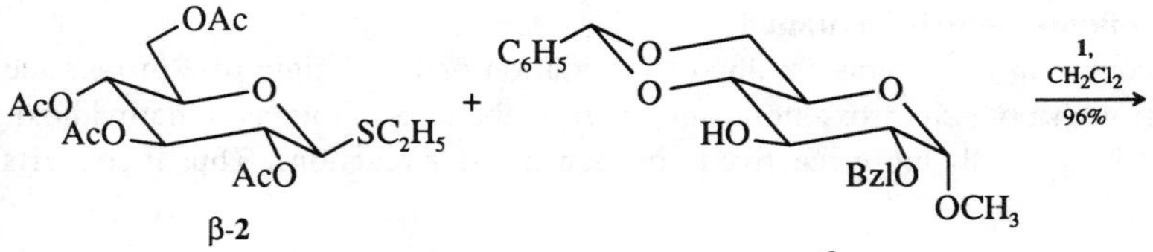

β-2

α-3

3-(Tetraacetyl-β-D-glucopyranosyl)-α-D-glucopyranoside (**4**)

protected derivative (α – **3**) of glucose to give the β-linked disaccharide (**4**) almost exclusively because of the 2-acyl substituent.

[1] F. Dasgupta and P. J. Garegg, *Carbohydrate Res.*, **177**, C13 (1988).

Methyl *p*-tolyl sulfoxide, 14, 115.

α,β-Sulfinyl epoxides. The γ-chloro β-keto sulfoxide **2**, prepared by reaction of methyl chloroacetate with the anion of methyl (R)-*p*-tolyl sulfoxide (**1**) in 83%

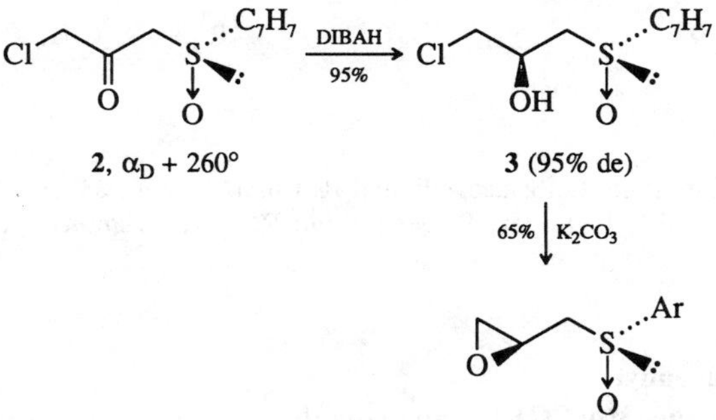

2, α_D + 260°

3 (95% de)

(SR)-**4**, α_D + 239°

yield, can be converted in two steps into the (SR)-α,β-sulfinyl epoxide **4**. These epoxides undergo ready elimination with basic reagents, but do react with cyanocuprates to provide homoallyl β-hydroxysulfoxides (**5**), which undergo Pummerer rearrangement to homoallylic α-hydroxy aldehydes.

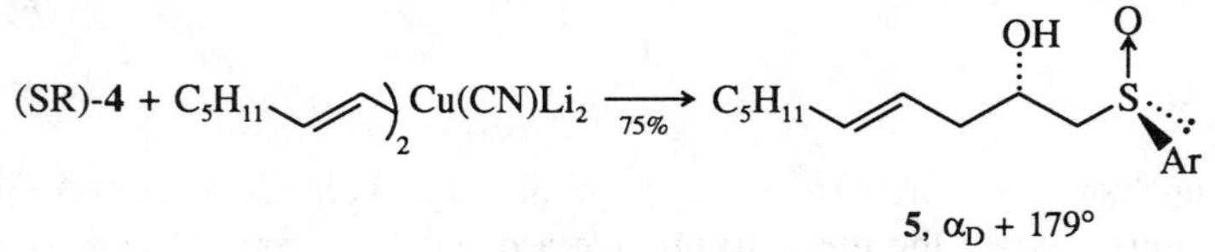

5, α_D + 179°

[1] G. Solladié, C. Hamdouchi, and M. Vicente, *Tetrahedron Letters*, **29**, 5929 (1988).

Methyl(trifluoromethyl)dioxirane (1).

This compound has been obtained by oxidation of 1,1,1-trifluoro-2-propanone (Aldrich) with potassium oxomonosulfate. It is about as stable as dimethyldioxirane, but is markedly more reactive in oxygen transfer reactions. Thus it converts

(1)

phenanthrene to the 9,10-oxide within 5 minutes at −20°. It can also oxidize C—H bonds. Thus cyclohexane is oxidized by **1** to cyclohexanone (−20°, 95% yield).[1]

Epoxidation of enol ethers.[2] This dioxirane can effect epoxidation of such hindered enol ethers as **2**. A similar reaction of 2,3-dihydropyran with **1** affords the epoxide in >92% yield.

$$R_2C{=}C(OCH_3)C_6H_5 \xrightarrow[94\%]{1,\ CH_2Cl_2,\ CF_3COCH_3} \mathbf{3}$$

R = adamantyl

[1] R. Mello, M. Fiorentino, O. Sciacovelli, and R. Curci, *J. Org.*, **53**, 3890 (1988).
[2] L. Troisi, L. Cassidei, L. Lopez, R. Mello, and R. Curci, *Tetrahedron Letters*, **30**, 257 (1989).

Molybdenum carbonyl.

Cyclohexadiene–Mo(CO)$_2$Indenyl complexes. The reaction of 1,4-dimethyl-1,3-cyclohexadiene with the dimer[1] of InMo(CO$_3$) (In = indenyl) provides the complex **1** (*cf.* Mo(CO)$_3$Cp complexes with cyclohexadiene, **12**, 332). Reaction of

$$\xrightarrow[52\%]{[InMo(CO)_3]_2,\ AgBF_4,\ CH_2Cl_2} \mathbf{1}\ (\overset{+}{Mo}(CO)_2In,\ BF_4^-) \xrightarrow{RM} \mathbf{2}$$

1 with nucleophiles (Grignard reagents, enolates of ketones or esters) results in substitutions *trans* to the metal to provide a quaternary center.[2] This technology has been used to prepare the *cis*-bicyclic lactone **3** in high yield.

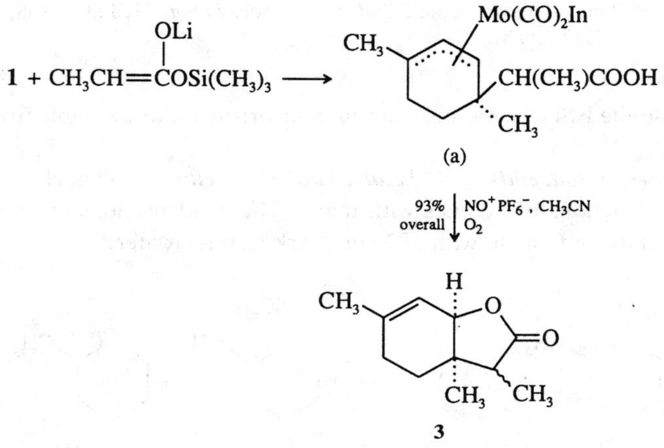

$$1 + CH_3CH{=}C(OLi)OSi(CH_3)_3 \longrightarrow$$

(a)

93% overall; $NO^+PF_6^-$, CH_3CN; O_2

3

[1] M. Bottrill and M. Green, *J.C.S. Dalton*, 2365 (1977).
[2] A. J. Pearson and V. D. Khetani, *J. Org.*, **53**, 3395 (1988).

Montmorillonite.

Michael addition to α,β-enoates. Aluminum ion-exchanged montmorillonite (Al-Mont[1]) is a very effective catalyst for Michael addition of silyl ketene acetals or silyl enol ethers to α,β-enoates. In fact this heterogeneous catalyst is more effective than Lewis acids, which are generally required in a stoichiometric amount. It also facilitates Michael addition to α,β-enones.[2]

Examples:

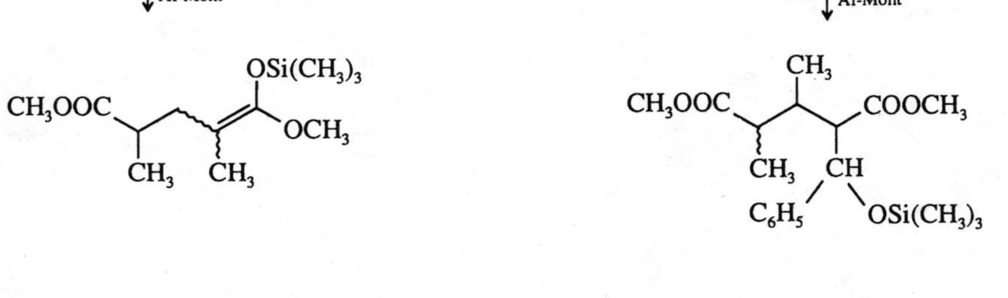

Al-Mont, −78°; 84%

syn/anti = 27:73

79%; $CH_2{=}C(CH_3)COOCH_3$, Al-Mont

80%; C_6H_5CHO, Al-Mont

[1] M. Kawai, M. Onaka, and Y. Izumi, *Bull. Chem. Soc. Japan*, **61**, 1237 (1988).
[2] *Idem, ibid.*, **61**, 2157 (1988).

Montmorillonite KSF. This is an acidic but noncorrosive clay available from Aldrich and Fluka.

Dithianes, thioacetals and -ketals, enol thiol ethers. This clay is a useful catalyst for reactions of ketones with thiols. The condensations are usually conducted in refluxing toluene with a Dean–Stark trap for water.[1]

SC_6H_5 SC_6H_5

CH_3 CH_3 CH_3

+ C_6H_5SH $\xrightarrow{\text{M-KSF}}$ +

(58%) (14%)

[1] B. Labiad and D. Villemin, *Syn. Comm.*, **19**, 31 (1989); *idem*, *Synthesis*, 143 (1989).

N

1-(1-Naphthyl)ethylamine, $C_{10}H_7CH(CH_3)NH_2$ (**1**).

Alkylation of α-nitro keto imines. The key step in a conversion of cyclic α-nitro ketones to cyclic 3-substituted 1-nitroalkenes (**4**) involves alkylation of the dianion of an α-nitro imine (**2**) or hydrazone. The dianions are easily generated,

$$\text{2-nitrocyclohexanone} \xrightarrow[65-80\%]{H_2NR^1} \mathbf{2} \xrightarrow[55-85\%]{1)\ \text{base (2 equiv.)};\ 2)\ R^2I} \mathbf{3} \xrightarrow[20-40\%\ \text{overall}]{NaBH_4,\ CeCl_3,\ CH_3OH} \mathbf{4} + H_2NR^1$$

particularly with *sec*-BuLi in THF/HMPT at −78°, and alkylation occurs at the position adjacent to the imine or hydrazone group. The products undergo facile reduction and elimination to **4** with $NaBH_4$ and $CeCl_3$ (2 equiv. of each) in an alcohol, usually at room temperature.[1]

Asymmetric alkylation of the intermediate can be effected by use of a chiral amine in the first step.[2] Best results are obtained with (S)-**1** even though it lacks a coordinating group. However, use of only *sec*-BuLi as the base results in low yields and moderate diastereoselectivity but use of LDA (1 equiv.) and *sec*-BuLi (2 equiv.) increases the optical yield to 79% de and to 87% de if HMPT is also present. Evidently an amide base plays an important role in the diastereoselectivity, probably by serving as a ligand for the dianion.

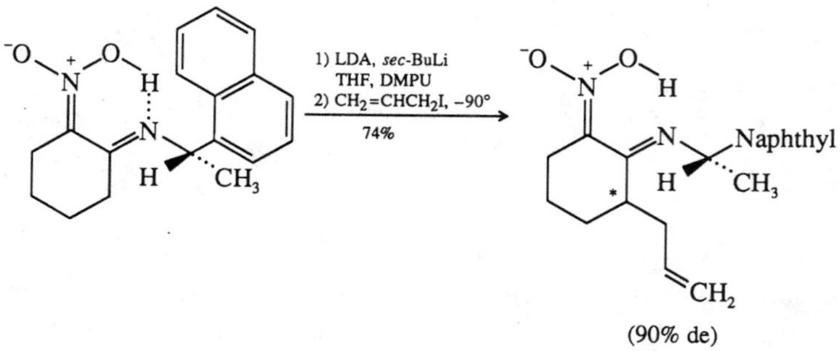

1) LDA, *sec*-BuLi
THF, DMPU
2) $CH_2=CHCH_2I$, –90°
74%

(90% de)

[1] S. E. Denmark, J. A. Sternberg, and R. Lueoend, *J. Org.*, **53**, 1251 (1988).
[2] S. E. Denmark and J. J. Ares, *Am. Soc.*, **110**, 4432 (1988).

Nickel boride.

Deselenation.[1] C–Se bonds are cleaved reductively by nickel boride in methanol in generally high yield. This reduction can even be more facile than reduction of C=C bonds.

[1] T. G. Back, V. I. Birss, M. Edwards, and M. V. Krishna, *J. Org.*, **53**, 3815 (1988).

Nickel carbonyl.

Aziridine → β-lactam.[1] This transformation can be effected by insertion of CO into the N–C bond of an aziridine. Thus reaction of N-benzyl-2-methylaziridine (**1**) with lithium iodide in refluxing THF provides a lithium dialkylamide (a) that reacts with $Ni(CO)_4$ to form the β-lactam **2** in 51% yield. Both reactions probably

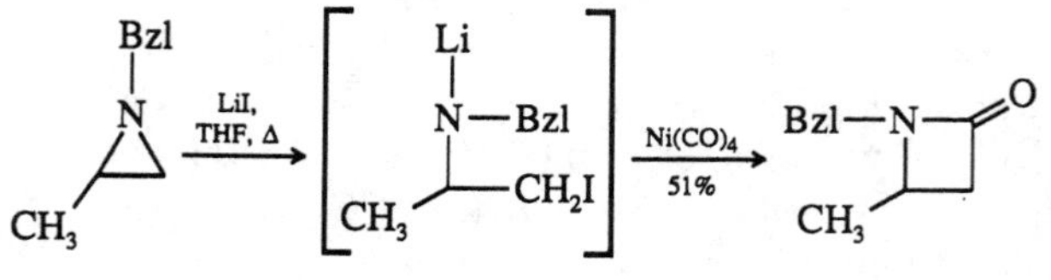

LiI, THF, Δ → a → $Ni(CO)_4$, 51% → 2

1 **a** **2**

proceed with inversion; and, indeed, 1-benzyl-*cis*-2,3-dimethylaziridine is converted by this procedure into 1-benzyl-*cis*-3,4-dimethylazetidinone (37% yield).

Cyclocarbonylation; cyclopentenones.[2] The adducts of allyl halides and alkynes,[3] particularly those with *cis*-configuration, can be converted into mono- or disubstituted cyclopentenones by reaction with $Ni(CO)_4$ in CH_3CN containing CH_3OH and $N(C_2H_5)_3$.

$$HC{\equiv}CSi(CH_3)_3 + CH_2{=}CHCH_2Br \xrightarrow[62\%]{Cl(CH_3CN)_2Pd(II),\ 25°} \text{(Z)-}(CH_3)_3SiC(Br){=}CHCH_2CH{=}CH_2$$

$$\xrightarrow[65\%]{Ni(CO)_4,\ N(C_2H_5)_3,\ CH_3OH,\ CH_3CN} \text{2-}Si(CH_3)_3\text{-4-}(CH_2COOCH_3)\text{-2-cyclopentenone}$$

[1] W. Chamchaang and A. R. Pinhas, *J.C.S. Chem. Comm.*, 710 (1988).
[2] F. Camps, J. Coll, A. Llebaria, and J. M. Moretó, *Tetrahedron Letters*, **29**, 5811 (1988).
[3] K. Kaneda, T. Uchiyama, Y. Fujiwara, T. Imanaka, and S. Teranishi, *J. Org.*, **44**, 55 (1979).

Nickel(II) chloride/Zinc/Pyridine.

Conjugate addition of RBr to α,β-enoates.[1] A Ni(0) catalyst formed from $NiCl_2{\cdot}6H_2O$, zinc (2 equiv.), and pyridine can effect addition of alkyl bromides(iodides) to electron-deficient C–C double bonds. A wide range of organic halides can undergo this reaction. A trace of water is necessary for turnover numbers of 20 or more. A number of bases can replace pyridine: α-picoline, 2,4-lutidine, DBU; but not TMEDA or 2,4,6-collidine.

$$\underset{\text{(E or Z)}}{C_6H_5CH{=}CHBr} + CH_2{=}CHCOOC_2H_5 \xrightarrow[72\text{–}75\%]{NiCl_2\cdot 6H_2O,\ Zn,\ Py,\ THF} C_6H_5CH{=}CHCH_2CH_2COOC_2H_5$$

$$\underset{\text{(Z)}}{CH_3CH{=}CHBr} + CH_2{=}CHCOCH_3 \xrightarrow[58\%]{} CH_3CH{=}CHCH_2CH_2COCH_3$$

[1] R. Sustmann, P. Hopp, and P. Holl, *Tetrahedron Letters*, **30**, 689 (1989).

***o*-Nitrobenzenesulfenyl chloride,** *o*-$O_2NC_6H_4SCl$ (**1**), **10**, 277.

Protection of amino groups. The *o*-nitrobenzenesulfenyl (NPS) group is useful for protection of terminal amino groups of peptides and the exocyclic amino groups of bases such as adenine. The group also increases the stability of an N-glycoside bond. It is removable by reduction with Bu_3P in aqueous dioxane.[1,2] It is also stable to phosphorylation. These properties have been used to synthesize nucleopeptides, in which a peptide is joined via a phosphodiester bond to the 5′-end of a nucleic acid.

[1] J. Heikkilä, N. Balgobin, and J. Chattopadhyaya, *Acta Chim. Scand. B.*, **37**, 857 (1983).
[2] E. Kuyl-Yeheskiely, C. M. Tromp, A. W. M. Lefeber, G. A. van der Marel, and J. H. van Boom, *Tetrahedron*, **44**, 6515 (1988).

2-Nitrobenzenesulfinyl chloride, $2\text{-}O_2NC_6H_4SOCl$.

$R^1SR^2 \rightarrow R^1S(O)R^2$. This sulfinyl chloride is oxidized by potassium superoxide to a sulfinyl peroxy intermediate that oxidizes sulfides to sulfoxides at −25° in >90% yield with only traces of the corresponding sulfones.[1]

$$2\text{-}O_2NC_6H_4SOCl + KO_2 + C_6H_5SCH_3 \xrightarrow[92\%]{CH_3CN,\ -25^\circ} C_6H_5\overset{O}{\overset{\|}{S}}CH_3$$

[1] Y. H. Kim and D. C. Yoon, *Tetrahedron Letters*, **29**, 6453 (1988).

***p*-Nitrobenzenesulfonyl peroxide,** $(p\text{-}O_2NC_6H_4SO_2O)_2$ (**1**), m.p. 122° dec.

Halogenation of arenes.[1] Chlorination or bromination of arenes can be effected with $(C_2H_5)_3NHCl$ or $(C_2H_5)_3NHBr$ in CH_3CN or CH_2Cl_2 in the presence of *p*- or *m*-nitrobenzenesulfonyl peroxide. The effective reagents evidently are sulfonyl hypohalites.

$$C_6H_5CH_3 + (C_2H_5)_3NHCl \xrightarrow[78\%]{\mathbf{1},\ CH_3CN,\ 0^\circ} ClC_6H_4CH_3 \quad (p/o = 58:42)$$

$$C_6H_5OCH_3 + (C_2H_5)_3NHBr \xrightarrow[76\%]{} p\text{-}BrC_6H_4OCH_3 \quad (p/o = 100:0)$$

[1] M. Yoshida, H. Mochizuki, and N. Kamigata, *Chem. Letters*, 2017 (1988).

***N*-(*p*-Nitrobenzylidene)benzenesulfonamide,** $C_6H_5SO_2N{=}CHC_6H_4NO_2\text{-}p$, **13**, 23. Preparation.[1]

Catalytic oxidation of R_2S to $R_2S{=}O$.[2] This sulfonamide can function as a catalyst for oxidation of sulfides selectively to sulfoxides by Oxone® (potassium peroxymonosulfate). The net effect involves transfer of oxygen from an N-sulfonyloxaziridine to the sulfide. This catalytic system is useful because sulfones are formed if at all only in traces.

[1] L. C. Vishwakarma, O. D. Stringer, and F. A. Davis, *Org. Syn.*, **66**, 203 (1987).
[2] F. A. Davis, S. G. Lal, and H. D. Durst, *J. Org.*, **53**, 5004 (1988).

Nitrogen dioxide, N_2O_4.

N-Nitrosation.[1] N-Nitroso- and N-nitrolactams are known to decompose on warming to lactones. A review of this method for conversion of lactams to lactones concludes that the route via nitrosolactams is preferable because nitrosation of lactams proceeds in almost quantitative yield with N_2O_4 and sodium acetate in CH_2Cl_2 at $-10°$.

$(CH_2)_n$ C=O / NH (**1**, n = 4–14) $\xrightarrow[92-100\%]{N_2O_4,\ CH_3COONa}$ $(CH_2)_n$ C=O / N—NO (**2**) $\xrightarrow[75-85\%]{CCl_4,\ \Delta}$ $(CH_2)_n$ C=O / O (**3**) + N≡N→O

[1] N. Torra, F. Urpi, and J. Vilarrasa, *Tetrahedron*, **45**, 863 (1989).

1-Nitro-2-(phenylthio)ethylene, $C_6H_5SCH{=}CHNO_2$ (**1**). Preparation.[1]

Indolizines.[2] The reagent reacts with the pyridinium salts (**2**) in the presence of $N(C_2H_5)_3$ to give the indolizine derivatives **3** and **4**. When R = CH_3, **4** is not formed.

2, R = H, CH_3 (Br^-, N-$CH_2COOC_2H_5$) + **1** $\xrightarrow{N(C_2H_5)_3,\ C_2H_5OH}$ **3** (56%) ($C_2H_5O_2C$) + **4** (0–39%) ($C_2H_5O_2C$, NO_2)

[1] N. Ono, A. Kamimura, and A. Kaji, *J. Org.*, **51**, 2139 (1986).
[2] Y. Tominaga, Y. Ichihara, and A. Hosomi, *Heterocycles*, **27**, 2345 (1988).

(1S,2R)-Norephedrine, $C_6H_5CH(OH)CH(CH_3)NH_2$ (*erythro*) (**1**).

Enantioselective alkylation of ω-formyl esters.[1] In the presence of a catalytic amount of the N,N-dibutyl derivative (**2**) of **1**, dialkylzincs add to the formyl group of esters such as **3** or **4** to form (S)-4- or -5-hydroxyalkanoic esters. These esters cyclize to optically active alkyl-substituted lactones.

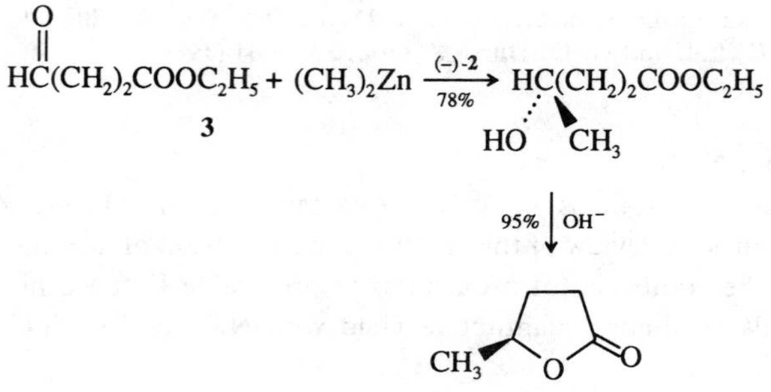

$$\underset{\mathbf{3}}{\mathrm{HC(=O)(CH_2)_2COOC_2H_5}} + \mathrm{(CH_3)_2Zn} \xrightarrow[78\%]{(-)\text{-}\mathbf{2}} \mathrm{HC(OH)(CH_3)(CH_2)_2COOC_2H_5} \xrightarrow[95\%]{OH^-} \text{(S)-5-methyl-γ-butyrolactone}$$

S, 90% ee

$$\underset{\mathbf{4}}{\mathrm{HC(=O)(CH_2)_3COOCH_3}} + \mathrm{(C_2H_5)_2Zn} \xrightarrow[83\%]{(-)\text{-}\mathbf{2}} \xrightarrow[97\%]{OH^-} \text{(S)-6-ethyl-δ-valerolactone}$$

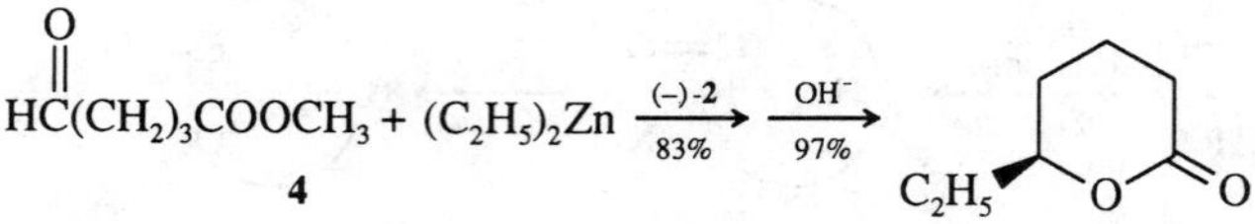

S, 88% ee

[1] K. Soai, S. Yokoyama, T. Hayasaka, and K. Ebihara, *Chem. Letters*, 843 (1988).

O

Organoaluminum reagents.

Review.[1] Use of these reagents in synthesis since 1984 has been reviewed (130 references).

[1] K. Maruoka and H. Yamamoto, *Tetrahedron*, **44**, 5001 (1988).

Organoboranes, chiral.

Synthesis of pure enantiomers. Brown and Singaram[1] have reviewed the chiral organoboranes obtained from (+)- and (−)-pinene and their use for synthesis of optically pure amines, halides, hydrocarbons, lithium alkylborohydrides, ketones, aldehydes, α-chiral acids, esters, nitriles, alkynes, and alkenes.

[1] H. C. Brown and B. Singaram, *Acct. Chem. Res.*, **21**, 287 (1988).

Organocerium compounds.

$RCeCl_2$. The reagents, prepared *in situ* from RLi + $CeCl_3$,[1] are more reactive than RLi in 1,2-addition to carbonyl groups. An example is the reaction with 2-alkoxypropionoyl-1,3-dithianes (**1**), prepared from methyl (S)-lactate.[2] The diastereoselectivity obtains even when the protecting group is incapable of chelation.

R^1O, S, CH_3, S, O $\xrightarrow{R^2M}$ R^1O, S, CH_3, S, HO, CH_3 + *anti*-**2**

1, $R^1 = SiMe_2$-*t*-Bu → **2** (*syn*)

CH_3Li	40%	80:20
CH_3CeCl_2	67%	94:6

[1] T. Imamoto, T. Kusumoto, Y. Tawarayama, Y. Sugiura, T. Mita, Y. Hatanaka, and M. Yokoyama, *J. Org.*, **49**, 3904 (1984).
[2] T. Sato, R. Kato, K. Gokyu, and T. Fujisawa, *Tetrahedron Letters*, **29**, 3955 (1988).

Organocopper reagents.

Role of $BF_3 \cdot O(C_2H_5)_2$ in reactions of R_2CuLi.[1] It is generally accepted that the beneficial effect of BF_3 etherate on cuprate reactions results from substrate activation. Lipschutz *et al.*[1] conclude that BF_3 etherate also modifies the cuprate. Spectroscopic studies of the effect of BF_3 etherate on $(CH_3)_2CuLi$ indicate that the

Lewis acid converts the homocuprate into a mixture of four components: $[(CH_3)_2CuLi]_2$, $(CH_3)_3Cu_2Li$, $CH_3Li \cdot BF_3$, and BF_3. In general, $(CH_3)_3Cu_2Li$ is more reactive than $(CH_3)_2CuLi$, and can be preformed from 3 CH_3Li + 2 CuI, then 2 BF_3. It can also be generated by warming $(CH_3)_2CuLi$ with BF_3 etherate to about 10° before addition of the substrate.

RCu/S(CH₃)₂. Various alkylcoppers are more thermally stable and also more reactive in dimethyl sulfide than in ether or tetrahydrofuran. This solvent can improve yields in conjugate additions and in alkylation of acid chlorides.[2]

Coupling with ROCl.[3] The lithium naphthalenide reduction of the CuI-triphenylphosphine complex results in a highly active Cu(0). This species reacts with alkyl bromides to give an alkylcopper reagent that reacts with acid chlorides to furnish ketones. This synthesis tolerates functional groups such as esters, nitriles, halide, and even some epoxides. Homocoupling can be a competing reaction when alkyl iodides are used in place of the bromides and with phosphines other than $P(C_6H_5)_3$.

$$C_2H_5OOC(CH_2)_3Br \xrightarrow[81\%]{1)\ Cu(0)\quad 2)\ C_6H_5COCl} C_6H_5CO(CH_2)_3COOC_2H_5$$

$$\overset{O}{CH_2\text{—}CH}(CH_2)_6Br \xrightarrow[58\%]{} C_6H_5CO(CH_2)_6\overset{O}{CH\text{—}CH_2}$$

***Cleavage of ketals* (12, 347–348).** Certain chiral cyclic ketals are cleaved by $R_2CuLi/BF_3 \cdot O(C_2H_5)_2$ diastereoselectively to chiral secondary alcohols.[4]

$$\xrightarrow[93\%]{(CH_3)_2CuLi,\ BF_3 \cdot O(C_2H_5)_2}$$

(91% de)

Conjugate addition.[5] The key step in a stereocontrolled synthesis of the Lythraceae alkaloid lasubine II (**4**) is the conjugate addition of an alkylcopper complexed with BF_3 to the N-acyl-2,3-dihydro-4-pyridone **1** to give the *cis*-product **2** in 56% yield and >96% stereoselectivity. Hydrogenation in the presence of Li_2CO_3 effects cyclization and deprotection of nitrogen to give the ketone **3**, which is reduced stereoselectively by lithium trisiamylborohydride to the desired alcohol **4**.

1, Ar = $C_6H_4(OCH_3)_2$-3,4

$Cl(CH_2)_4MgBr$, $CuBr \cdot S(CH_3)_2$, $BF_3 \cdot O(C_2H_5)_2$, 56%

2 (*cis*, >96% ee)

82% H_2, Pd/C, Li_2CO_3

$LiB[CH(CH_3)CH(CH_3)_2]_3H$, 81%

4 **3**

***1,6-Addition to an enynone* (1).**[6] Cuprates, particularly higher-order ones, undergo 1,6-addition to **1** to provide the less stable (Z)-dienone **2** as the major product. The (Z)-stereoselectivity increases with the bulk of the transferred R group. The addition may involve an allenyl enol. Addition of C_6H_5SLi to **1**, however, provides (E)-dienones, possibly via an intermediate 1,2-adduct.

1 + t-$Bu_2Cu(CN)Li_2$ $\xrightarrow{93\%}$ **2** (Z/E = 34:1)

1,3-Dienes.[7] A new route to (E,Z)-1,3-dienes involves reaction of (Z)-alkenylcuprates with *cis*- or *trans*-epoxysilanes in the presence of BF_3 etherate to give *erythro*- or *threo*-β-hydroxysilanes, respectively. These products are converted to either (Z,E)- or (Z,Z)-1,3-dienes by basic or acidic elimination of $(CH_3)_3SiOH$.

(Z)-1 + $(C_2H_5CH{=}CH)_2CuLi$ (2) $\xrightarrow{BF_3,\ ether,\ -70°}$ 3

$C_2H_5CH{=}CHCH{=}CH$Heptyl (Z,Z > 97%) $\xleftarrow[50\%\ from\ 1]{NaH}$ 3 $\xrightarrow[52\%\ from\ 1]{BF_3 \cdot O(C_2H_5)_2}$ $C_2H_5CH{=}CHCH{=}CH$Heptyl (Z,E > 97%)

(E)-4 + 2 ⟶ 5

$C_2H_5CH{=}CHCH{=}CH$Pentyl (Z,E > 97%) $\xleftarrow[65\%\ from\ 4]{NaH}$ 5 $\xrightarrow[70\%\ from\ 4]{BF_3 \cdot O(C_2H_5)_2}$ $C_2H_5CH{=}CHCH{=}CH$Pentyl (Z,Z > 97%)

Reaction with β-propiolactones.[8] These lactones are cleaved to 3-substituted propionic acids by a number of organocopper reagents including R_2CuMgX, R_2CuLi, $R_2Cu(CN)Li_2$, $RCu \cdot Bu_3P$. The reaction involves selective cleavage of the β-carbon–oxygen bond.

$\xrightarrow[>80\%]{R_2CuLi,\ RCu,\ etc.}$ $R(CH_2)_2COOH$

Polypropionate 1,3-diols.[9] The method for synthesis of *syn*-1,3-polyols (**12**, 139) of the polyacetate type has been extended to polypropionates merely by use of an (E)- or (Z)-propenylcuprate, $(CH_3CH{=}CH)_2Cu(CN)Li_2$ (**1**), in place of a vinylcuprate.

$$\text{BzlO}\text{-epoxide} \xrightarrow[90\%]{(E)\text{-}\mathbf{1},\ \text{ether}} \text{BzlO}\ldots\text{CH(OH)}\ldots\text{CH=CH}\text{CH}_3 \xrightarrow[84\%]{1)\ \text{VO(acac)}_2;\ 2)\ \text{ClSiR}_3} \text{BzlO}\ldots\text{CH(OSiR}_3)\ldots\text{epoxide-CH}_3$$

$$\xrightarrow[68\%]{(E)\text{-}\mathbf{1}} \text{BzlO}\ldots\text{CH(OSiR}_3)\ldots\text{CH(OH)CH(CH}_3)\text{CH=CH}\text{CH}_3 \xrightarrow{74\%} \text{BzlO}\ldots\text{CH(OSiR}_3)\ldots\text{CH(OH)CH(CH}_3)\text{-epoxide-CH}_3$$

$$\xrightarrow[91\%]{(\text{CH}_2\text{=CH})_2\text{Cu(CN)Li}_2} \text{BzlO}\ldots\text{CH(OSiR}_3)\ldots\text{CH(OH)CH(CH}_3)\text{CH(OH)CH(CH}_3)\text{CH=CH}_2$$

Vinyllithiums **(*cf.* 13,** 142).[10] An improved route to vinyllithiums involves coupling of vinyl and aryl triflates with tributyl- or trimethylstannylcuprates of the formula $(R_3Sn)_2Cu(CN)Li_2$, prepared from LDA, R_3SnH, and CuCN in THF at −50° to −60°. The higher-order cuprates couple with vinyl and aryl triflates at −20° to provide a vinyltin intermediate. Work-up of the intermediate with AgOAc provides a vinyltin, with bromine a vinyl bromide, or with CF_3COOH an aryltin.

$$\text{1-OTf-cyclohexene} \xrightarrow[97\%]{1)\ (\text{Bu}_3\text{Sn})_2\text{CuCNLi}_2;\ 2)\ \text{AgOAc}} \text{1-SnBu}_3\text{-cyclohexene} \xrightarrow{\text{RLi}} \text{1-Li-cyclohexene} + \text{RSnBu}_3$$

Benzylic organocopper reagents, $C_6H_5CH_2Cu(CN)ZnBr$ **(1).** Preparation:

$$\text{(I)}\quad \text{C}_6\text{H}_5\text{CH(R)Br} \xrightarrow[>90\%]{\text{Zn(BrCH}_2\text{CH}_2\text{Br), THF}} \text{C}_6\text{H}_5\text{CH(R)ZnBr} \xrightarrow{\text{CuCN}\cdot 2\text{LiCl, THF, }-20^\circ} \text{C}_6\text{H}_5\text{CH(R)Cu(CN)ZnBr}\ (\mathbf{1})$$

$(R = H, CH_3)$

These organocopper reagents (equation I) react with various electrophiles, aldehydes, acid chlorides, enones, allylic bromides, in 85–95% yield.[11]

$$2,6\text{-}Cl_2C_6H_3CH_2Cu(CN)ZnBr + CH_2{=}CHCH_2Br \xrightarrow[96\%]{} 2,6\text{-}Cl_2C_6H_3CH_2CH_2CH{=}CH_2$$

$$C_6H_5CH_2Cu(CN)ZnBr + PrCHO \xrightarrow[94\%]{BF_3\cdot O(C_2H_5)_2} C_6H_5CH_2CH(OH)Pr$$

$$3\text{-}(C_6H_{11}CO)C_6H_4CH_2Cu(CN)ZnBr + C_6H_5COCl \xrightarrow[90\%]{} 3\text{-}(C_6H_{11}CO)C_6H_4CH_2COC_6H_5$$

$$3\text{-}NCC_6H_4CH_2Cu(CN)ZnBr + \text{3-iodo-2-cyclohexenone} \xrightarrow[84\%]{ClSi(CH_3)_3(cat.)} \text{3-(3-cyanobenzyl)-2-cyclohexenone}$$

Lithium bis(trimethylsilyl)cuprate.[12] This cuprate, prepared from $(CH_3)_3SiLi$ and CuCN, converts acyl chlorides into acylsilanes in moderate to high yield.

***Methyloxidocopperlithium* (1).**[13] In the presence of traces of water, lithium dimethylcuprate is converted into a bright yellow reagent, tentatively formulated

$$3(CH_3)_2CuLi + H_2O \longrightarrow 2CH_4 + CH_3Cu + \begin{array}{ccc} CH_3{-}Li{-}O{-}Li \\ | \quad\quad\quad | \\ Cu \quad\quad\quad Cu \\ | \quad\quad\quad | \\ CH_3{-}Li{-}CH_3 \end{array}$$

1

as **1**, and a polymeric methylcopper. This new reagent undergoes rapid and diastereoselective addition to enones.

$$C_6H_5CH(OMOM)COCH{=}CHC_6H_5 \xrightarrow[93-100\%]{1} MOMOCH(C_6H_5)COCH_2CH(CH_3)C_6H_5 + MOMOCH(C_6H_5)COCH_2CH(CH_3)C_6H_5 \quad (16:1)$$

Alkylbis(t-butylphosphido)cuprates, RCuP(t-Bu)$_2$Li.[14] These cuprates are obtained by reaction of RLi with $[CuP(t\text{-}Bu)_2]_4$. The cuprates are generally stable in THF at 25° for 4 hours, but can be used for conjugate addition, halide displacement (particularly of primary halides), and epoxide cleavage.

α-Alkoxyorganocuprates.[15] α-Alkoxyorganocuprates such as **1**, prepared as shown, provide a novel synthesis of tetrahydrofurans. Thus conjugate addition to

$(CH_3)_2CHCH(OCH_2OCH_3)(SnBu_3)$ —1) BuLi, 2) CuCN→ $[(CH_3)_2CHCH(OMOM)]_2Cu(CN)Li_2$ (**1**)

2 + **1** —$ClSi(CH_3)_3$→ [**a**: $OSi(CH_3)_3$, OCH_2OCH_3, $CH(CH_3)_2$] —$TiCl_4$, CH_2Cl_2, 77%→ **3** (69:31)

2 in the presence of TMSCl gives a single silyl enol ether (**a**), which on treatment with $TiCl_4$ at −78° cyclizes to a *cis*-fused annelated tetrahydrofuran. The reaction can also show diastereoselectivity at the exocyclic position of the tetrahydrofuran ring.

α-Alkoxyorganocuprates when combined with chlorotrimethylsilane also undergo preferential 1,4-addition to α,β-enals. Of greater interest, the addition is *syn*-selective (45–250:1). The 1,2-addition shows no diastereoselectivity.[16]

$[BuCH(OMOM)]_2Cu(CN)Li_2$ + BuCH=CHCHO —$ClSi(CH_3)_3$, THF, −78→20°→

(1,2-adduct, 17%) + *syn*-**1** (30%) + *anti*-**1** (250:1)

syn-**1** —18%; 1) H_3O^+, 2) CrO_3→ (*cis*)

Dilithium(trimethylstannyl)(2-thienyl)cyanocuprate, $(CH_3)_3SnCu(2\text{-}Th)CNLi_2$ (**1**)[17]. The cuprate is prepared by addition of CH_3Li (2 equiv.) in ether to a THF solution of $[(CH_3)_3Sn]_2$ (1 equiv.) and thiophene (1 equiv.); the reagent is formed after addition of CuCN (1 equiv.).

The reagent transfers the trimethyltin group to the β-position of a variety of α,β-unsaturated esters, including acetylenic esters. It is generally superior to the lower-order cuprate $[(CH_3)_3SnCuSC_6H_5]Li$ (**10**, 289).

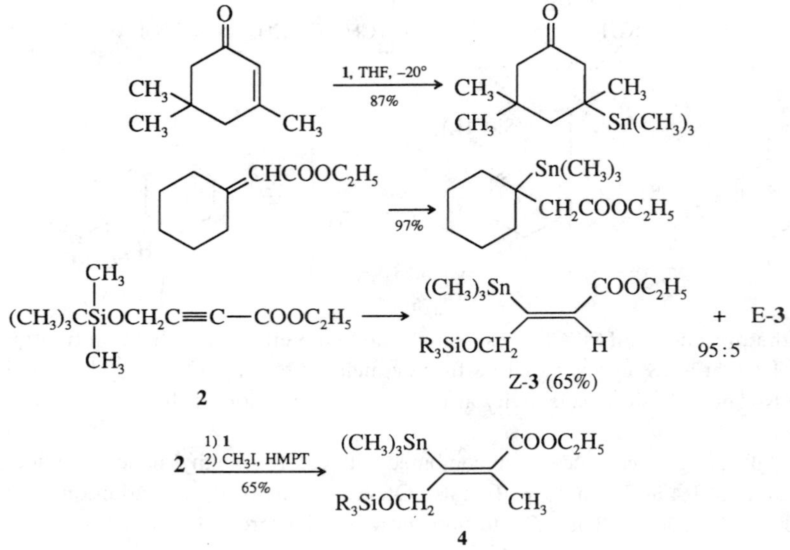

[2-Thienyl(cyano)copper]lithium, (2-thienyl)Cu(CN)Li (**1**).[18] This cuprate is prepared by reaction of 2-lithiothiophene with CuCN in THF at −78 → −40°. It is relatively stable, but can be converted to higher-order cuprates by reaction with RLi or RMgX at −78°. Thus reaction of **1** with vinyllithium provides $(CH_2{=}CH)(2\text{-thienyl})Cu(CN)Li_2$(**2**), which is more useful than $(CH_2{=}CH)Cu(CN)Li$ for cleavage of epoxides and for conjugate addition to enones.[19]

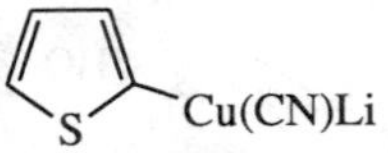

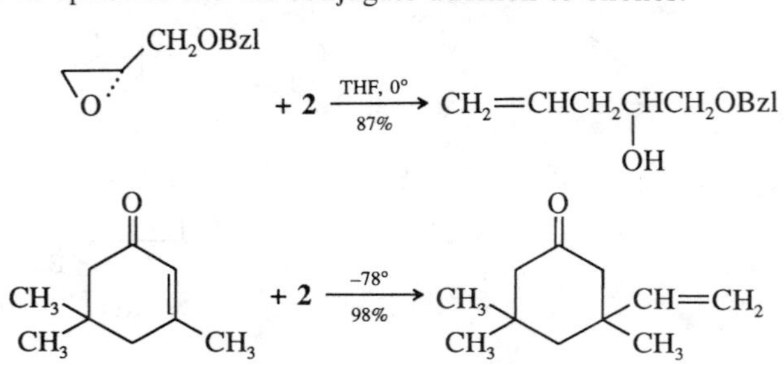

RCu–ISi(CH$_3$)$_3$.[20] This reagent, when solubilized in PBu_3 is an excellent reagent for 1,4-addition to enones. (Z)-Silyl enol ethers are formed preferentially.

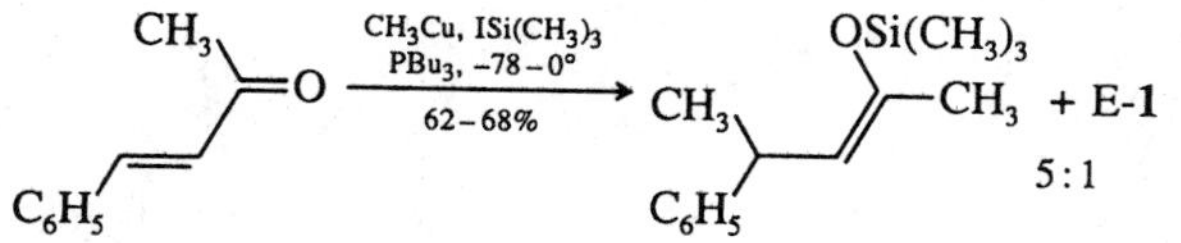

CH_3Cu, $ISi(CH_3)_3$, PBu_3, −78 – 0°; 62–68%

Z-1 + E-1 5:1

95%

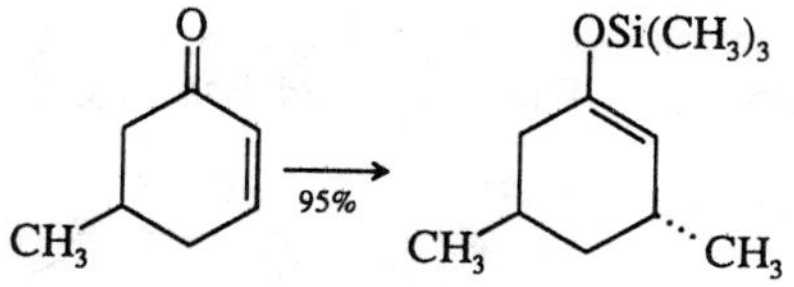

(*trans/cis* = 5:1)

[1] B. H. Lipshutz, E. L. Ellsworth, and T. J. Siahaan, *Am. Soc.*, **111**, 1351 (1989).
[2] S. H. Bertz and G. Dabbagh, *Tetrahedron*, **45**, 425 (1989).
[3] R. M. Wehmeyer and R. D. Rike, *Tetrahedron Letters*, **29**, 4513 (1988).
[4] J. F. Normant, A. Alexakis, A. Ghribi, and P. Mangeney, *Tetrahedron*, **45**, 507 (1989).
[5] J. D. Brown, M. A. Foley, and D. L. Comins, *Am. Soc.*, **110**, 7445 (1988).
[6] M. Hulce, *Tetrahedron Letters*, **29**, 5851 (1988).
[7] A. Alexakis and D. Jachiet, *Tetrahedron*, **45**, 381 (1989).
[8] M. Kawashima, T. Sato, and T. Fujisawa, *Tetrahedron*, **45**, 403 (1989).
[9] B. H. Lipshutz and J. C. Barton, *J. Org.*, **53**, 4495 (1988).
[10] S. R. Gilbertson, C. A. Challener, M. E. Bos, and W. D. Wulff, *Tetrahedron Letters*, **29**, 4795 (1988).
[11] S. C. Berk, P. Knochel, and M. C. P. Yeh, *J. Org.*, **53**, 5789 (1988).
[12] A. Capperucci, A. Degl'Innocenti, C. Faggi, P. Dembech, G. Seconi, and A. Ricci, *J. Org.*, **53**, 3612 (1988).
[13] E. J. Corey, F. J. Hannon, and N. W. Boaz, *Tetrahedron*, **45**, 545 (1989).
[14] S. F. Martin, J. R. Fishpaugh, J. M. Power, D. M. Giolando, R. A. Jones, C. M. Nunn, and A. H. Cowley, *Am. Soc.*, **110**, 7226 (1988).
[15] R. J. Linderman and A. Godfrey, *ibid.*, **110**, 6249 (1988).
[16] R. J. Linderman and J. R. McKenzie, *Tetrahedron Letters*, **29**, 3911 (1988).
[17] E. Piers and R. D. Tillyer, *J. Org.*, **53**, 5366 (1988).
[18] B. H. Lipshutz, M. Koerner, and D. A. Parker, *Tetrahedron Letters*, **28**, 945 (1987).
[19] B. H. Lipshutz, R. Moretti, and R. Crow, *Org. Syn.*, submitted (1988).
[20] M. Bergdahl, E.-L. Lindstedt, M. Nilsson, and T. Olsson, *Tetrahedron*, **45**, 535 (1989).

Organocopper/Zinc reagents, RCu(CN)ZnI. These reagents, in which R can be substituted by a chloride, ester, or enoate group, are prepared by reaction of RZnI with CuCN and LiCl (2 equiv.).[1] These reagents undergo 1,4-addition to α,β-enones, even β,β-disubstituted ones, in high yield when activated by BF_3 etherate (equation I).[2]

(I) [3-iodocyclohexenone] $\xrightarrow[84\%]{Cl(CH_2)_4Cu(CN)ZnI,\ THF,\ -30°}$ [3-$(CH_2)_4Cl$-cyclohexenone] $\xrightarrow[86\%]{PivO-CH(CH_3)(CH_2)_3Cu(CN)ZnI,\ BF_3 \cdot O(C_2H_5)_2,\ -13 \rightarrow 0°}$ [3-$(CH_2)_3CH(CH_3)OPiv$-3-$(CH_2)_4Cl$-cyclohexanone]

In some cases an intermediate bicyclic difluoroboron enolate such as **1** can be isolated from the BF_3-promoted reaction and converted into a bicyclic ketone (equation II).

(II) [3-methylcyclohexenone] + $IZn(CN)Cu(CH_2)_3CN$ $\xrightarrow[82\%]{BF_3 \cdot O(C_2H_5)_2}$ **1** $\xrightarrow[71\%]{NaOCH_3,\ CH_3OH}$ [bicyclic hydroxy enone, CH_3]

[1] P. Knochel, M. C. P. Yeh, S. C. Berk, and J. Talbert, *J. Org.*, **53**, 2390 (1988).
[2] *Idem, Tetrahedron Letters*, **29**, 6693 (1988).

Organolithium reagents.

Generation in situ.[1] A synthesis (**12**, 310–311) of 1,1,2-trisubstituted dihydronaphthalenes (**3**) involves as the first step addition of alkyllithiums to the 1-naphthyloxazoline **1** followed by trapping with an electrophile. The products (**2**) can be converted to formyl-1,2-dihydronaphthalenes (equation I). Addition of RLi

(I) **1** $\xrightarrow{1)\ RLi;\ 2)\ CH_3I}$ **2** (>100:1) $\longrightarrow$ **3** proceeds in 90–99% yield at −45° with *n*-, *sec*-, and *t*-BuLi, but allyl-, benzyl-, and vinyllithiums fail to add to **1**. However, these organolithiums when generated *in situ* from the corresponding tetrasubstituted stannane with CH_3Li[2] undergo this addition also in high yield. The reactive species may be an ate complex of R_4Sn and CH_3Li, or it may be nonaggregated reagent. In some cases addition of HMPT also is helpful. The R_4Sn reagents are prepared in about 70–90% yield by reaction of Mg with RCl in THF to form RMgCl, which is then added to $SnCl_4$ in benzene.

Magnesium 2-ethoxyethoxide, $Mg(OCH_2CH_2OC_2H_5)_2$. Preparation.[3] This metal alkoxide (**1**) decreases the activity of RLi reagents to such an extent that they can be prepared in THF in the presence of 0.5 equiv. of **1**. The advantages of this fact are that some unstable organolithiums can be generated by this technique and that directed *ortho*-metallation can be suppressed if desired.[4]

(Dialkoxymethyl)lithium, $(RO)_2CHLi$.[5] Both cyclic and acyclic reagents of this type have been prepared by lithiation of $(RO)_2CHSC_6H_5$ at −95° or by transmetalation of $(RO)_2CHSnBu_3$ at −110°. The acyclic reagents decompose rapidly at −95°, but cyclic ones are relatively stable at −45 to −78°. Typical reactions of this type are shown for the lithiodioxane reagent **1**, which reacts with a number of electrophiles to give acetals. It forms 1,2-adducts with ketones and enones in high

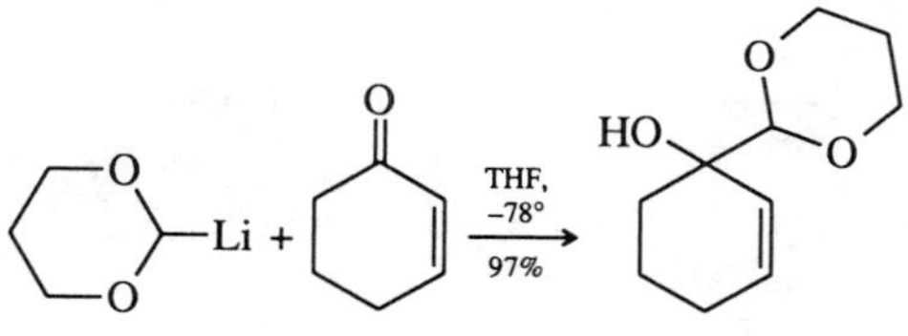

1 + **2** $\xrightarrow[97\%]{THF,\ -78°}$

1 + CuI•PBu_3 + **2** $\xrightarrow[92\%]{BF_3 \cdot O(C_2H_5)_2}$

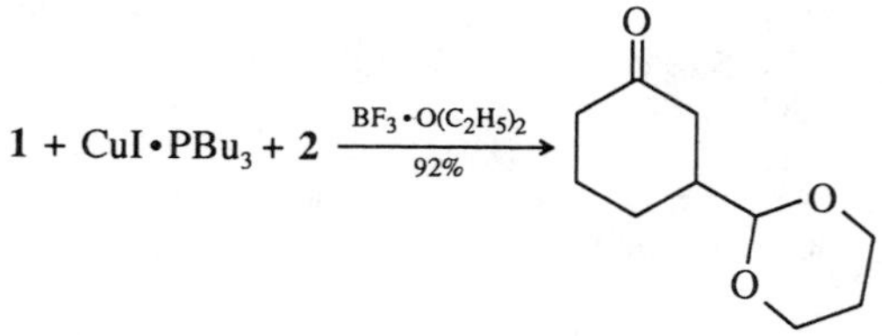

yield. The derived copper reagent when activated by BF_3 etherate reacts with enones by conjugate addition. Unlike most organolithium reagents, these new reagents react with alkyl halides, often in high yield in the case of primary bromides.

(ω-Haloalkenyl)lithiums, $[X(CH_2)_nCR{=}CZ]Li$.[6] The I/Li exchange of ω-halo-1-iodo-1-trimethylsilyl-1-alkenes (**1**) with BuLi results in three- to seven-membered cycloalkenylsilanes (**2**) in 65–85% yield. The presence of the trimethylsilyl group is not essential for the cyclization, but facilitates isomerization to the required *cis*-relationship between Li and the ω-halo alkyl group for cyclization.

Cyclization is also possible with (ω-halo-1-silyl-1-alkyl)metals in which the metal is Al, Zn, Zr, or Si. In this cyclization a silyl group is essential. In this reaction,

$R(X(CH_2)_n)C{=}C(Si(CH_3)_3)I \xrightarrow{\text{BuLi}}$ 2 (cyclic R, $Si(CH_3)_3$ alkene with $(CH_2)_n$)

1, n = 2, R = H — 80%
n = 4, R = CH_3 — 70%

cyclization to three- and four-membered rings is more facile than that to five- or six-membered rings.

$$X(CH_2)_2C{\equiv}CSi(CH_3)_3 \xrightarrow[Cl_2ZrCp_2]{(CH_3)_3Al} \left[X(CH_2)_2(CH_3)C{=}C(Si(CH_3)_3)Al(CH_3)_2 \right] \xrightarrow[92\%]{25^\circ} CH_3, Si(CH_3)_3\text{-cyclobutene}$$

$$Br(CH_2)_4C{\equiv}CSi(CH_2)_3 \xrightarrow[Cl_2ZrCp_2]{(CH_3)_3Al} CH_3, Si(CH_3)_3\text{-cyclohexene } (53\%) + CH_2{=}CH(CH_2)_2C(CH_3){=}CHSi(CH_3)_3 \ (20\text{–}25\%)$$

[1] A. I. Meyers, K. A. Lutomski, and D. Laucher, *Tetrahedron*, **44**, 3107 (1988).
[2] D. Seyferth and M. A. Weiner, *Am. Soc.*, **83**, 3583 (1961).
[3] C. G. Screttas and M. J. Micha-Screttas, *J. Organometal. Chem.*, **290**, (1985).
[4] C. G. Screttas and B. R. Steele, *J. Org.*, **54**, 1013 (1989).
[5] C. S. Shiner, T. Tsunoda, B. A. Goodman, S. Ingham, S. Lee, and P. E. Vorndam, *Am. Soc.*, **111**, 1381 (1989).
[6] E. Negishi, L. D. Boardman, H. Sawada, V. Bagheri, A. T. Stoll, J. M. Tour, and C. L. Rand, *Am. Soc.*, **110**, 5383 (1988).

Organomagnesium/zinc reagents.

These reagents can be obtained by reaction of vinyl Grignard reagents with allylic zinc bromides in THF at −35 to 0°. Reaction of **1** with $ClSn(CH_3)_3$ provides an α-trimethyltin–zinc reagent (**2**) that is oxidized by air to an aldehyde or ketone (**3**).[1]

This sequence can be used to obtain β-alkoxy aldehydes with high stereoselectivity from 3-ethoxyallylzinc bromide (**4**) (**6**, 607), equation (I).[2] The reaction involves almost complete transfer of the stereochemistry of the vinyl Grignard reagent to the aldehyde **5**.

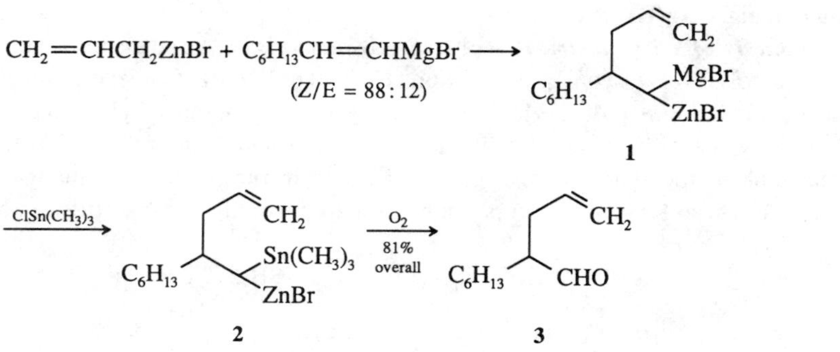

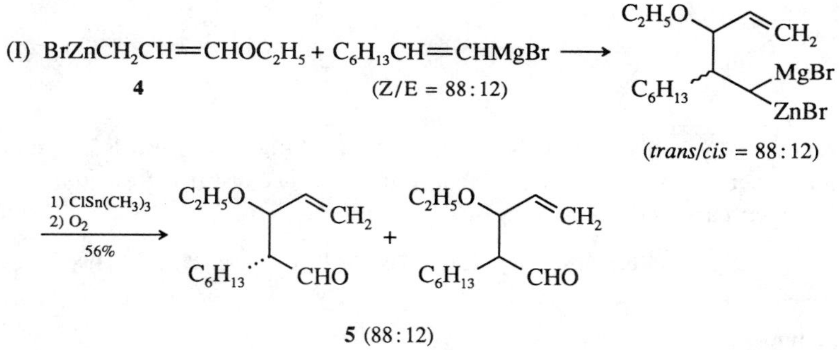

$CH_2{=}CHCH_2ZnBr + C_6H_{13}CH{=}CHMgBr \longrightarrow$ C_6H_{13}, CH_2, $MgBr$, $ZnBr$ **1**

(Z/E = 88:12)

$\xrightarrow{ClSn(CH_3)_3}$ C_6H_{13}, CH_2, $Sn(CH_3)_3$, $ZnBr$ **2** $\xrightarrow[\text{81\% overall}]{O_2}$ C_6H_{13}, CH_2, CHO **3**

(I) $BrZnCH_2CH{=}CHOC_2H_5 + C_6H_{13}CH{=}CHMgBr \longrightarrow$ C_2H_5O, CH_2, $MgBr$, C_6H_{13}, $ZnBr$

4 (Z/E = 88:12)

(*trans*/*cis* = 88:12)

$\xrightarrow[\text{56\%}]{\text{1) ClSn(CH}_3)_3 \text{ 2) O}_2}$ C_2H_5O, CH_2, C_6H_{13}, CHO + C_2H_5O, CH_2, C_6H_{13}, CHO

5 (88:12)

This mild oxidation of diorganometallics has been applied to oxidation of silylamines.[3] Thus conversion of a primary or secondary silylamine to a lithium or zinc salt followed by air-oxidation (−60 to −20°) provides an aldehyde or ketone.

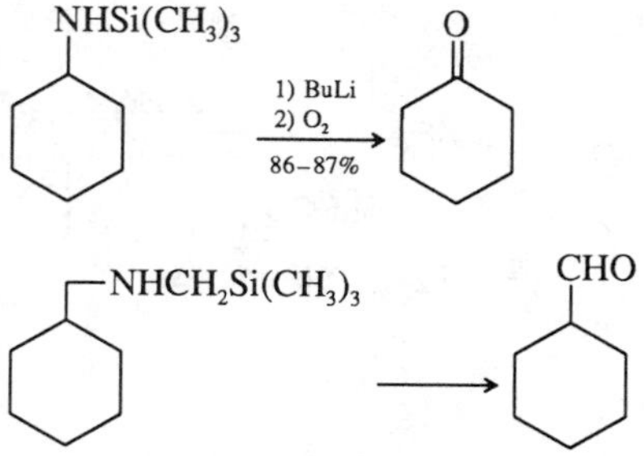

$NHSi(CH_3)_3$ $\xrightarrow[\text{86–87\%}]{\text{1) BuLi 2) O}_2}$ O

$NHCH_2Si(CH_3)_3$ $\longrightarrow$ CHO

[1] P. Knochel and J. F. Normant, *Tetrahedron Letters*, **27**, 1039 (1986).
[2] P. Knochel, C. Xiao, and M. C. P. Yeh, *ibid.*, **29**, 6697 (1988).
[3] H. G. Chen and P. Knochel, *ibid.*, **29**, 6701 (1988).

Organomanganese(II) halides.

***Synthesis of unsymmetrical* sec- *and* tert-*alcohols*.**[1] An attractive route to these alcohols involves a one-pot acylation (**13**, 210) of RMnBr followed by reaction with an alkyllithium (**14**, 229). The overall yield is generally higher than that obtained by reaction of the alkyllithium with the isolated, intermediate ketone. An example is the synthesis of α-bisabolol (**2**) from the homoallylic manganese bromide **1**. These RMnBr reagents can be prepared from RLi using $MnBr_2$, which

$(CH_3)_2C{=}CH(CH_2)_2MnBr$ (**1**) → 1) CH_3–(cyclohexenyl)–COCl; 2) CH_3Li; 80% → **2**

is commercially available and cheaper than MnI_2. Although $MnBr_2$ is insoluble as such in ether, it dissolves in this solvent in the presence of LiBr, formed *in situ* on reaction with an alkyllithium reagent.

[1] G. Cahiez, J. Rivas-Enterrios, and P. Clery, *Tetrahedron Letters*, **29**, 3659 (1988).

Organomanganese reagents.

***Benzylmanganesepentacarbonyl, BzlMn(CO)$_5$ (1); indenols*.**[1] Aryl ketones react with **1** selectively at the *ortho*-position to form a manganacycle (**2**), which after oxidative decarbonylation with $(CH_3)_3NO$ in CH_3CN is converted into a product (**a**) that reacts regioselectively with alkynes to form substituted 1-alkyl-1*H*-inden-*1*-ols (**3**). This reaction is regioselective; a terminal alkyne reacts to form 2-sub-

$C_6H_5COCH_3$ + 1 → (heptane, 76%) → **2** (m.p. 114–116°) [$Mn(CO)_4$] → ($(CH_3)_3NO$, CH_3CN) → [**a**, $Mn(CO)_3$] → ($R^1C{\equiv}CR^2$) → **3**

stituted indenols in 60–75% yield. Reactions with 1-trimethylsilyl-1-alkynes results in 2-trimethylsilyl-3-alkylindenols. The regioselectivity of reactions with unsymmetrical internal alkynes is determined by steric factors; the larger group is attached to the 2-position. Reactions with ethoxycarbonylalkynes, however, lead to 2-ethoxycarbonyl substituted indenols regardless of the other substituent on the alkyne.

***Pentacarbonyl(trimethylsilyl)manganese (1)*, 11,** 401. Improved preparation:

$$(CH_3)_3SiOTf + NaMn(CO)_5 \xrightarrow[60-70\%]{-50°} (CO)_5MnSi(CH_3)_3$$

1

The complex reacts with epoxides to yield a silyloxyalkylmanganese complex (**a**) that undergoes insertion with alkenes or alkynes to furnish manganacycles. Demetallation results in β-hydroxy carbonyl compounds, spiroketal lactones, and cyclopentenones.[2]

[1] L. S. Liebeskind, J. R. Gasdaska, and J. S. McCallum, *J. Org.*, **54**, 669 (1989).
[2] P. De Shong and D. R. Sidler, *ibid.*, **53**, 4892 (1988).

Organoosmium compounds.

Dialkylnitriloosmium chromates, $\overset{+}{N}Bu_4[N{\equiv}OsR_2](CrO_4)^-$ **(1)**. Preparation.[1] These metal chromates are effective catalysts for oxygenation of primary and secondary alcohols.[2] Primary alcohols are oxidized exclusively to aldehydes and no esters are formed from secondary alcohols. Primary alcohols are oxidized more rapidly than secondary ones.

[1] P. A. Shapley, H. S. Kim, S. R. Wilson, *Organometallics*, **7**, 928 (1988).
[2] N. Zhang, C. M. Mann, and P. A. Shapley, *Am. Soc.*, **110**, 6591 (1988).

Organosilanes.

Optically active binaphthylsilacycloheptanes. Jung and Hogan[1] have prepared the optically active binaphthylic cyclic silanes **1** and **2**, in which the chirality

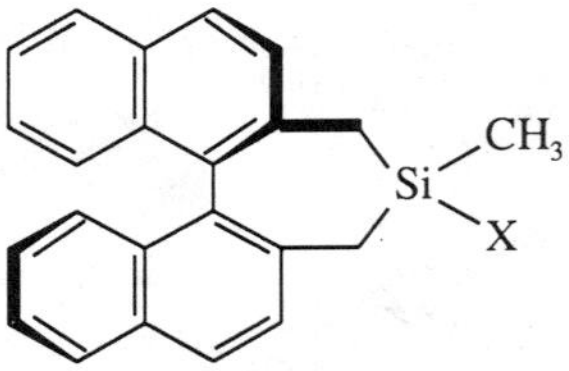

(R)-**1** 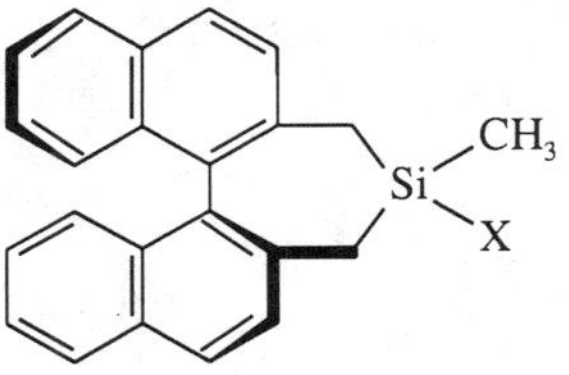 (S)-**2**

resides in the binaphthyl groups. These silanes (X = OCH_3) can effect partial asymmetric catalytic hydrogenation of enones and ketones. The highest enantioselectivity (35% ee) is observed with (S)-**3** in an aldol reaction (equation I).

(I) (S)-**3** + $C_6H_5CH(OCH_3)_2$ $\xrightarrow[78\%]{(CH_3)_3SiOTf,\ CH_2Cl_2,\ -78°}$

(17% ee) + (35% ee), 4.4 : 1

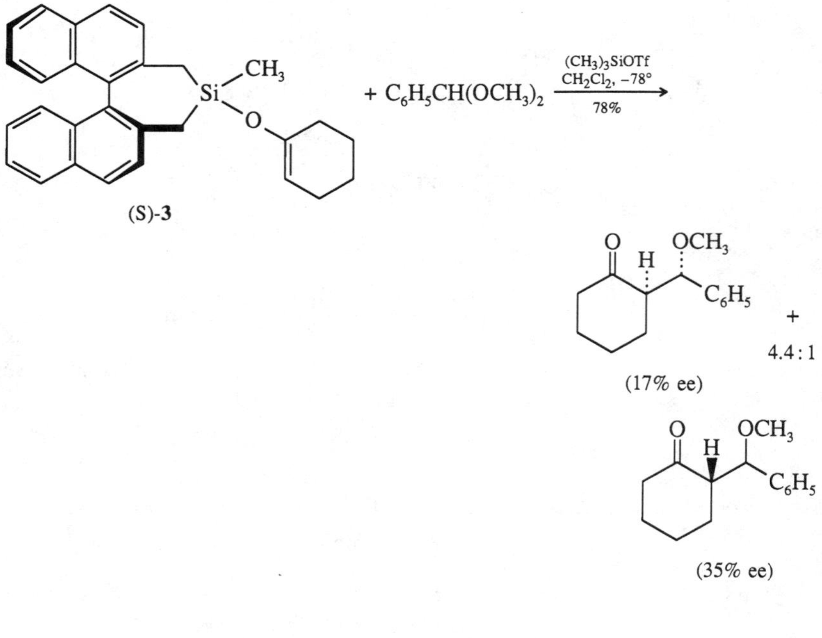

[1] M. E. Jung and K. T. Hogan, *Tetrahedron Letters*, **29**, 6199 (1988).

Organotin reagents.

α-Alkoxyallyltributyltins, $Bu_3SnCH(OR)CH{=}CH_2$ (**1**).[1] These reagents undergo Pd(0)-catalyzed coupling with ArBr to give a mixture of (E)- and (Z)-vinyl ethers (equation I).

$$\text{(I)}\ C_6H_5Br + \mathbf{1}\ (R = C_2H_5) \xrightarrow[83\%]{Pd(O),\ C_6H_6,\ 110^\circ} C_6H_5CH_2CH{=}CHOC_2H_5$$

(E/Z = 95:5)

In the presence of BF_3 etherate, **1** isomerizes to γ-alkoxyallyltins, which react with aryl aldehydes to form mainly *syn*-α-glycol monoethers (equation II). However, thermal reactions of **1** with aryl aldehydes result in mixtures of *syn*- and *anti*-vinyl ethers.

$$\text{(II)}\ \mathbf{1} \xrightarrow{BF_3 \cdot O(C_2H_5)_2} Bu_3SnCH_2CH{=}CHOC_2H_5 \xrightarrow[95\%]{C_6H_5CHO,\ -78^\circ} C_6H_5CH(OH)CH(OC_2H_5)CH{=}CH_2$$

(Z/E = 85:15) (*syn*, 93%)

1,1-Dimethoxy-3-trimethylstannylpropane, $(CH_3)_3Sn(CH_2)_2CH(OCH_3)_2$ (**1**). The reagent is prepared by reaction of $(CH_3)_3SnMgCl$ with

1-trimethylsilyloxycyclohexene + **1** $\xrightarrow{\text{Lewis acid}}$ [2-(1-methoxy-3-trimethylstannylpropyl)cyclohexanone, $(CH_2)_2Sn(CH_3)_3$, OCH_3] → hydroxy(methoxy)hydrindanes (OH, H, OCH_3) + epimer

Lewis acid	Yield	Ratio
TMSOTf	64%	90:10
BF_3	27%	66:34
$TiCl_4$	60%	60:40

$Br(CH_2)_2CH(OCH_3)_2$. It is useful for [3 + 2]annelation to *cis*-fused cyclopentanes and to spirocyclic compounds (equation I).[2]

(I) $\xrightarrow{\text{1, TMSOTf}}$ [ ] $\xrightarrow[\text{83\%}]{\text{PDC}}$

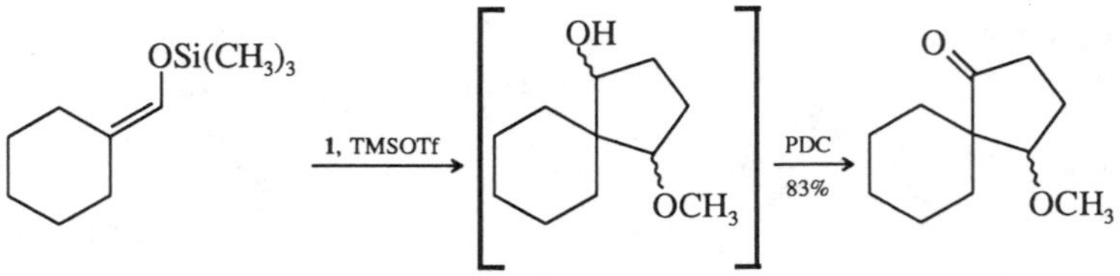

[1] J.-P. Quintard, G. Dumartin, B. Elissondo, A. Rahm, and M. Pereyre, *Tetrahedron*, **45**, 1017 (1989).
[2] T. V. Lee, K. A. Richardson, K. L. Ellis, and N. Visani, *ibid.*, **45**, 1167 (1989).

Organotitanium reagents.

Trichloromethyltitanium (**10**, 270; **12**, 355; **13**, 216). Diastereoselective addition to the chiral β-keto sulfoxide (**1**) is difficult because of ready enolization in the presence of a base. However, this titanium reagent reacts with marked diastereoface selectivity, which is the opposite of that observed with $(CH_3)_3Al$. Similar

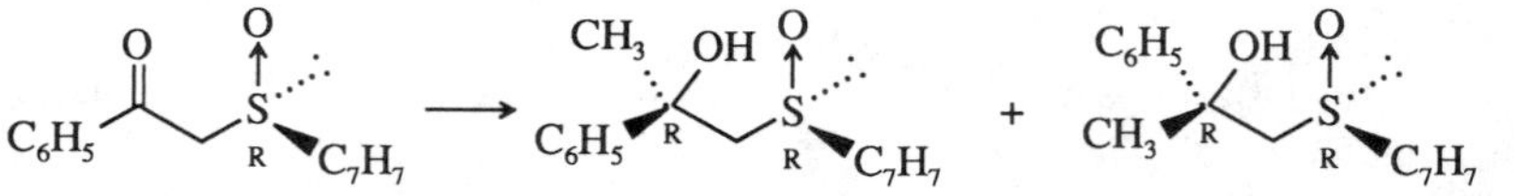

1		**2a** + **2b**
CH_3TiCl_3	79%	82:18
$(CH_3)_3Al$	66%	26:74

selectivity with both these organometallics is observed when the C_6H_5 group of **1** is replaced by other aromatic groups.

[1] T. Fujisawa, A. Fujimura, and Y. Ukaji, *Chem. Letters*, 1541 (1988).

Organozinc reagents.

$RSO_2C_6H_5 \rightarrow RAr$.[1] The organozinc reagents prepared from ArMgBr and $ZnBr_2$ can displace phenylsulfonyl groups at the 2-position of cyclic ethers. The reaction succeeds with reagents prepared from vinyl Grignard reagents, but not with those from alkyl Grignard reagents.

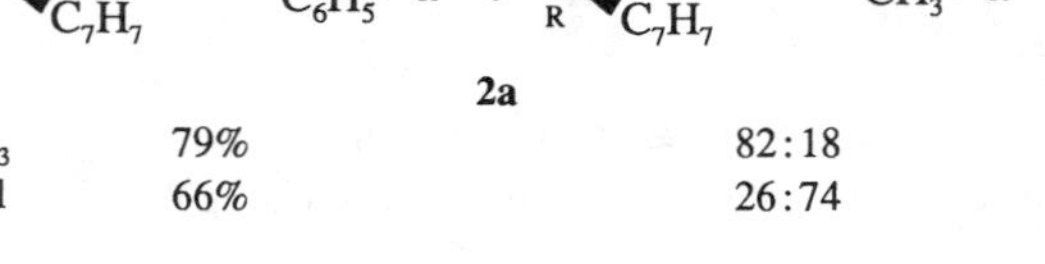

$\xrightarrow[\text{90\%}]{C_6H_5MgBr,\ ZnBr_2}$

$$C_6H_5\text{-(tetrahydropyranyl)-}SO_2C_6H_5 \xrightarrow[82\%]{CH_2{=}CHMgBr,\ ZnBr_2} C_6H_5\text{-(tetrahydropyranyl)-}CH{=}CH_2$$

These phenylsulfones can be converted into acetals by a reaction with alcohols in the presence of $MgBr_2 \cdot O(C_2H_5)_2$ and $NaHCO_3$.[2]

$$\text{(tetrahydropyranyl)-}SO_2C_6H_5 + ROH \xrightarrow[70-100\%]{MgBr_2 \cdot O(C_2H_5)_2,\ NaHCO_3,\ THF} \text{(tetrahydropyranyl)-}OR$$

***3-Methyl-2-butenylzinc bromide,* $(CH_3)_2C{=}CHCH_2ZnBr$ (1).**[3] This reagent is prepared from 1-bromo-3-methyl-2-butene and zinc (activated with 1,2-dibromoethane). It is a sticky gray solid, soluble in CH_2Cl_2. This allylic zinc bromide, unlike allylzinc bromide, reacts in the presence of $NiBr_2(PBu_3)_2$ with the α,β-unsaturated acetal (**2**) to give the α-alkylated acetal exclusively. The intermediate zinc homoenolate can be trapped with H_2O or RX to afford α,β-dialkylated acetals (**3**).

$$\mathbf{1} + CH_2{=}CH\text{-(5,5-dimethyl-1,3-dioxan-2-yl)}\ (\mathbf{2}) \xrightarrow[50\%]{1)\ Ni(II)\quad 2)\ CH_3I,\ HMPT,\ 25^\circ} (CH_3)_2C{=}CHCH_2CH(CH_2CH_3)\text{-(5,5-dimethyl-1,3-dioxan-2-yl)}\ (\mathbf{3})$$

2

3

When heated above 25° the intermediate zinc homoenolate undergoes an intramolecular alkylation, particularly in the presence of $ClSi(CH_3)_3$, to afford a cyclopropanol ether (equation I).

$$\text{(I)}\quad \mathbf{1} + \mathbf{2} \xrightarrow[53\%]{1)\ Ni(II)\quad 2)\ ClSi(CH_3)_3,\ 40^\circ} (CH_3)_2C{=}CHCH_2\overset{CH_2}{\overline{CH{-}CH}}{-}OCH_2C(CH_3)_2CH_2OSi(CH_3)_3$$

Lithium trialkylzincates (**8**, 515–516).[4] 1,1-Dibromoalkenes react with LiR_3Zn by a Br/Zn exchange at −85° to form a 1-bromoalkenylzincate (**a**), which undergoes an intramolecular alkylation in the presence of acetic acid at 0°. The alkylation involves at least partial inversion at the carbenoid carbon.

$$\text{RHC}=\text{CBr}_2 + \text{LiBu}_3\text{Zn} \xrightarrow{-85^\circ} \text{RHC}=\text{C(ZnBu}_3\text{Li)Br}\ (\mathbf{a}) \xrightarrow[82\%]{-85^\circ,\ \text{HOAc}} \text{RHC}=\text{CHBr}\quad E/Z = 2.6:1$$

R= $C_6H_5(CH_3)CH$

$$\mathbf{a} \xrightarrow[61\%]{0^\circ,\ \text{HOAC}} \text{RHC}=\text{CHBu}\quad E/Z = 1:1.8$$

$$(CH_3)_3C\text{-}C_6H_{10}{=}CBr_2 + \text{LiBu}_3\text{Zn} \xrightarrow[64\%]{1)\ \text{THF},\ -85^\circ;\ 2)\ H^+,\ 0^\circ} (CH_3)_3C\text{-}C_6H_{10}{=}\text{CHBu}$$

[1] D. S. Brown and S. V. Ley, *Tetrahedron Letters*, **29**, 4869 (1988).
[2] D. S. Brown, S. V. Ley, and S. Vile, *ibid.*, **29**, 4873 (1988).
[3] A. Janagisawa, S. Habaue, and H. Yamamoto, *Am. Soc.*, **111**, 366 (1989).
[4] T. Harada, D. Hara, K. Hattori, and A. Oku, *Tetrahedron Letters*, **29**, 3821 (1988).

Osmium tetroxide.

Asymmetric dihydroxylation of alkenes (**14**, 235–239). Further study[1] of this reaction reveals that the optical yields of products can be markedly improved by slow addition (5–26 hours) of the alkene to the catalyst in acetone–water at 0° with stirring. The enantioselectivity can also be increased by addition of tetraethylammonium acetate, which facilitates hydrolysis of osmate esters. The report suggests that the first product (**1**) of osmylation can undergo a second osmylation to provide **2**, with reverse enantioselectivity of the first osmylation.

RCH=CHR $\xrightarrow{OsO_4,\ \text{NMO}}$ **1** (osmate ester · NMO) $\xrightarrow{\text{RCH=CHR}}$ **2** (bis-osmate ester)

1 $\xrightarrow{H_2O}$ RCH(OH)CH(OH)R (chiral diol); **2** $\xrightarrow{H_2O}$ RCH(OH)CH(OH)R

By use of this simple technique the enantioselectivity can be increased as much as tenfold. Thus the ee for dihydroxylation of 1-phenylcyclohexene can be increased from 8 to 78% in a reaction of 78 hours.

[1] J. S. M. Wai, I. Markó, J. S. Svendsen, M. G. Finn, E. N. Jacobsen, and K. B. Sharpless, *Am. Soc.*, **111**, 1123 (1989).

Osmium tetroxide–Dihydroxyphenylborane, OsO_4–$C_6H_5B(OH)_2$.

Dihydroxylation of alkenes. $C_6H_5B(OH)_2$ is useful for capture of the diols formed on osmylation with OsO_4 and N-methylmorpholine N-oxide, particularly unstable or water-soluble diols. Osmylation with the borane can also be conducted in a nonaqueous medium with enhanced rates.[1]

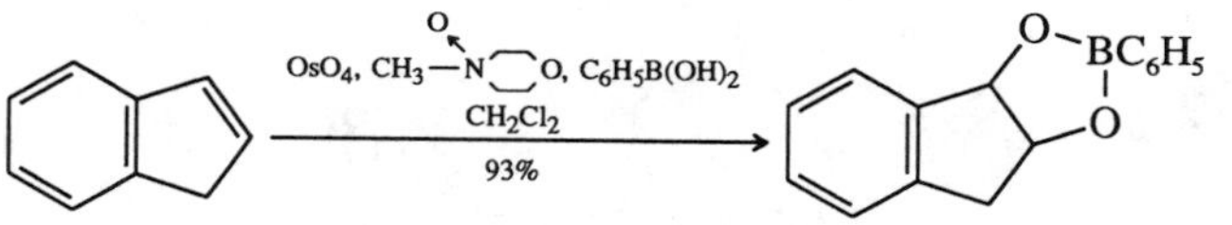

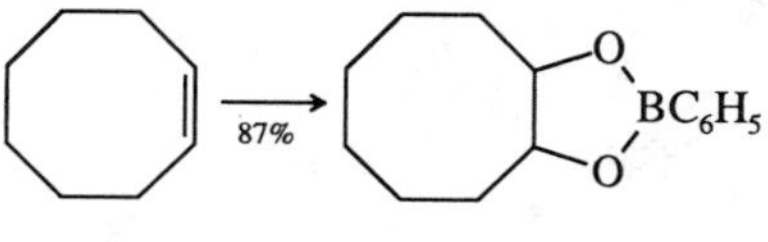

[1] N. Iwasawa, T. Kato, and K. Narasaka, *Chem. Letters*, 1721 (1988).

Oxazaborolidines, chiral, **14**, 239–242.

(S)-2-Hydroxymethylindoline (**1**)[1] reacts with borane·dimethyl sulfide to form the oxazaborolidine **2**, which functions as a catalyst for reduction of ketones by

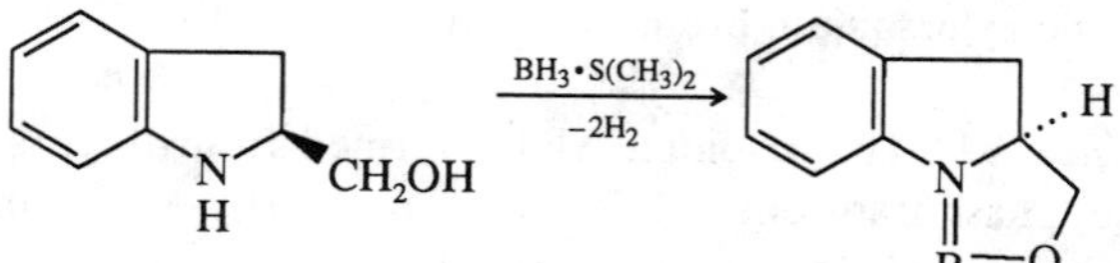

1 **2**

$BH_3 \cdot S(CH_3)_3$, but which is required in stoichiometric amount to achieve high optical yields. Alkyl aryl ketones can be reduced in this way in 84–93% ee to (R)-alcohols, but dialkyl ketones are reduced in only 45–60% ee.

[1] I. K. Youn, S. W. Lee, and C. S. Pak, *Tetrahedron Letters*, **29**, 4453 (1988).

Oxazolidines, chiral.

3-Arylsulfonyloxazolidines.[1] The optically active oxazolidine **1**, prepared from (R)-(N-toluenesulfonyl)phenylglycinol and 2-hydroxymethylenecyclohexanone,

1 $\xrightarrow[89\%]{CH_3MgCl}$ **2**, $\alpha_D - 113°$ $\xrightarrow{HS(CH_2)_3SH}$ **3** **4**

undergoes highly diastereoselective addition of metal nucleophiles to the carbonyl group. On treatment of the adduct with 1,3-propanedithiol, the chiral auxiliary is released with formation of the dithiane (**3**) of the aldehyde **4**.

[1] I. Hoppe, D. Hoppe, C. Wolff, E. Egert, and R. Herbst, *Angew. Chem. Int. Ed.*, **28**, 67 (1989).

2-(Oxoalkyl)triphenylarsonium bromides, $(C_6H_5)_3\overset{+}{As}CH_2CORBr^-$ (1), 13, 137; 14, 170.

(E)-α,β-Enones.[1] The arsonium ylides formed from these salts react with aldehydes under phase-transfer conditions to form (E)-α,β-enones in 71–99% yield (equation I). This Wittig olefination is particularly useful for reactions of unstable aldehydes because of the mild conditions.

(I) $\mathbf{1}[R = CH(CH_3)_2, n\text{-}C_5H_{11}] + R'CHO \xrightarrow[71-99\%]{K_2CO_3,\ CH_2Cl_2,\ H_2O,\ 25°} R'CH{=}CHCOR$

[1] Y.-Z. Huang, L.-L. Shi, and S.-W. Li, *Synthesis*, 975 (1988).

Oxygen, singlet.

Endoperoxidation. Conduritol-A (**5**) can be prepared[1] stereospecifically from cyclohexa-1,4-diene via **2** by singlet oxygenation (tetraphenylporphyrine), which provides the *anti*-endoperoxide **3**. Reduction of the endoperoxide with thiourea[2] followed by deketalization gives the cyclohexenetetraol **5**.

1) $KMnO_4$ 2) $(CH_3)_2C(OCH_3)_2$; $^1O_2$, TPP 95%; H_2NCSNH_2, CH_3OH, 25° 80%; HCl, CH_3OH 100%

1 2 3 4 5

Deoxygenation of cyclic peroxides.[3] α-Methylene-β-lactones (**3**) have been obtained for the first time by deoxygenation with $(C_6H_5)_3P$ of the α-methylene-β-peroxylactones (**2**) obtained by photosensitized oxygenation of methacrylic acid derivatives (**1**).

1) $^1O_2$ 2) H_3O^+ 60–70%; $(C_6H_5)_3P$ 30–50%

1, R = CH_3 or C_2H_5 2 3

[1] Y. Sütbeyaz, H. Seçen, and M. Balci, *J.C.S. Chem. Comm.*, 1330 (1988).
[2] *Idem*, *J. Org.*, **53**, 2312 (1988).
[3] W. Adam, L. Hasemann, and F. Prechtl, *Angew. Chem. Int. Ed.*, **27**, 1536 (1988).

Ozone

α-Hydroperoxy carbonyl compounds.[1] α-Hydroperoxy aldehydes are probably too unstable to prepare directly, but can be obtained in 15–25% yield via the more stable *t*-butyldimethylsilyl peroxides. Thus ozonation followed by reduction of the trialkylsilylated allylic hydroperoxides **1**, obtained by sensitized oxygenation of an appropriate alkene, provides α-silylperoxy aldehydes **2** albeit in low yield.

$$\text{1 (R = H, CH}_3\text{)} \xrightarrow[50-70\%]{t\text{-Bu(CH}_3)_2\text{SiCl}} \text{1 (OOSi(CH}_3)_2\text{-}t\text{-Bu)} \xrightarrow[15-24\%]{O_3,\ CH_3OH;\ S(CH_3)_2} \mathbf{2}$$

Desilylation of **2** results in decomposition attended with chemiluminescence. α-Silylperoxy ketones can be prepared (and in higher yield) by the same sequence and these compounds can be desilylated by methanol.

[1] W. Adam, L. H. Catalani, C. R. Saha-Möller, and B. Will, *Synthesis*, 121 (1989).

P

Palladium

Exocyclic enol ethers.[1] A route to (E)-cyano- or -methoxycarbonyl substituted exocyclic enol ethers involves hydrogenation and isomerization catalyzed by Pd/C of an endocyclic dienol ether. Thus hydrogenation of **1a** or **1b** in toluene at −35° catalyzed by Pd/C gives (E)-**2a** or **2b** as the major product in about 80% yield.

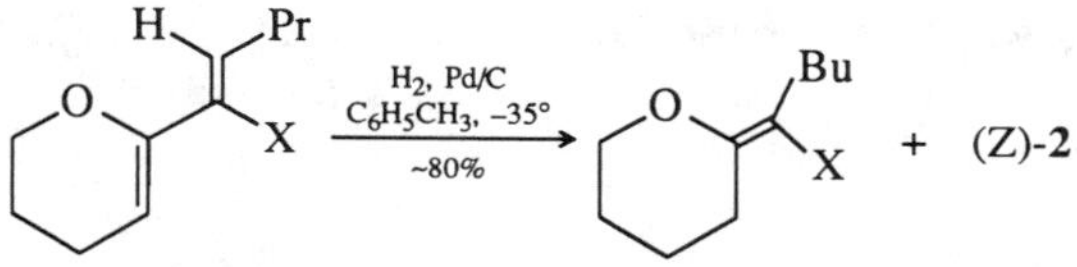

H_2, Pd/C, $C_6H_5CH_3$, −35°, ~80%

+ (Z)-**2**

1a, X = CN
1b, X = $COOCH_3$

(E)-**2a** 11 : 1
(E)-**2b** 18 : 1

No other hydrogenation catalyst shows comparable stereoselectivity. This hydrogenation was useful for a synthesis of the antitumor agent nileprost (**3**).

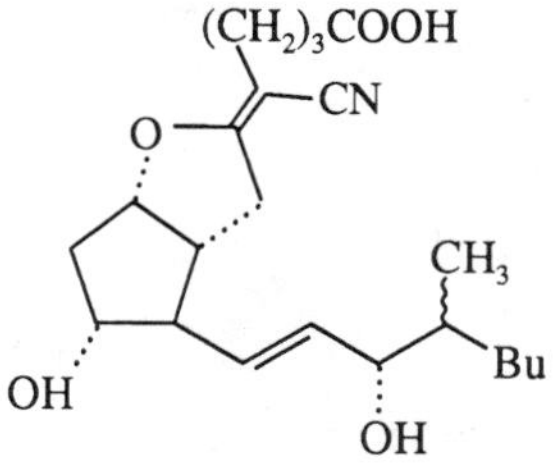

$(CH_2)_3COOH$, CN, CH_3, Bu, OH, OH

3

RN_3 → RNHBoc.[2] Vicinal azido alcohols are converted directly into *vic*-N-*t*-butoxycarbonylamino alcohols by Pd/C-catalyzed hydrogenation in the presence of di-*t*-butyl dicarbonate in ethyl acetate at 25°. Yields are 71–93%.

[1] A. Takahashi, Y. Kirio, M. Sodeoka, H. Sasai, and M. Shibasaki, *Am. Soc.*, **111**, 643 (1989).
[2] S. Saito, H. Nakajima, M. Inaba, and T. Moriwake, *Tetrahedron Letters*, **30**, 837 (1989).

Palladium(II) acetate.

Lactonization of dienes by 1,4-functionation.[1] Reaction of the diene **1** with acetic acid and benzoquinone catalyzed by $Pd(OAc)_2$ results in the isomeric lactones **2** and **3**. The stereoselectivity of this lactonization can be controlled in part by

CH_2COOH HOAc, quinone, $(CH_3)_2C=O$, Pd(II) 77% AcO ... =O + 99:1 AcO ... =O

1 + LiOAc, LiCl 50–57% 2 ~40:60 3

addition of LiOAc or LiCl in catalytic amounts. Chlorine-substituted lactones are obtained by addition of 2 equiv. of LiCl.

1,3,5-Trienes.[2] These trienes can be prepared in moderate yield by Pd-catalyzed reaction of fumaryl chloride (**1**) with 1-alkenes in the presence of N-ethylmorpholine (**2**), which is superior to Bu_3N as the base. (E,E)-5-Phenyl-2,4-pentadienyl chloride (**3**) can also be used in the condensation (last example), but this reaction is not (E,E,E)-specific.

Cl ... Cl + $C_6H_5CH{=}CH_2$ $\xrightarrow[44\%]{Pd(II),\ 2,\ 120°}$ C_6H_5 ... C_6H_5 (E,E,E)

1

1 + $CH_3OOCC(CH_3){=}CH_2$ $\xrightarrow[56\%]{}$ CH_3OOC ... $COOCH_3$ (E,E,E)

C_6H_5 ... Cl + $C_2H_5O_2CCH{=}CH_2$ ⟶

3

C_6H_5 ... $COOC_2H_5$ + (Z,E,E)-**4**

(E,E,E)-**4**, 38% 16%

Cyclization of o-iodophenyl allyl ethers to benzofurans.[3] This cyclization can be effected in DMF with HCOONa (1 equiv.), Bu_4NCl (1 equiv.), and Na_2CO_3 (2 equiv.) catalyzed by $Pd(OAc)_2$. The function of sodium formate is not known, but the salt improves the yield.

O I CH_3 $\xrightarrow[83\%]{Pd(II),\ HCOONa,\ Bu_4NCl,\ Na_2CO_3}$ O C_2H_5

Coupling of $CH_2{=}CHSi(CH_3)_3$ with vinyl iodides.[4] The reaction can be effected with $Pd(OAc)_2$ as catalyst. Added silver nitrate accelerates the reaction and has the additional virtue of suppressing desilylation and increasing the yield of 1-trimethylsilyl-1,3-dienes.

$$\text{1-iodocyclohexene} + CH_2{=}CHSi(CH_3)_3 \xrightarrow[64\%]{Pd(II),\ AgNO_3,\ DMSO,\ 50^\circ} \text{1-(2-}Si(CH_3)_3\text{-vinyl)cyclohexene}$$

[1] J.-E. Bäckvall, P. G. Andersson, and J. O. Vägberg, *Tetrahedron Letters*, **30**, 137 (1989).
[2] A. Kasahara, T. Izumi, and N. Kudou, *Synthesis*, 704 (1988).
[3] R. C. Larock and D. E. Stinn, *Tetrahedron Letters*, **29**, 4687 (1988).
[4] K. Karabelas and A. Hallberg, *J. Org.*, **53**, 4909 (1988).

Palladium(II) acetate–1,3-Bis(diphenylphosphino)propane, $Pd(OAc)_2$-dppp (**1**).

Carbonylation of allylamines.[1] This is the most effective catalyst for carbonylation of allylamines to give β,γ-unsaturated amides, which are reducible to homoallylic amines.

$$CH_3CH{=}CHCH_2NBu_2 + CO \xrightarrow[77\%]{1,\ C_6H_5CH_3,\ 110^\circ} CH_3CH{=}CHCH_2\overset{O}{\overset{\|}{C}}NBu_2 \xrightarrow[95\%]{LiAlH_4} CH_3CH{=}CHCH_2CH_2NBu_2$$

Methoxycarbonylation. Aryl triflates or iodides are converted into alkyl benzoates by reaction with CO and an alcohol at 70° catalyzed by $Pd(OAc)_2$ in combination with dppp; yields are 72–95%. The method is useful for esterification of phenols under mild conditions.[2]

This procedure was used for the synthesis of the aromatic segment (**1**) of cali-

$$\text{(}CH_3\text{, I, I, BzlOCH}_2\text{O, OCH}_3\text{, OCH}_3\text{-benzene)} \xrightarrow[44\%]{CO,\ cat.,\ N(C_2H_5)_3,\ DMSO/CH_3OH} \text{(}CH_3\text{, I, COOCH}_3\text{, BzlOCH}_2\text{O, OCH}_3\text{, OCH}_3\text{-benzene)} \xrightarrow[89\%]{HCl,\ CH_3OH,\ 25^\circ} \text{(}CH_3\text{, I, COOCH}_3\text{, HO, OCH}_3\text{, OCH}_3\text{-benzene)}\ \mathbf{1}$$

chemicins,[3] a new class of natural antibiotic and antitumor agents. Crystallization of **1** is accompanied by spontaneous resolution to give the enantiomers of **1a**.

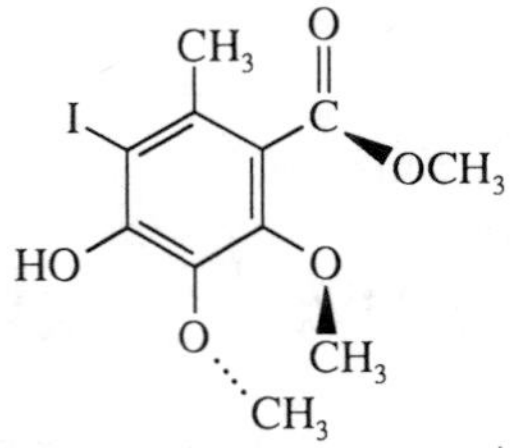

1a

[1] S. Murahashi, Y. Imada, and K. Nishimura, *J.C.S. Chem. Comm.*, 1578 (1988).
[2] R. E. Dolle, S. J. Schmidt, and L. I. Kruse, *ibid.*, 904 (1987).
[3] K. C. Nicolaou, T. Ebata, N. A. Stylianides, R. D. Groneberg, and P. J. Carroll, *Angew. Chem. Int. Ed.*, **27**, 1097 (1988).

Palladium(II) acetate–Triphenylphosphine.

Heck cyclization.[1] Vinylation of halides in the presence of a hydride source (formic acid) can give cyclic products.

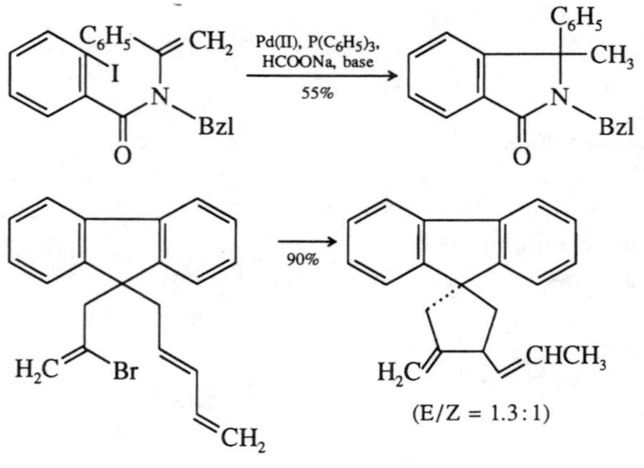

Pd(II), $P(C_6H_5)_3$, HCOONa, base; 55%

90%

(E/Z = 1.3:1)

[1] B. Burns, R. Grigg, V. Sridharan, T. Worakun, and P. Stevenson, *Tetrahedron Letters*, **29**, 4325, 4329 (1988).

Palladium(II) chloride.

Pyranosides → cyclohexanones.[1] The C_5-exomethylene pyranosides (**1**) derived from D-glucosamine can be transformed into cyclohexanones (**2**) by treatment with $PdCl_2$ or $Pd(OAc)_2$ (10–20 mole %) in acidic aqueous dioxane or acetone at 60° for 1 hour. The transformation may be related to the Ferrier rearrangement.[2]

H_2C O OBzl BzlO OR NRBzl $\xrightarrow{Pd(II),\ H_3O^+,\ (CH_3)_2C{=}O,\ 60°}$ O OBzl BzlO OH NRBzl

1, R = CH_3, $COCH_3$, Bzl **2**

[1] S. Adam, *Tetrahedron Letters*, **29**, 6589 (1988).
[2] R. J. Ferrier, *J.C.S. Perkin I*, 1455 (1979).

Palladium(II) chloride–Benzoquinone.

Oxidation of 1-alkene-4-ols. These substrates are oxidized by $PdCl_2$ (catalytic) and *p*-benzoquinone (3 equiv.) or by $CuCl/O_2$ at 40° to γ-butyrolactols (cyclic hemiacetals), precursors to γ-butyrolactones.[1]

$CH_2{=}CHCH(COOCH_3)CH(OH)C_6H_{13}$ $\xrightarrow[64\%]{oxid.}$ HO–(O)–C_6H_{13}, $COOCH_3$ (lactol)

$\downarrow$ 95%, CrO_3, H_2SO_4, $(CH_3)_2C{=}O$

O=(O)–C_6H_{13}, $COOCH_3$ (lactone)

[1] J. Nokami, H. Ogawa, S. Miyamoto, T. Mandai, S. Wakabayashi, and J. Tsuji, *Tetrahedron Letters*, **29**, 5181 (1988).

Palladium(II) chloride–Copper(II) chloride.

Oxidation–carbonylation of 1-alkenes. These alkenes can undergo an aerobic palladium-catalyzed oxidation–carbonylation in acetic acid–acetic anhydride (4:1) with $CuCl_2$ as a reoxidant and a trace of NaCl, required for satisfactory yields.[1]

$$RCH{=}CH_2 + CO + Ac_2O \xrightarrow[O_2,\ 80^\circ]{PdCl_2,\ CuCl_2,\ NaCl} RCH(OAc)CH_2COOAc$$

R	Yield
$R = C_4H_9$	84%
$= CH_2C_6H_5$	86%
$= AcO$	53%

$$HOCH_2CH{=}CH_2 + CO + C_2H_5OH \xrightarrow[57\%]{PdCl_2,\ CuCl_2,\ NaCl,\ CH(OC_2H_5)_3,\ O_2,\ 50^\circ} C_2H_5OCH_2CH(OC_2H_5)CH_2COOC_2H_5$$

[1] H. Urata, A. Fujita, and T. Fuchikami, *Tetrahedron Letters*, **29**, 4435 (1988).

Palladium(II) chloride–Bis(salicylidene)ethylenediamine (salen).

Hydrogenation catalyst.[1] $PdCl_2$ and salen can form two different types of a palladium(II) complex, a yellow complex **1** similar to complexes of Cu(II) and an insoluble green complex **2**, best formed from K_2PdCl_4, salen, and $N(C_2H_5)_3$ as base.

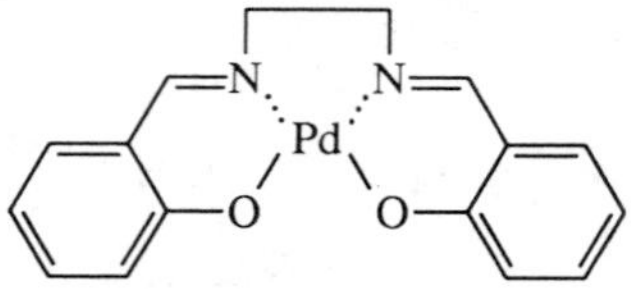

1, yellow

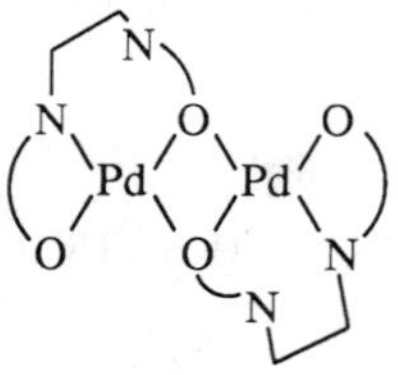

2, green

Impure samples of **1** containing a green impurity were originally believed to catalyze hydrogenation, but this activity is really due to the green complex, formulated as **2**. This complex is more effective than Pd/C for hydrogenation of alkenes and alkynes, and can reduce nitroarenes to anilines directly. Reduction of internal alkynes affords *cis*-alkenes with slight contamination with an alkane.

$$HO(CH_2)_2C{\equiv}CC_2H_5 \xrightarrow[100\%]{H_2,\ \mathbf{2}} (Z)\text{-}HO(CH_2)_2CH{=}CHC_2H_5$$

$$BuC{\equiv}CH \xrightarrow{H_2,\ \mathbf{2}} \underset{(98\%)}{BuCH{=}CH_2} + \underset{(2\%)}{C_6H_{14}}$$

[1] J. M. Kerr, C. J. Suckling, and P. Bamfield, *Tetrahedron Letters*, **29**, 5545 (1988).

(2R,4R)-2,4-Pentanediol, $CH_3CH(OH)CH_2CH(OH)CH_3$ **(1).**

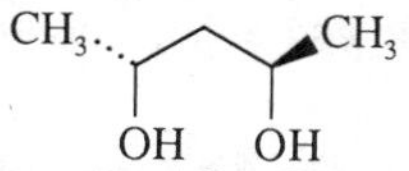

Asymmetrization of* meso*-ketones.[1] The reaction of tri-*i*-butylaluminum in CH_2Cl_2 at 0° with the homochiral acetal **3** of the ketone **2** provides the vinyl ether **4** with high diastereoselectivity (9:1). Somewhat higher selectivity obtains with the dialkylaluminum amide **6** (94:6).

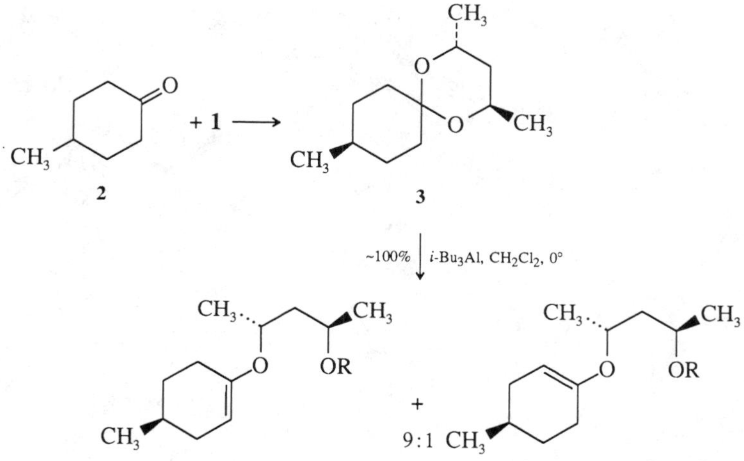

2 + 1 ⟶ **3**

~100% ↓ *i*-Bu_3Al, CH_2Cl_2, 0°

4a, R = H + 9:1 **5a**, R = H

c-$C_5H_{11}CH(Bu$-$t)NAl(Bu$-$i)_2$ with $C(CH_3)_3$ on N

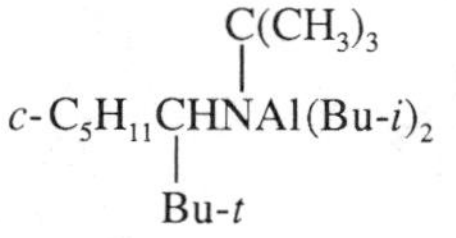

6

[1] Y. Naruse and H. Yamamoto, *Tetrahedron*, **44**, 6021 (1988).

4-Pentenyloxymethyl chloride, $CH_2{=}CH(CH_2)_3OCH_2Cl$ (POMCl), **1**. The reagent is prepared by reaction of 4-pentenol with formaldehyde and hydrogen chloride.

Protection of alcohols.[1] Primary and secondary alcohols are converted into POM ethers by reaction with **1** and Hünig's base in CH_2Cl_2. The alcohols are regenerated by oxidative hydrolysis in aqueous acetonitrile (79–81% yield).

Vicinal diols are converted into acetals by reaction with 4-pentenal catalyzed by camphorsulfonic acid. The acetals are converted on reaction with NIS into an iodo ether, which is reduced by zinc and a catalytic amount of NH_4Cl in alcohol to the diol.

[1] Z. Wu, D. R. Mootoo, and B. Fraser-Reid, *Tetrahedron Letters*, **29**, 6549 (1988).

Periodinane of Dess-Martin (1), 12, 378–379.

Oxidation.[1] The diol **2**, prepared from 3-hexene-1,5-dyne as shown, undergoes aromatization to 1,2-dibenzoylbenzene (**3**) on treatment with methanesulfonyl

Li
Li
C_6H_5CHO
74%
OH
CHC_6H_5
CHC_6H_5
OH
2
MsCl, THF
$N(C_2H_5)_3$
4–6%
COC_6H_5
COC_6H_5
3
1 (equiv.)
30–40%
MsCl, THF
$N(C_2H_5)_3$
O
CC_6H_5
CC_6H_5
O
4
\+ **2** +
1:1:1
O
C_6H_5C
C_6H_5CH
OH
5

chloride and a base, but in poor yield. Oxidation of **2** with 1 equiv. of periodinane gives a 1:1:1 mixture of the expected diketone (**4**), the starting material (**2**), and an isomeric hydroxy ketone (**5**) which aromatizes to **3** in reasonable yield under the same conditions as for **2**.

α-Keto esters.[2] This periodinane (suspended in CH_3CN or CH_2Cl_2) can be used for oxidation of α-hydroxy esters to α-keto esters at 25°.

[1] S. J. Danishefsky, D. S. Yamashita, and N. B. Mantlo, *Tetrahedron Letters*, **29**, 4681 (1988).
[2] J. P. Burkhart, N. P. Peet, and P. Bey, *ibid.*, **29**, 3433 (1988).

Phase-transfer catalysts.

ArI → ArCOOH.[1] Carbonylation of ArI to form ArCOOH can be effected by catalysis with nickel cyanide and a phase-transfer catalyst, cetyltrimethylammonium bromide (equation I). Carbonylation of ArI catalyzed by $Co_2(CO)_8$ requires irradiation (350 nm.).

$$\text{(I) ArI + CO} \xrightarrow[40-80\%]{Ni(CN)_2,\ C_{16}H_{33}\overset{+}{N}(CH_3)_3Br^-,\ C_6H_5CH_3,\ H_2O,\ NaOH} \text{ArCOOH}$$

1-Chloroalkanes.[2] The conversion of primary alcohols to 1-chloroalkanes with aqueous hydrochloric acid is not useful because of mediocre yields (30–60%). The rate and yield of this reaction are improved considerably by use of a surfactant such as cetyltrimethylammonium bromide or cetylpyridinium bromide.

Ester interchange.[3] Exchange of methyl or ethyl esters can be conducted with 1 equiv. of an alcohol and a base (K_2CO_3) in the presence of Aliquat or Bu_4NHSO_4 and in the absence of an additional solvent. Indeed it can be carried out at 70° in the absence of a phase-transfer catalyst.

RX → RCN.[4] Nitriles are obtained in 75–93% yield by heating a suspension of an alkyl halide, NaCN (2 equiv.), and cetyltrimethylammonium bromide (1.5 mole %) in water (26 mole %) at 100° overnight.

[1] I. Amer and H. Alper, *J. Org.*, **53**, 5147 (1988).
[2] B. Juršić, *Synthesis*, 868 (1988).
[3] J. Barry, G. Bram, and A. Petit, *Tetrahedron Letters*, **29**, 4567 (1988).
[4] B. Juršić, *J. Chem. Res.* (*S*), 336 (1988).

Phenethylamines.

Alkaloid synthesis from* vic*-tricarbonyl compounds. Wasserman's group[1] has used the strongly electrophilic character of the central carbonyl of a *vic*-tricarbonyl system for synthesis of several alkaloids. Thus several isoquinoline alkaloids can be prepared by reaction of a phenethylamine with the tricarbonyl **1**.

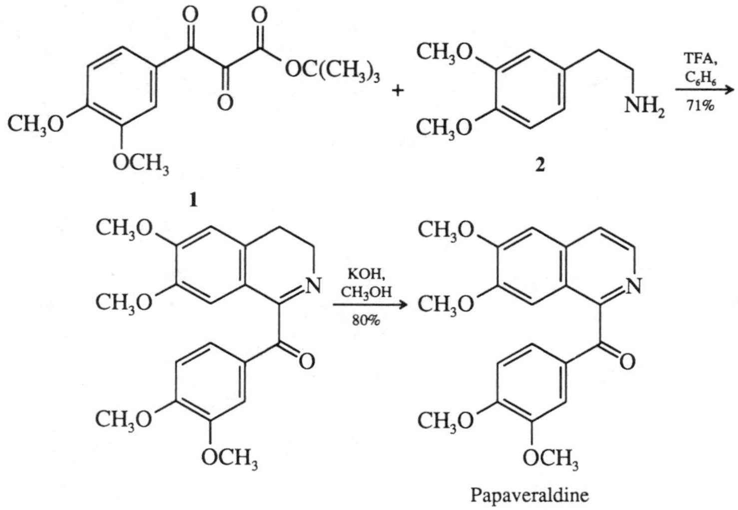

$$\mathbf{1} + \mathbf{2} \xrightarrow[71\%]{TFA,\ C_6H_6} \text{dihydroisoquinoline} \xrightarrow[80\%]{KOH,\ CH_3OH} \text{Papaveraldine}$$

Papaveraldine

The reaction of tryptamine (**4**) with the tricarbonyl compound **3** provides a route to indole alkaloids.

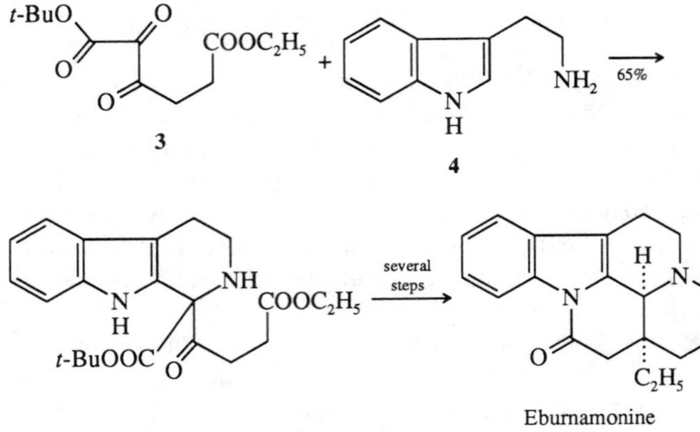

t-BuO O COOC$_2$H$_5$ O O **3** + NH$_2$ N H **4** 65%

NH N H COOC$_2$H$_5$ t-BuOOC O several steps H N N O C$_2$H$_5$

Eburnamonine

[1] H. H. Wasserman, R. Amici, R. Frechette, and J. H. van Duzer, *Tetrahedron Letters*, **30**, 869 (1989).

Phenylboric acid, $C_6H_5B(OH)_2$.

Phenanthrenequinone from benzoin. The photocyclization of stilbene to phenanthrene requires a *trans* to *cis* isomerization. This step can be circumvented by reaction of benzoin with phenylboric acid to form the 1,3-dioxoborole **2**. This adduct can be converted to phenanthrenequinone by photocyclization in the presence of diphenyl diselenide as the oxidant.

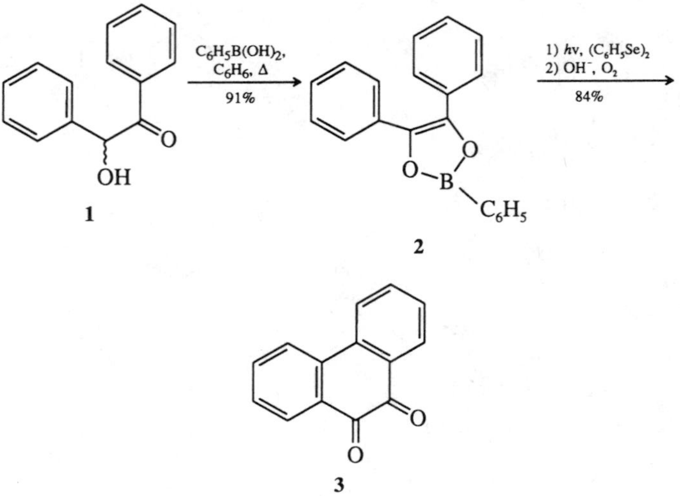

O OH **1** $C_6H_5B(OH)_2$, C_6H_6, Δ 91% O O B C$_6$H$_5$ **2** 1) $h\nu$, $(C_6H_5Se)_2$ 2) OH^-, O_2 84%

O O **3**

[1] J. G. de Vries, and S. A. Hubbard, *J.C.S. Chem. Comm.*, 1172 (1988).

S-Phenyl carbonochloridothioate, $C_6H_5S\overset{O}{\overset{\|}{C}}Cl$ (**1**).

1,4-Keto aldehydes or 1,4-diones. Reaction of this acid chloride with the Grignard reagent **2a** or **2b** catalyzed by $Cl_2Ni(dppe)$ (**12**, 171) results in coupling to **3** in 80–85% yield. Coupling of a second Grignard reagent catalyzed by $Fe(acac)_3$

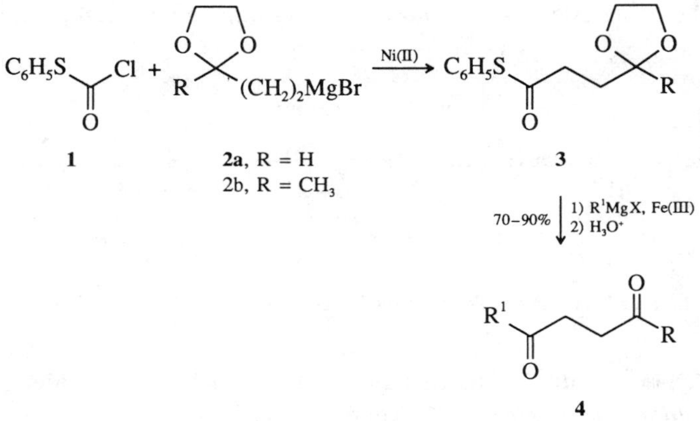

(**12**, 557) followed by deacetalization results in a 1,4-keto aldehyde (R = H) or a 1,4-dione (R = CH_3).

[1] V. Fiandanese, G. Marchese, and F. Naso, *Tetrahedron Letters*, **29**, 3587 (1988).

Phenyl dichlorophosphate–Sodium iodide–Dimethylformamide.

Dethioacetalization.[1] The combination of phenyl dichlorophosphate, $C_6H_5OPOCl_2$, and NaI converts acetals to the corresponding aldehyde or ketone, but has little, if any, effect on thioacetals. However, the reagent obtained by addition of DMF converts thioketals into the corresponding carbonyl compound at room temperature in 1–17 hours in 70–95% yield.

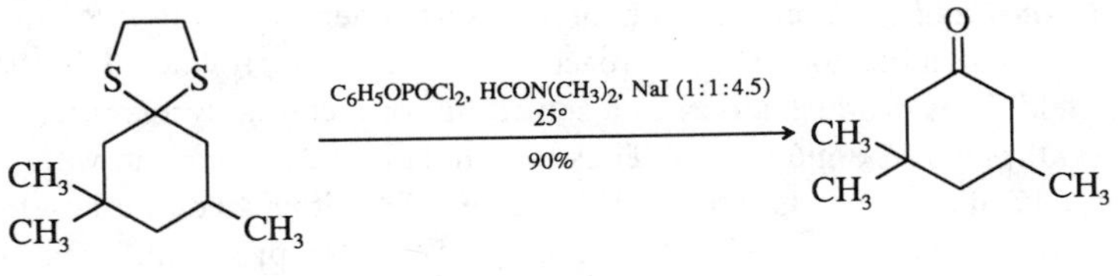

[1] H.-J. Liu and V. Wiszniewski, *Tetrahedron Letters*, **29**, 5471 (1988).

Phenyl dichlorophosphite, $C_6H_5OPCl_2$.

Phospholipids. This phosphite in combination with ethyldiisopropylamine can be used to prepare phosphite triesters, which on oxidation (H_2O_2) and deprotection

$$\text{(I) ROH + R}^1\text{OH} \xrightarrow[\text{base}]{C_6H_5OPCl_2} \text{ROP(OC}_6\text{H}_5\text{)OR}^1 \xrightarrow[90\%]{1)\ H_2O_2\ 2)\ H_2,\ PtO_2} \text{ROP(=O)(OH)OR}^1$$

furnish phosphate esters. This protocol provides a route to typical phospholipids:

$$\text{RCO}_2\text{CH}_2\text{CH(O}_2\text{CR}^1\text{)CH}_2\text{OH} + \text{HOCH}_2\text{CH(NHCbo)COOBzl} \xrightarrow[65-70\%\ \text{overall}]{1)\ C_6H_5OPCl_2\ 2)\ H_2O_2\ 3)\ H_2,\ PtO_2} \text{RCO}_2\text{CH}_2\text{CH(O}_2\text{CR}^1\text{)CH}_2\text{OP(=O)(OH)OCH}_2\text{CH(NH}_2\text{)COOH}$$

[1] S. F. Martin and J. A. Josey, *Tetrahedron Letters*, **29**, 3631 (1988).

Phenylethyl isocyanate, $OCN\overset{*}{C}H(CH_3)C_6H_5$ (**1**). Available from Aldrich or Fluka.

Resolution of sec-amines.[1] Reaction of *sec*-amines with (+)- or (−)-**1** results in two diasteriomeric ureas, $R_2NCONH(CH_3)C_6H_5$, which can usually be resolved by either chromatography or crystallization. The resolved *sec*-amine is obtained by alcoholysis of the optically pure urea. The method is particularly valuable where both optical isomers of the amine are desired; *see also* (R)-(−)-(1-naphthyl)ethyl isocyanate (**6**, 416) for a similar reagent.

[1] B. Schönenberger and A. Brossi, *Helv.*, **69**, 1486 (1986); L. A. Chrisey and A. Brossi, *Org. Syn.*, submitted 1989.

D-2-Phenylglycinol, $C_6H_5CH(NH_2)CH_2OH$ (**1**). both D- and L-**1** are available from Aldrich.

α-Amino acids.[1] Condensation of D-**1** with phenyl 2-bromoacetate (**2**) in ethyldiisopropylamine followed by reaction with Boc_2O provides **3** in 40–70% overall yield. This product serves as a substitute for a chiral glycine in enantioselective synthesis of α-amino acids. Thus the enolate of **3**, generated with sodium hexamethyldisilazide, reacts with alkyl halides at −78° with high diastereoselectivity to give (3R)-derivatives (**4**). Note that change of the Boc protecting group of **3** to

Bzl results in opposite diastereoselectivity on alkylation. The alkylated products are readily converted to derivatives of α-amino acids.

$$\mathbf{1} + \underset{\mathbf{2}}{BrCH_2COOC_6H_5} \xrightarrow[40-70\%]{1)\ (i\text{-}Pr)_2NC_2H_5;\ 2)\ Boc_2O} \mathbf{3}$$

Boc

3

$$\mathbf{3} \xrightarrow[85\%]{1)\ NaN[Si(CH_3)_3]_2;\ 2)\ CH_3I,\ -78^\circ} \mathbf{4} \xrightarrow[\Delta]{HCl,\ C_2H_5OH}$$

C_6H_5 N CH_3 Boc 3

4 (3R/3S > 200 : 1)

OH CH_3 C_6H_5 $\overset{+}{N}$ $COOC_2H_5$ H_2Cl^-

$\downarrow H_2$, Pd/C

CH_3 $Cl^-H_3\overset{+}{N}$ $COOC_2H_5$

$$\xrightarrow[83\%]{1)\ NaN[Si(CH_3)_3]_2;\ 2)\ BzlBr}$$

C_6H_5 N Bzl → C_6H_5 N $CH_2C_6H_5$ Bzl

[1] J. F. Dellaria and B. D. Santarsiero, *Tetrahedron Letters*, **29**, 6079 (1988).

Phenyliodine(III) bis(trifluoroacetate) $C_6H_5I(O_2CCF_3)_2$ (**1**).

***Oxidation of p*-alkoxyphenols.** These substrates are oxidized by $C_6H_5I(OCOCF_3)_2$[1] or $C_6H_5I(OAc)_2$[2] to *p*-quinones in high yield, probably via hemiacetal intermediates. CAN can also be used as oxidant but yields are lower.

Hofmann rearrangement.[3] Two laboratories[4,5] have reported that this hypervalent iodine reagent is useful for rearrangement of primary amides to primary amines, without effect on a secondary sulfonamide group. In the example cited here, the Hofmann rearrangement failed completely under usual conditions.

$$ArSO_2NH(CH_2)_5CONH_2 \xrightarrow[38-75\%]{\mathbf{1},\ CH_3CN/H_2O} ArSO_2NH(CH_2)_5NH_2$$

Dethioacetalization.[6] This reaction can be effected with $C_6H_5I(OCOCF_3)_2$ (**1**) in aqueous methanol (9:1) or acetonitrile at 25° without effect on ester, nitrile, hydroxyl groups, or unsaturated bonds.

$$p\text{-}(CH_3)_2NC_6H_4\text{-(1,3-dithian-2-yl)} \xrightarrow[99\%]{1,\ CH_3OH,\ H_2O} p\text{-}(CH_3)_2NC_6H_4CHO$$

$$\text{2-}CH_3\text{-2-}C_6H_5\text{-1,3-dithiane} \xrightarrow[85\%]{} \begin{matrix}CH_3\\C_6H_5\end{matrix}\!\!>\!C{=}O$$

$$C_6H_5SC(=O)(CH_2)_4CH(SC_2H_5)_2 \xrightarrow[97\%]{1,\ CH_3OH} C_6H_5SC(=O)(CH_2)_4CH(OCH_3)_2$$

[1] Y. Tamura, T. Yakura, H. Tohma, K. Kikuchi, and Y. Kita, *Synthesis*, 126 (1989).
[2] A. Pelter and S. Elgendy, *Tetrahedron Letters*, **29**, 677 (1988).
[3] E. S. Wallis and J. F. Lane, *Org. React.*, **3**, 267 (1949).
[4] G. M. Loudon, A. S. Radhakrishna, M. R. Almond, J. K. Blodgett, and R. H. Boutin, *J. Org.*, **49**, 4272 (1984).
[5] V. H. Pavlidis, E. D. Chan, L. Pennington, M. McParland, M. Whitehead, and I. G. C. Coutts, *Syn. Comm.*, **18**, 1615 (1988).
[6] G. Stork and K. Zhao, *Tetrahedron Letters*, **30**, 287 (1989).

Phenyliodine(III) diacetate.

Iododecarboxylation. Primary and secondary acids undergo this reaction when treated with $C_6H_5I(OAc)_2$ and iodine (equation I).[1]

$$\text{(I)}\ RCOOH + C_6H_5I(OAc)_2 \xrightarrow[h\nu]{I_2,\ CCl_4} [RCO_2I] \xrightarrow[70\text{–}94\%]{} RI + CO_2$$

The reaction can be extended to aromatic acids that are soluble in carbon tetrachloride.[2]

$$R\text{–}C_6H_4\text{–}COOH + C_6H_5I(OAc)_2 + I_2 \xrightarrow[25\text{–}55\%]{CCl_4,\ h\nu} R\text{–}C_6H_4\text{–}I$$

[1] J. I. Concepción, C. G. Francisco, R. Freire, R. Hernández, J. A. Salazar, and E. Suárez, *J. Org.*, **51**, 402 (1986).
[2] R. Singh and G. Just, *Syn. Comm.*, **18**, 1327 (1988).

Phenyliodonium bis(phenylsulfonyl)methylide, $C_6H_5I^+C^-(SO_2C_6H_5)_2$ (**1**), m.p. 129–131.°.

Preparation:

$$C_6H_5I(OAc)_2 + CH_2(SO_2C_6H_5)_2 \xrightarrow[92\%]{\text{KOH, } CH_3OH,\ -10^\circ} \mathbf{1}$$

gem-*Bis(phenylsulfonyl)cyclopropanes.*[1] The reagent reacts with simple alkenes thermally or photochemically to give these cyclopropanes in 25–70% yield. The products decompose slowly on standing with loss of SO_2. Complex products are formed from alkynes.

[1] L. Hatjiarapoglou, A. Varvoglis, N. W. Alcock, and G. A. Pike, *J.C.S. Perkin I*, 2839 (1988).

Phenyl isocyanate, $C_6H_5N{=}C{=}O$.

cis-*Oxazolidin-2-ones.*[1] Aryl isocyanates, with the exception of *p*-toluenesulfonyl isocyanate, react with vinylepoxides in the presence of a Pd(0) catalyst to form the less stable *cis*-oxazolidinones, regardless of the stereochemistry of the vinylepoxide.

$$C_6H_5(CH_2)_2\text{-epoxide-}CH{=}CH_2 + C_6H_5N{=}C{=}O \xrightarrow[88\%]{\text{Pd(0), (RO)}_3\text{P, THF}} \text{3-}C_6H_5\text{-4-}(CH{=}CH_2)\text{-5-}(C_6H_5(CH_2)_2)\text{-oxazolidin-2-one}$$

cis/trans = 1:2 *cis*, 3:1

If 2-methoxy-1-naphthyl isocyanate is used, the N-aryloxazolidinone can be cleaved by CAN to furnish the corresponding N-unsubstituted oxazolidinone in 70–85% yield.

[1] B. M. Trost and A. R. Sudhakar, *Am. Soc.*, **110**, 7933 (1988).

Phenyl isocyanide dichloride, $C_6H_5N{=}CCl_2$ (**1**), **6**, 458. Supplier: Aldrich.

Dialkynyl ketones.[1] In the presence of $Cl_2Pd[P(C_6H_5)_3]_2$ **1** couples with alkynyltin compounds (**2**) to give (N-phenyl)dialkynylimines (**3**) in 30–75% yield. These products are hydrolyzed to dialkynyl ketones (**4**) by aqueous acetone (BF_3 etherate) in ~60% yield. Under the same conditions, Grignard reagents undergo coupling to alkanes.

$$1 + 2\ Bu_3SnC{\equiv}CR \xrightarrow[30-75\%]{Pd(II),\ C_6H_6} C_6H_5N{=}C(C{\equiv}CR)_2 \xrightarrow[60\%]{H_3O^+} O{=}C(C{\equiv}CR)_2$$

2 **3** **4**

$$1 + 2\ RMgBr \xrightarrow[75-90\%]{Pd(II),\ THF,\ 25^\circ} R{-}R + C_6H_5NC$$

[1] Y. Ito, M. Inouye, and M. Murakami, *Tetrahedron Letters*, **29**, 5379 (1988).

(−)-8-Phenylmenthol, R*OH.

Asymmetric Wittig–Horner reaction; chiral olefinations.[1,2] Reaction of the chiral ketone **1** with the ylide from (−)-8-phenylmenthyl phosphonoacetate (**2**) at −30 to −60° gives the (E)-olefin in a 90:10 ratio. The geometry of the alkene is determined mainly by the chiral auxiliary. Use of *ent*-**2** results in **3** with the E/Z ratio of about 20:80 in 90% yield. This chiral olefination is limited to enantiomerically pure ketones.

$$\mathbf{1} + (CH_3O)_2\overset{O}{\overset{\|}{P}}CH_2COOR^* \ (\mathbf{2}) \xrightarrow[\sim 95\%]{KOC(CH_3)_3,\ -30\ to\ -60^\circ} (E)\text{-}\mathbf{3} + (Z)\text{-}\mathbf{3} \quad (\sim 90:10)$$

H, H, CH_2OSiR_3, OH, O

1

$\overset{*}{R}OOC$, H, H, H, CH_2OSiR_3, OH

(E)-**3**

Asymmetric intramolecular double Michael reaction.[3] Treatment of the 8-phenylmenthyl α,β-unsaturated amide ester (**2**) with *t*-butyldimethylsilyl triflate and triethylamine effects this Michael reaction with almost complete diastereoselectivity to give the indolizidine **3**, which was used for synthesis of (−)tylophorine (**4**).

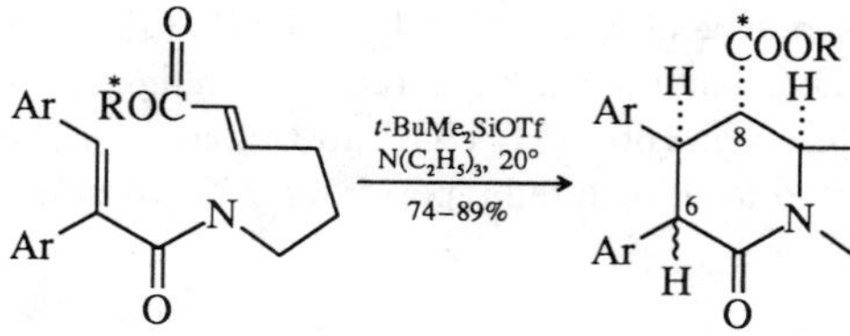

O
Ar ROC*
Ar N
O

t-BuMe$_2$SiOTf
N(C$_2$H$_5$)$_3$, 20°
74–89%

C*OOR
H H
Ar 8
Ar 6 N
H
O

3 (4:1)

2, Ar = $-C_6H_3(OCH_3)_2$ (OCH$_3$, OCH$_3$)

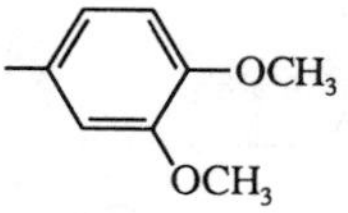

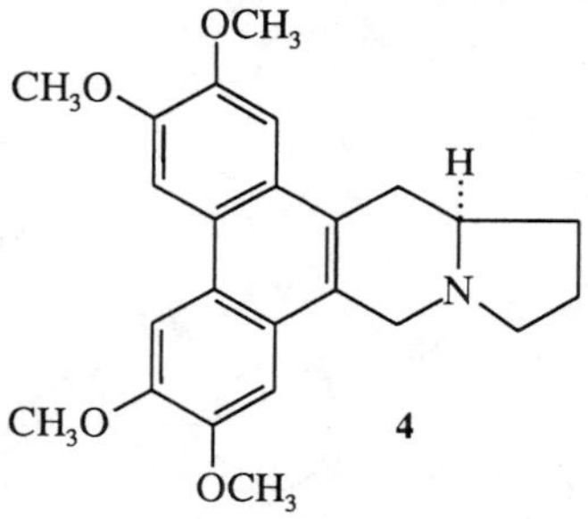

OCH$_3$
CH$_3$O
H
N
CH$_3$O
OCH$_3$
4

[1] H.-J. Gais, G. Schmiedl, W. A. Ball, J. Bund, G. Hellmann, and I. Erdelmeier, *Tetrahedron Letters*, **29**, 1773 (1988).
[2] H. Rehwinkel, J. Skupsch, and H. Vorbrüggen, *ibid.*, **29**, 1775 (1988).
[3] M. Ihara, Y. Takino, and K. Fukumoto, and T. Kametani, *Heterocycles*, **28**, special issue, 63 (1989).

Phenylselenophosphonic dichloride, $C_6H_5P(=Se)Cl_2$. The reagent is prepared by reaction of phenyldichlorophosphine with selenium at 170–175°.[1]

>C=O → >C=Se.[2] The reagent reacts with enones and several masked carbonyl compounds to form selenocarbonyl compounds.

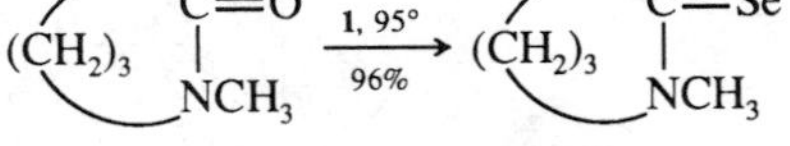

$$(CH_2)_3\langle C{=}O,\ NCH_3\rangle \xrightarrow[96\%]{\mathbf{1},\ 95^\circ} (CH_2)_3\langle C{=}Se,\ NCH_3\rangle$$

[1] H. W. Roesky and W. Kloker, *Z. Naturforsch*, **28B**, 697 (1973).
[2] J. P. Michael, D. H. Reid, B. G. Rose, and R. A. Speirs, *J.C.S. Chem. Comm.*, 1494 (1988).

4-(Phenylsulfonyl)butanoic acid, $C_6H_5SO_2(CH_2)_3COOH$ (**1**).

2-Piperidones. The dianion (**2**) of **1** reacts with imines activated by BF_3 etherate to afford, after cyclization (TFAA), substituted 2-piperidones (**3**). These lactams can be converted to substituted piperidines (**5**) by desulfonation and LAH reduction.[1]

$CH_3N{=}CHCH{=}CHC_6H_5$ + 2 → (1) $BF_3 \cdot O(C_2H_5)_2$, 2) TFAA; 87%) → **3**

3 → (Na/Hg, 93%) → **4** → ($LiAlH_4$, 71%) → **5**

[1]C. M. Thompson, D. L. C. Green, and R. Kubas, *J. Org.*, **53**, 5389 (1988).

1-Phenylsulfonylpropanol-2, $C_6H_5SO_2CH_2CH(OH)CH_3$ (**1**).

Reaction with electrophiles. The dianion of this and other β-hydroxy sulfones shows unexpected stereoselectivity. Thus the dianion of (S)-**1** is alkylated at the position α to the sulfonyl group with high *syn*-stereoselectivity, which can be markedly affected by additives.[1]

(S)-**1** → (1) 2BuLi, 2) $CH_2{=}CHCH_2Br$, THF; 83%) → *syn*-**2** + *anti*-**2** 73:27

+ HMPT	79%	58:42
+ TMEDA	82%	85:15

The reaction of the dianion with aldehydes is similar to that with alkyl halides and results in formation mainly of the 1,3-*syn* isomer.[2]

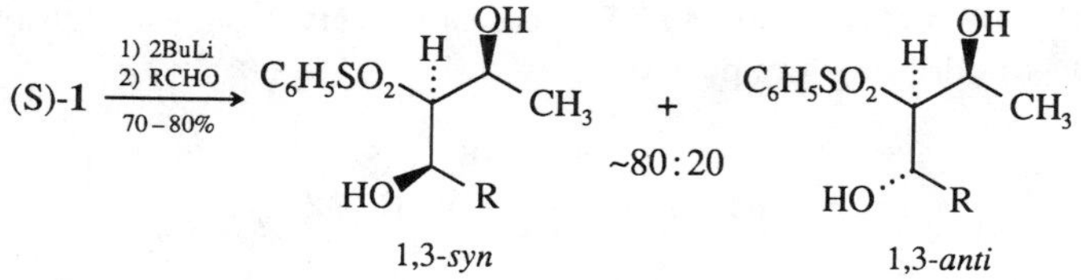

[1] K. Tanaka, K. Ootake, K. Imai, N. Tanaka, and A. Kaji, *Chem. Letters*, 633 (1983).
[2] R. Tanikaga, K. Hosoya, and A. Kaji, *J.C.S. Perkin I*, 2397 (1988).

Phenylthiomethylenetriphenylphosphorane, $(C_6H_5)_3P{=}CHSC_6H_5$ (**1**). Preparation.[1]

Lactone ring expansion.[2] The γ-lactone **2** can be expanded to the protected γ-lactol **4**, characteristic of mevinic acids, by reaction of the lactol of **2** with this Wittig reagent to provide a vinyl sulfide **3**. Conversion of the vinyl sulfide to **4** was carried out by alkoxymercuration followed by reduction and concomitant ring closure.

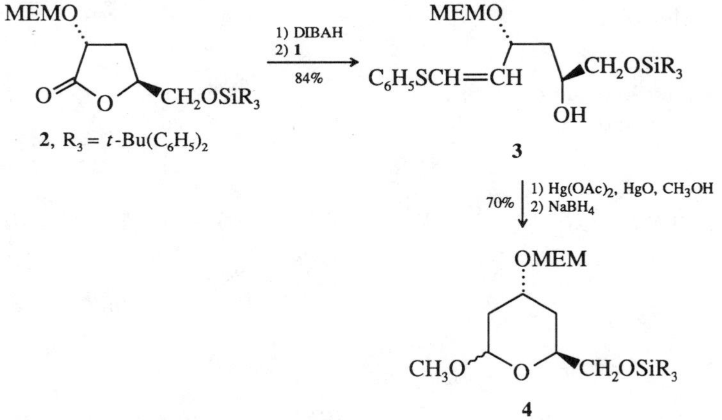

[1] T. J. Lee, *Tetrahedron Letters*, 4995 (1985).
[2] A. H. Davidson, C. D. Floyd, C. N. Lewis, and P. L. Myers, *J.C.S. Chem. Comm.*, 1417 (1988).

Phenylthiomethyl(trimethyl)silane, $C_6H_5SCH_2Si(CH_3)_3$ (**1**), **10**, 313–314.

β-Hydroxyalkyl phenyl sulfides.[1] In the presence of Bu_4NF (0.1 equiv.) this silane can effect phenylthiomethylation of aldehydes or ketones to provide silyl ethers of β-hydroxyalkyl phenyl sulfides in moderate yield. The solution of Bu_4NF in THF should be dried with 5-Å molecular sieves before use. Yields are usually improved by use of 1 equiv. of Bu_4NF in the case of aryl aldehydes substituted by an electron-withdrawing group.

$$ArCHO + \mathbf{1} \xrightarrow[50-96\%]{Bu_4NF,\ 25°,\ THF} ArCH(OH)CH_2SC_6H_5$$

$$C_6H_5COCH_3 + \mathbf{1} \xrightarrow[35\%]{} C_6H_5C(CH_3)(OH)CH_2SC_6H_5$$

$$(CH_2)_5C{=}O + \mathbf{1} \xrightarrow[36\%]{} (CH_2)_5C(OH)CH_2SC_6H_5$$

[1] A. Hosomi, K. Ogata, K. Hoashi, S. Kohra, and Y. Tominaga, *Chem. Pharm. Bull.*, **36**, 3736 (1988).

Phenylthionitrile oxide (1).

Generation and trapping with an alkene:

$$C_6H_5SCH_2NO_2 + C_6H_5NCO \xrightarrow{C_6H_6,\ N(C_2H_5)_3} [C_6H_5SC{\equiv}N{\rightarrow}O]\ (\mathbf{1}) + C_6H_5NHCONHC_6H_5$$

$$\mathbf{1} \xrightarrow[80\%]{CH_3(CH_2)_3CH{=}CH_2} \mathbf{2}\ \text{(3-}C_6H_5S\text{-5-}C_4H_9\text{-isoxazoline)} \xrightarrow[97\%]{ClC_6H_4CO_3H} \mathbf{3}\ \text{(3-}C_6H_5SO_2\text{-5-}C_4H_9\text{-isoxazoline)}$$

This route provides a two-step synthesis of 3-(phenylsulfonyl)-Δ^2-isoxazolines (**3**). The phenylsulfonyl group can be displaced by various nucleophiles.[1]

[1] D. P. Curran and J.-C. Chao, *J. Org.*, **53**, 5369 (1988).

(Phenylthio)nitromethane, $C_6H_5SCH_2NO_2$ (**1**).

Preparation[1]:

$$O(CH_2CH_2)_2NSC_6H_5 + NO_2CH_2COOC_2H_5 \xrightarrow[92\%]{1)\ KOH;\ 2)\ NH_2OH} \mathbf{1}$$

Ring-enlarged α-phenylthiocycloalkanones.[2] The adducts (**2**) of the dianion of **1** (2BuLi, THF/HMPT) with cycloalkanones rearrange in the presence of $AlCl_3$ (2 equiv.) to α-phenylthiocycloalkanones.

$$\text{2-methyl-3-phenylcyclobutanone} + C_6H_5S\bar{C}{=}NO_2^- \xrightarrow{85\%} \mathbf{2}\ (\text{1-HO-1-}CH(SC_6H_5)NO_2\text{-2-}CH_3\text{-3-}C_6H_5\text{-cyclobutane}) \xrightarrow[74\%]{2AlCl_3} \text{2-}SC_6H_5\text{-3-}CH_3\text{-4-}C_6H_5\text{-cyclopentanone}$$

[1] F. G. Bordwell and J. E. Bartmess, *J. Org.*, **43**, 3101 (1978).
[2] S. Kim and J. H. Park, *Chem. Letters*, 1323 (1988).

Phosgene.

cis-*Oxyamination of alkenes.*[1] *trans*-1,2-Bromohydrins on consecutive reaction with phosgene and then benzylamine are converted into a bromo carbamate, which cyclizes on treatment with NaH (2 equiv.) to a *cis*-oxazolidinone. N-Debenzylation followed by hydrolysis provides a *cis*-1,2-amino alcohol.

$$\text{trans-2-bromocyclohexanol} \xrightarrow[73\%]{1)\ COCl_2,\ N(C_2H_5)_3;\ 2)\ C_6H_5CH_2NH_2} \text{Br, OCONHBzl} \xrightarrow[93\%]{2NaH,\ THF} \text{N-Bzl oxazolidinone} \xrightarrow[84\%]{Li,\ NH_3} \text{N-H oxazolidinone} \xrightarrow[quant.]{HCl,\ THF} \text{cis-2-aminocyclohexanol}\ (NH_2, OH)$$

[1] J. Das, *Syn. Comm.*, **18**, 907 (1988).

Phosphines, chiral.

Hydrogenation catalysts. A number of new optically active phosphine ligands have been developed for selective enantioselective reduction of N=C bonds of amino acid precursors, including DPCB(**1**)[1], the methylene homolog (**2**)[2] of dipamp, and the novel ligand **3**, in which the source of chirality is a Re atom.[3] A ligand **4** has been used as a Rh(I) complex for hydrosilylation.[4]

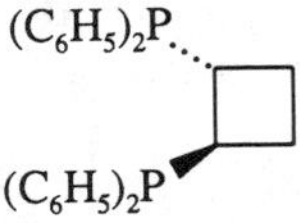

$(C_6H_5)_2P$ $(C_6H_5)_2P$ **1** (DPCB)

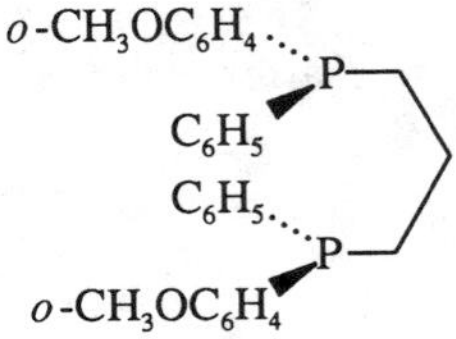

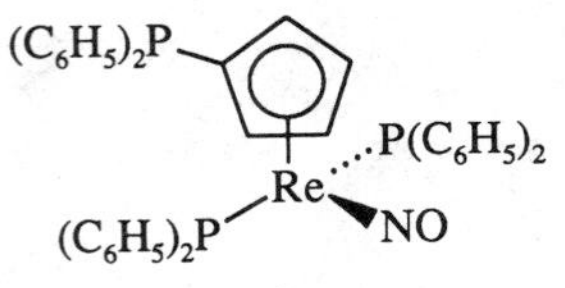

o-$CH_3OC_6H_4$ P C_6H_5 C_6H_5 P o-$CH_3OC_6H_4$ **2**

$(C_6H_5)_2P$ Re $P(C_6H_5)_2$ NO $(C_6H_5)_2P$ **3**

CH_3 As C_6H_5 CH_3 As C_6H_5 **4**

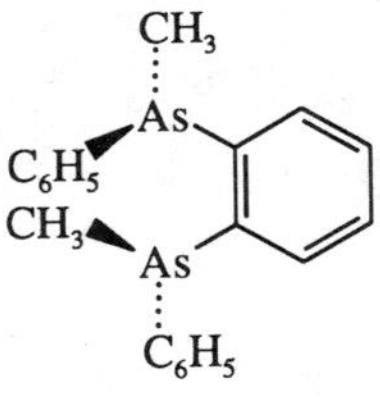

[1] T. Minami, Y. Okada, R. Nomura, S. Hirota, Y. Nagahara, and K. Fukuyama, *Chem. Letters*, 613 (1986).
[2] C. R. Johnson and T. Imamoto, *J. Org.*, **52**, 2170 (1987).
[3] B. D. Zwick, A. M. Arif, A. T. Patton, and J. A. Gladysz, *Angew. Chem., Int. Ed.*, **26,** 910 (1987).
[4] A. F. F. M. M. Rahman and S. B. Wild, *ibid.*, **39**, 155 (1987).

Phosphoric acid.

Cationic olefin cyclization of aldehyde enol acetates.[1] The enol acetate (**2**) of citronellal (**1**) undergoes cyclization to dihydrocyclocitral (**3**) in the presence of

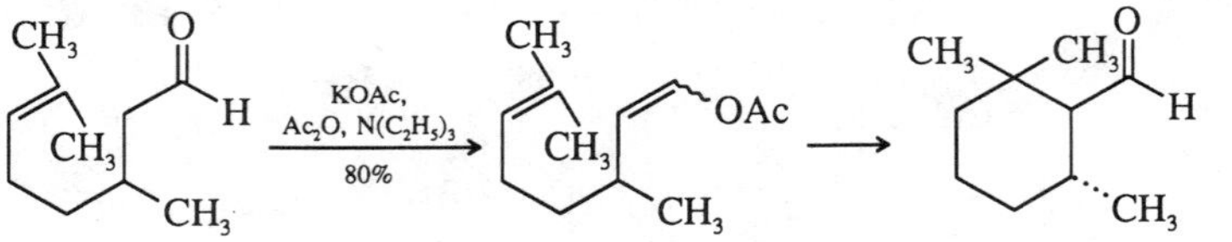

1 (CH_3, CH_3, CH_3, CHO) → (KOAc, Ac_2O, $N(C_2H_5)_3$, 80%) → **2** (E/Z = 2:1) (CH_3, CH_3, CH_3, OAc) → **3** (CH_3, CH_3, CH_3, CHO)

$SnCl_4$, CH_2Cl_2,	49–52%	(*trans/cis*) = 18:1
CH_3SO_3H, $C_6H_5CH_3$, 0°	42%	" = 16:1
H_3PO_4, $C_6H_5CH_3$, 100°	80%	" = 8:1

either Lewis or Brönsted acids. Cyclization is achieved in the highest yield with 85% aqueous phosphoric acid. Higher stereoselectivity obtains with Lewis acids, but the rate is slower, and the yield lower. Cyclization of **4** to the drimane **5** is best achieved with $SnCl_4$. In this case, use of H_3PO_4 provides **5** in 14% yield.

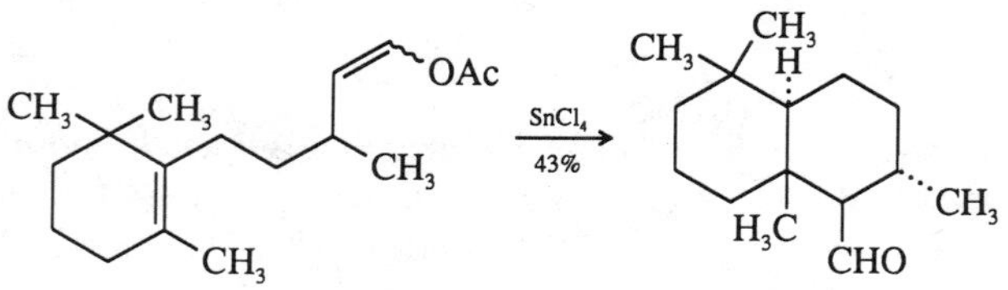

4 (E/Z = 2:1) (CH_3, CH_3, CH_3, CH_3, OAc) → ($SnCl_4$, 43%) → **5** (CH_3, CH_3, H, H_3C, CHO, CH_3)

[1] D. P. Simmons, D. Reichlin, and D. Skuy, *Helv.*, **71**, 1000 (1988).

Phosphorus oxychloride (Phosphoryl chloride), $POCl_3$.

Dehydration. Although early studies of dehydration of tertiary alcohols were shown to involve an *anti*-elimination, little use of this stereoselectivity has been applied to preparation of (E)- or (Z)-alkenes. This reaction is useful for preparation of (E)- and (Z)- Δ^{22}- or Δ^{23}-sterols, and in combination with *syn*-hydroboration of alkenes for isomerization of trisubstituted steroidal alkenes.[1]

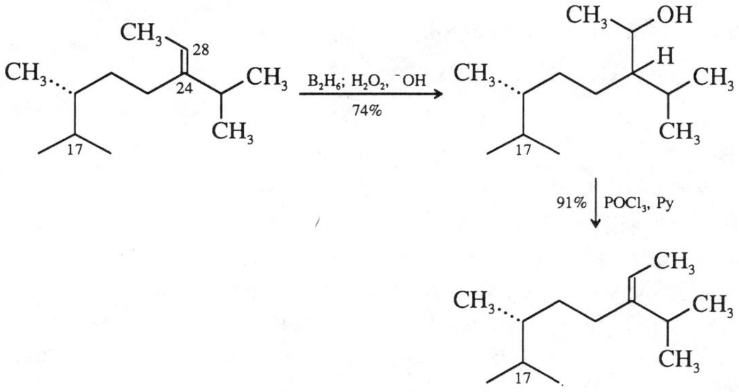

CH3 28 CH3 24 CH3 CH3 17 — B_2H_6; H_2O_2, ^-OH (74%) → CH3 OH H CH3 CH3 CH3 17 — 91% $POCl_3$, Py → CH3 CH3 CH3 CH3 17

[1] J.-L. Giner, C. Margot, and C. Djerassi, *J. Org.*, **54**, 369 (1989).

Phosphorus pentoxide, P_2O_5.

2-Amino-6-arylaminopurines. These interesting purines (**2**), many of which have useful biological effects, can be prepared directly from guanine (**1**) by reaction with an arylamine hydrochloride and P_2O_5 in the presence of $(C_2H_5)_3N{\cdot}HCl$ at 200°.[1]

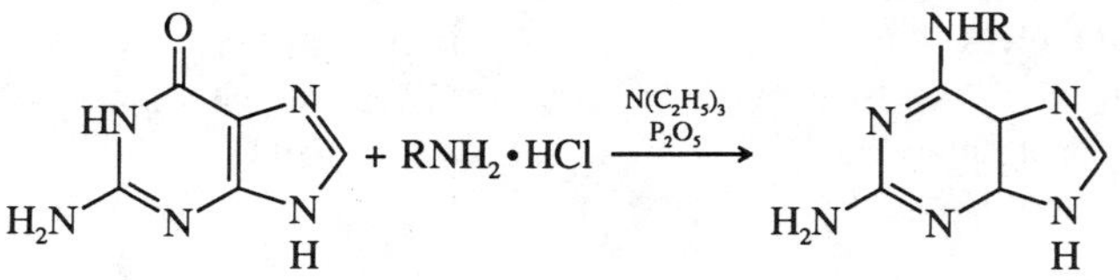

1 + $RNH_2{\cdot}HCl$ $\xrightarrow{N(C_2H_5)_3,\ P_2O_5}$ **2**

R	Yield
$R = C_6H_5$	80%
$R = p\text{-}FC_6H_4$	61%
$R = m\text{-}CH_3C_6H_4$	82%

[1] E.-S. M. A. Yakout, H. M. Soe, and E. B. Pederson, *Acta Chem. Scand.*, **B42**, 257 (1988).

Picric acid (2,4,6-trinitrophenol). Caution: explosive and toxic.

Picrates.[1] At one time organic bases were isolated and identified as the picrates, which are often highly crystalline. Picric acid is generally supplied as a moist solid containing water, which hinders crystallization of picrates. The water can be replaced by ethanol for storage. Picrates can be prepared by reaction in acetone, in which picric acid is fairly soluble and then allowed to crystallize by addition of ether.

[1] D. Lloyd and H. McNab, *J. Chem. Res.* (*S*), 18 (1989).

1-Piperidino-3,3-dimethyl-2-butanol (PDB), $(CH_3)_3C\overset{*}{C}H(OH)CH_2N$(piperidine ring) (**1**).

The β-amino alcohol is prepared by reaction of $(CH_3)_3CC(=O)CH_2Br$ with piperidine followed by reduction with $LiAlH_4$. Resolution with (−)-dibenzyl-L-tartaric acid provides (−)-**1**, α_D −72° (98% ee).

Enantioselective addition of $(C_2H_5)_2Zn$ to C_6H_5CHO.[1] In the presence of (−)-**1**, diethylzinc adds to C_6H_5CHO to provide the (R)-adduct. The optical purity of the catalyst (**1**) has a marked effect on the rate and also on the optical purity of the adduct. Thus (−)-PDB of only 10–20% ee provides (R)-1-phenylpropanol in 80–90% ee and about 95% chemical yield. This nonlinear effect in catalyzed asymmetric oxidations and aldolizations has been noted previously.[2] It may be a result of the molecularity of the reaction or of the aggregation or ligand exchange of the catalyst.

[1] N. Oguni, Y. Matsuda, and T. Kaneko, *Am. Soc.*, **110**, 7877 (1988).
[2] C. Puchot, O. Samuel, E. Dunach, S. Zhao, C. Agami, and H. B. Kagan, *Am. Soc.*, **108**, 2353 (1986).

Platinum(IV) oxide (Adam's catalyst, 1, 890).

5-Deoxyanthracyclines.[1] Hydrogenation of anthracyclines **1a** or **1b** catalyzed by $Pd/BaSO_4$ gives 7-deoxy-**1** (97% yield), which on further hydrogenation is converted into a 5,7-bisdeoxy product. However, hydrogenation of **1** catalyzed by Adam's catalyst in methanol containing chloroacetic acid affords the 5-deoxy product **2** in about 70% yield.

1a, R = H
1b, R = OH

2

[1] D. W. Cameron, G. I. Feutrill, and P. G. Griffiths, *Tetrahedron Letters*, **29**, 4629 (1988).

Polyphosphoric acid trimethylsilyl ester (PPSE), $(CH_3)_3SiO\left[\left(\overset{O}{\overset{\|}{\underset{OSi(CH_3)_3}{\underset{|}{P}}}}-O\right)_3\right]_4Si(CH_3)_3$

(**1**). Prepared as a mixture of four components by reaction of P_2O_5 with hexamethyldisiloxane in CH_2Cl_2. Soluble in most aprotic solvents.[1]

It effects a synthesis of amidines from simple carboxylic acids and amines[1]:

$$C_6H_5COOH + C_6H_5NH_2 \xrightarrow[80-90\%]{PPSE,\ 160^\circ} C_6H_5\overset{NC_6H_5}{\overset{\|}{C}}NHC_6H_5$$

This reagent is both a Lewis acid and a dehydrating agent. Thus it effects Pummerer rearrangement of sulfoxides and rearrangement of pinacol to pinacolone via an epoxide.[2]

$$(C_6H_5)_2\overset{OH}{\overset{|}{C}}H-\underset{OH}{\underset{|}{C}}H(C_6H_5)_2 \xrightarrow[(CH_2Cl)_2]{1} \text{(2,2,3,3-tetraphenyloxirane: } C_6H_5, C_6H_5 \text{ / O / } C_6H_5, C_6H_5) + C_6H_5\overset{O}{\overset{\|}{C}}-C(C_6H_5)_3$$

80°	61%	26%
150°	—	97%

[1] S. Ogata, A. Mochizuki, M. Kakimoto, and Y. Imai, *Bull. Chem. Soc. Japan*, **59**, 2171 (1986).
[2] M. Kakimoto, T. Seri, and Y. Imai, *ibid.*, **61**, 2643 (1988).

Potassium.

Demethoxylation.[1] 1-Alkyl-3,4,5-trimethoxybenzenes are converted into 1-alkyl-3,5-dimethoxybenzenes on treatment with potassium sand in THF. Either sodium or potassium can be used to convert 1,2,3-trimethoxybenzene into 1,3-dimethoxybenzene (85% yield).

$$\text{1-}CH_2R\text{-3,4,5-}(OCH_3)_3C_6H_2 \xrightarrow[72-87\%]{K,\ THF,\ 20^\circ} \text{1-}CH_2R\text{-3,5-}(OCH_3)_2C_6H_3$$

[1] U. Azzena, T. Denurra, E. Fenude, G. Melloni, and G. Rassu, *Synthesis*, 28 (1989).

Potassium–18-Crown-6.

Cleavage of β-lactones.[1] The blue solution of potassium–18-crown-6 in THF cleaves β-propiolactone at −20° to furnish a dianion (**1**) which on protonation gives methyl acetate. A similar reaction of the β-lactone **2** can be used to obtain isopropyl propionate (**3**).

$$\text{β-propiolactone} \xrightarrow{K^-(K^+)} K^+\ {}^-C(=O)(CH_2^-)-OCH_2^-\ (K^+)\ (\mathbf{1}) \xrightarrow[80\%]{H^+} CH_3C(=O)-OCH_3$$

$$\mathbf{1} \xrightarrow[70\%]{CH_3I} CH_3C(=O)-OCH_2CH_3$$

$$\mathbf{2} \xrightarrow[65\%]{1)\ K^-(K^+)\quad 2)\ CH_3I} C_2H_5C(=O)-OCH(CH_3)CH_3\ (\mathbf{3})$$

Cleavage of aryl ethers and esters.[2] Diphenyl ethers undergo C—O bond cleavage with this reagent in THF to give phenols and benzene. Phenyl benzoates are cleaved at the carbonyl–oxygen bond, and the rate of this ester cleavage is markedly increased by electron-donating groups, OCH_3 and OH, in either the acid or the phenol group.

[1] Z. Jedlinski, M. Kowalczuk, and A. Misiolek, *J.C.S. Chem. Comm.*, 1261 (1988).
[2] R. H. Fish and J. W. Dupon, *J. Org.*, **53**, 5230 (1988).

Potassium 3-aminopropylamide (KAPA).

$ROC{\equiv}CCH_3 \rightarrow ROCH_2C{\equiv}CH$. Propynyl ethers can be obtained from secondary alkoxides by reaction with trichloroethylene followed by dehydrochlorination and methylation. These can be rearranged to propargyl ethers by KAPA. This sequence is applicable to highly hindered secondary alcohols.[1]

$$c\text{-}C_6H_{11}OH \xrightarrow[87\%]{1)\ KH\quad 2)\ Cl_2C{=}CHCl\quad 3)\ BuLi,\ CH_3I} c\text{-}C_6H_{11}OC{\equiv}CCH_3 \xrightarrow[73\%]{KAPA} c\text{-}C_6H_{11}OCH_2C{\equiv}CH$$

[1] C. Almansa, A. Moyano, M. A. Pericas, and F. Serratosa, *Synthesis*, 707 (1988).

Potassium *t*-butoxide.

Haller–Bauer reaction.[1] A study of the base-induced cleavage of the optically active tertiary α-phenyl ketone (**1**)[2] shows that cleavage proceeds with retention with metal amides in benzene and alkoxides in an alcohol or benzene, but with

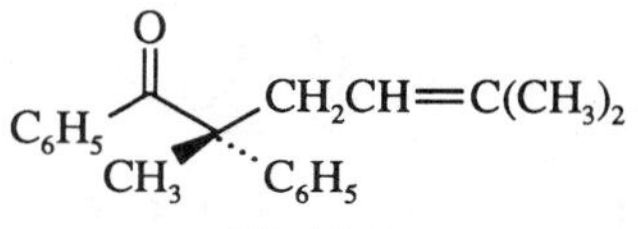

(R)-(−)-**1** 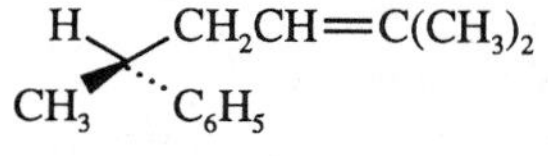 **2**

inversion in reactions with $KOCH_2CH_2OH$ in ethylene glycol. The highest stereoselectivity (88% retention) obtains with potassium *t*-butoxide. Lithium *t*-butoxide is ineffective. The cation reactivity thus is $K^+ > Na^+ > Li^+$.

Substantially the same results are observed in the cleavage of the cyclopentyl α-phenyl ketone (+)-**3**.[3] In this case the reaction with potassium *t*-butoxide in

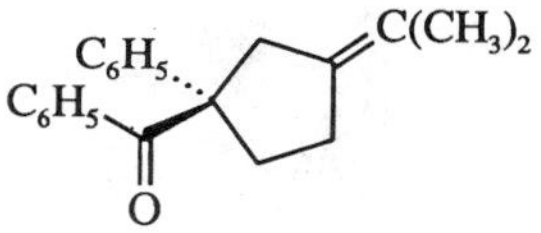

(R)-(+)-**3** 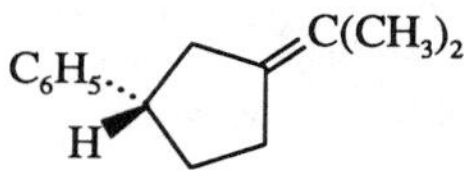 (S)-(+)-**4**

benzene proceeds with 98% retention. However, cleavage of the axial and equatorial isomers **5** and **6** is different from that of acyclic systems. Both are cleaved with retention (84–86%) by $KOC(CH_3)_3$, but only a 2-hour reaction time is nec-

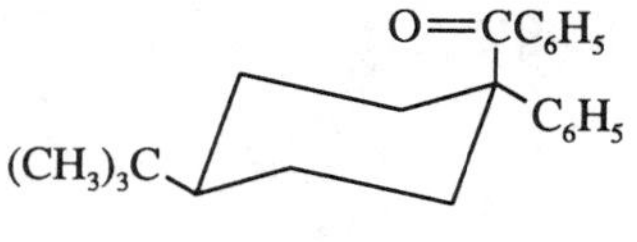

5 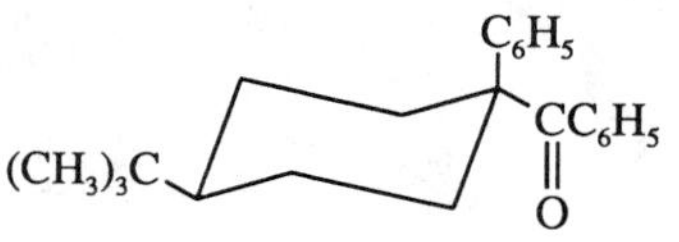 **6**

essary for cleavage of **6**, whereas 64 hours is required for cleavage of **5**. In the case of **5**, cleavage with $LiNH_2$ results in inversion (10%), but cleavage of **5** with the same base proceeds with retention (70%).

Cyclic enediynes.[4] A Ramburg–Backlund reaction of the cyclic diynes **1** furnishes cyclic enediynes.

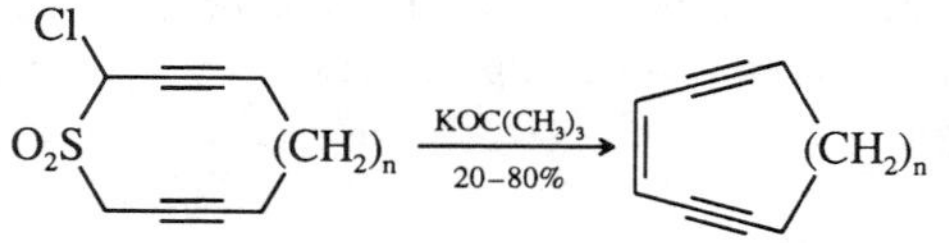

$KOC(CH_3)_3$, 20–80%

1, $n = 2–8$ **2**

[1] K. E. Hamlin and W. A. Weston, *Org. React.*, **9**, 1 (1957); E. M. Kaiser and C. D. Warner, *Synthesis*, 395 (1975).
[2] L. A. Paquette and J. P. Gilday, *J. Org.*, **53**, 4972 (1988).
[3] L. A. Paquette and C. S. Ra, *ibid.*, **53**, 4978 (1988).
[4] K. C. Nicolaou, G. Zuccarello, Y. Ogawa, E. Schweiger, and T. Kumazawa, *Am. Soc.*, **110**, 4866 (1988).

Potassium fluoride.

Fluorination. Freeze-dried KF can effect chloro–fluoro exchange in refluxing DMSO.[1]

$$\text{Cl-}C_6H_4NO_2 \xrightarrow[97.5\%]{\text{KF, DMSO}} \text{F-}C_6H_4NO_2$$

[1] Y. Kimura and H. Suzuki, *Tetrahedron Letters*, **30**, 1271 (1989).

Potassium fluoride–18-Crown-6.

[3+2]Annelation. Cleavage of 1-(trimethylsilyloxy)-2-alkoxycarbonylcyclopropanes (**1**) with KF–18-crown-6 generates γ-oxo-α-ester enolate anions, which can be trapped by electrophiles. Trapping with a vinyl phosphonium salt results in [3+2] annelation to provide annelated cyclopentanones (**2**).[1]

CH3, CH3, OSi(C2H5)3, COOC2H5, H — **1** $\xrightarrow[CH_3CN,\ \Delta]{\text{KF, crown ether}}$ [CH3, CH3, O, H, CHCOOC2H5, K] **a** $\xrightarrow[80\%]{(C_6H_5)_3\overset{+}{P}CH{=}CH_2Br^-}$

CH3, CH3, COOC2H5 — **2**

Octahydronapthalene synthesis.[2] An intramolecular version of this annelation using a Michael addition to a vinyl sulfone provides the octahydronapthalene unit (**1**) of compactin (**2**), a mevinic acid of interest as an inhibitor of cholesterol synthesis.

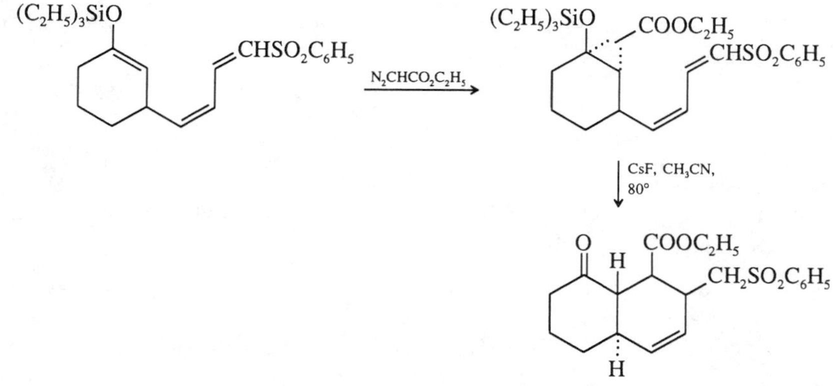

$(C_2H_5)_3SiO$ ... $CHSO_2C_6H_5$ $\xrightarrow{N_2CHCO_2C_2H_5}$ $(C_2H_5)_3SiO$ ... $COOC_2H_5$ ... $CHSO_2C_6H_5$

CsF, CH_3CN, 80°

O, H, $COOC_2H_5$, $CH_2SO_2C_6H_5$, H

1

HO, O, O, O, C_2H_5, O, $(CH_2)_2$, CH_3, H

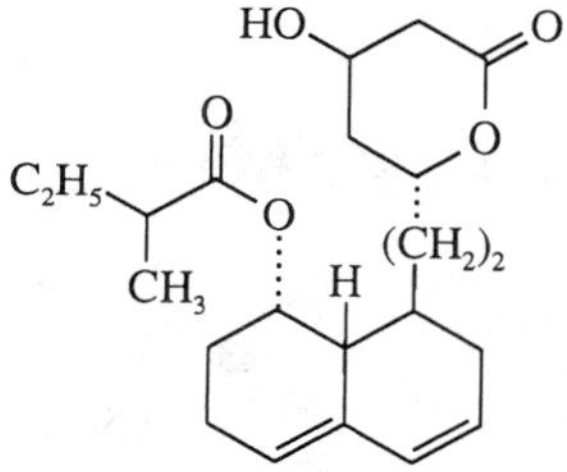

2

[1] J. P. Marino and E. Laborde, *J. Org.*, **52**, 1 (1987).
[2] J. P. Marino and J. K. Long, *Am. Soc.*, **110**, 7916 (1988).

Potassium fluoride–Hydrobromic acid.

Desilylation of phenolic ethers. Desilylation of *t*-butyldimethyl silyl ethers of phenols requires Bu_4NF or HF (for a base-sensitive phenol). Actually this desilylation can be effected at 25° with KF combined with a trace of hydrobromic acid.[1]

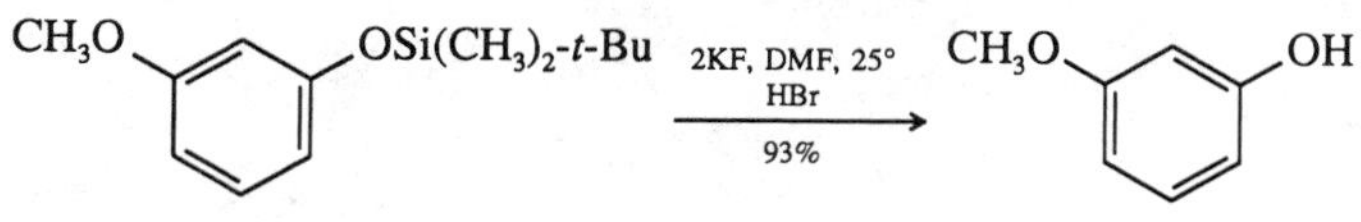

CH_3O–C_6H_4–$OSi(CH_3)_2$-*t*-Bu $\xrightarrow[93\%]{2KF,\ DMF,\ 25°,\ HBr}$ CH_3O–C_6H_4–OH

[1] A. K. Sinhababu, M. Kawase, and R. T. Borchardt, *Synthesis*, 710 (1988).

Potassium permanganate.

Imines → nitrones.[1] This oxidation is possible with $KMnO_4$ under phase-transfer conditions in aqueous CH_2Cl_2 in 15–90% yield. The best catalysts are Bu_4NCl or $(n\text{-}C_8H_{17})_4NBr$.

$$C_6H_5CH{=}NC(CH_3)_3 \xrightarrow[CH_2Cl_2,\ H_2O]{KMnO_4,\ Bu_4NCl,} C_6H_5CH{=}\overset{+}{N}(O^-)C(CH_3)_3 + C_6H_5\overset{O}{\overset{\|}{C}}NHC(CH_3)_3$$

(89%) (11%)

2-Hydroxy-3-oxocarboxylic esters.[2] β-Alkyl-α,β-unsaturated esters (**1**) are oxidized to these esters (**2**) by $KMnO_4$ in mildly acidic aqueous acetone at $-10°$ in 60–85% yield.

$$CH_3CH{=}C(CH_3){-}COOCH_3 \xrightarrow{80\%} CH_3\overset{O}{\overset{\|}{C}}{-}C(CH_3)(OH){-}COOCH_3$$

1 → **2**

[1] D. Christensen and K. A. Jorgensen, *J. Org.*, **54**, 126 (1989).
[2] D. H. G. Crout and D. L. Rathbone, *Synthesis*, 40 (1989).

Potassium persulfate, $K_2S_2O_8$.

Benzylic oxidation (**11**, 441).[1] Oxidation of the N-Boc-L-tyrosine **1** with $K_2S_2O_8$ with $CuSO_4$ as catalyst gives the *syn*-cyclic carbamate **2**, which can be converted to the β-hydroxy amino acid **4** as shown.

1 (p-$CH_3OC_6H_4CH_2CH(NHBoc)CH_2OAc$) $\xrightarrow[76\%]{K_2S_2O_8,\ CuSO_4}$ **2** (37:1) (cyclic carbamate, O—C=O, NH, CH_2OAc)

2 $\xrightarrow[100\%]{1)\ Ba(OH)_2;\ 2)\ Boc_2O}$ **3** (p-$CH_3OC_6H_4CH(OH)CH(NHBoc)COOH$)

3 $\xrightarrow[45\%]{1)\ PtO_2,\ O_2;\ 2)\ TFA}$ **4** (p-$CH_3OC_6H_4CH(OH)CH(NH_2)COOH$)

RCOOH → RCOOOH.[2] This oxidant converts water-insoluble carboxylic acids to the peracid in ether or CH_2Cl_2 under the influence of benzyltriethyl ammonium chloride or polyethylene glycol (PEG-400).

[1] K. Shimamoto and Y. Ohfune, *Tetrahedron Letters*, **29**, 5177 (1988).
[2] C. S. Pande and N. Jain, *Syn. Comm.*, **18**, 2123 (1988).

Potassium superoxide, KO_2.

Arene oxides.[1] KO_2 converts 2-nitrobenzenesulfonyl chloride into a peroxysulfur radical that effects epoxidation of polyarenes to give arene oxides in aprotic solvents at −30°. Anthracene is converted into the 9,10-quinone (70% yield).

SO_2Cl, NO_2, KO_2, CH_3CN, −30°, 50%, O

O, 95%

Desulfuration.[2] Thioamides are known to be converted into amides by potassium superoxide (**13**, 260). The reagent can also effect cyclodesulfuration in the case of N-(2-hydroxyphenyl)thioamides to provide benzoxazoles in generally high yield.

S, H, NCC_6H_5, OH, + $4KO_2$, CH_3CN, 20°, 93%, N, O, C_6H_5 + K_2SO_4

[1] H. K. Lee, K. S. Kim, J. C. Kim and Y. H. Kim, *Chem. Letters*, 561 (1988).
[2] Y. H. Kin, Y. I. Kim, H. S. Chang, and D. C. Yoon, *Heterocycles*, **29**, 213 (1989).

Pyrazinium chlorochromate, (N, $\overset{+}{N}$, H) $^-ClCrO_3$ (**1**). The reagent is obtained in 80% yield by reaction of pyrazine in 6*M* HCl with CrO_3.

Oxidation of alcohols. The reagent is comparable to PCC for oxidation of alcohols.[1]

[1] G. J. S. Doad, *J. Chem. Res.* (*S*), 270 (1988).

Pyridinium bromide perbromide.

α-Bromo-α,β-enones.[1] α-Bromo-α,β-cyclopentenones can be obtained in good yield by reaction of cyclopentenones with this reagent (about 3 equiv.) and pyridine (3 equiv.) in CH_2Cl_2. In the absence of pyridine, yields are lower because of polybromination.

$$\text{3-R-cyclopent-2-enone} \xrightarrow[\text{Py/CH}_2\text{Cl}_2]{\text{Py}\cdot\text{HBr}\cdot\text{Br}_2} \text{2-Br-3-R-cyclopent-2-enone}$$

R = CH_3 57%
R = CH_2OBzl 86%

[1] W. G. Dauben and A. M. Warshawsky, *Syn. Comm.*, **18**, 1323 (1988).

Pyridinium chlorochromate (PCC)

1,2-Diketones.[1] Alkyl benzyl ketones (**1**) are oxidized by PCC in combination with pyridine to 1,2-diketones (**2**) in 80–85% yield. This oxidation fails with dialkyl ketones.

$$C_6H_5CH_2COR + PCC \xrightarrow[60-85\%]{Py/CH_2Cl_2} C_6H_5\overset{O}{\overset{\|}{C}}-\overset{O}{\overset{\|}{C}}R$$

1 (R = CH_3, C_2H_5)

[1] F. Bonadies and C. Bonini, *Syn. Comm.*, **18**, 1573 (1988).

Pyridinium *p*-toluenesulfonate.

Selective cleavage of* t*-butyldimethylsilyl ethers. This reagent cleaves *t*-butyldimethylsilyl ethers at 22–55° in 80–92% yield in the presence of *t*-butyldiphenylsilyl ethers.

[1] C. Prakash, S. Saleh, and I. A. Blair, *Tetrahedron Letters*, **30**, 19 (1989).

Q

Quina alkaloids.

Chiral sulfinates, $(CH_3)_3CS(O)OR$.[1] Alkyl *t*-butylsulfinates can be prepared by reaction of *t*-butylmagnesium chloride with dialkyl sulfites. If the reaction is carried out in the presence of an optically active amine, the products can be optically active. The highest enantioselectivity is observed with amino alcohols, in particular with (−)-quinine (**1**).

Example:

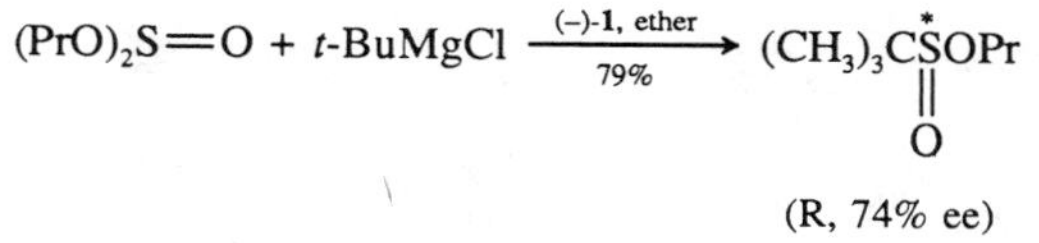

$$(PrO)_2S{=}O + t\text{-}BuMgCl \xrightarrow[79\%]{(-)\text{-}\mathbf{1},\ \text{ether}} (CH_3)_3C\overset{*}{S}(=O)OPr$$

(R, 74% ee)

Asymmetric reduction of benzil.[2] Reduction of benzil with SmI_2 in C_6H_6/THF with quinidine (**1**) as a proton source yields (R)-benzoin in about 40% ee. Addition of HMPT increases the rate and the enantioselectivity to 56% ee.

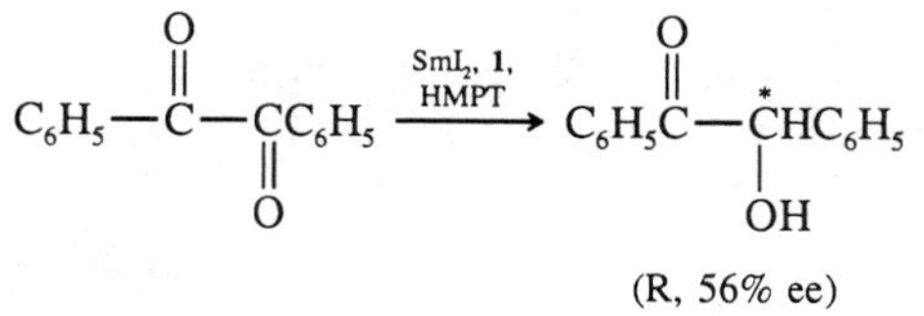

$$C_6H_5{-}C(=O){-}C(=O)C_6H_5 \xrightarrow{SmI_2,\ \mathbf{1},\ \text{HMPT}} C_6H_5C(=O){-}\overset{*}{C}H(OH)C_6H_5$$

(R, 56% ee)

[1] J. Drabowicz, S. Legedź, and M. Mikołajczyk, *Tetrahedron*, **44**, 5243 (1988).
[2] S. Takeuchi and Y. Ohgo, *Chem. Letters*, 403 (1988).

R

Raney nickel (1).

Transfer hydrogenation of nitroarenes.[1] Transfer hydrogenation of the nitroarene **2** with Raney nickel and isopropanol as the hydride donor results in the

$$C_6H_5NH-C_6H_4-NO_2 \xrightarrow[]{\mathbf{1},\ (CH_3)_2CHOH} C_6H_5NH-C_6H_4-NH_2 + (CH_3)_2C{=}O$$

2 → **3**

$$\left[C_6H_5NH-C_6H_4-N{=}C(CH_3)_2\right] \xrightarrow[95\%\ overall]{\mathbf{1},\ (CH_3)_2CHOH} C_6H_5NH-C_6H_4-NHCH(CH_3)_2$$

4 → **5**

amine **5**, formed via the unstable imine **4**. Other alcohols can be used, but increase in the bulk results in lower yields of the final product.

A similar result obtains on transfer hydrogenation of **2** with Pd–HCOOH in the presence of isobutyl methyl ketone (equation I) to provide **6**.

$$\text{(I)}\quad \mathbf{2} \xrightarrow{Pd/C,\ HCOOH} \mathbf{3} \xrightarrow{(CH_3)(i\text{-}Bu)C{=}O,\ Pd/C} \left[C_6H_5NH-C_6H_4-N{=}C(CH_3)(Bu\text{-}i)\right] \xrightarrow[100\%]{Pd/C,\ HCOOH} C_6H_5NH-C_6H_4-NH-CH(CH_3)(Bu\text{-}i)$$

6

[1] A. A. Banerjee and D. Mukesh, *J.C.S. Chem. Comm.*, 1275 (1988).

Rhodium(II) carboxylates.

Cyclopropanation of furanyl diazo ketones.[1] The reaction of α-diazo ketones of furanylpropionic acids with $Rh_2(OAc)_4$ involves formation of an unstable cyclopropane, which rearranges readily to a cyclic ketone. The intermediate cyclopropane (**2**) can be isolated in the case of the diazo benzofuran **1** and can be isomerized by a 1% H_2SO_4 solution to the ketone **3** or rearranged thermally to **4**.

$(CH_2)_2CCHN_2$ **1** —$Rh_2(OAc)_4$, C_6H_6, 25°, 71%→ **2**; **2** —H_3O^+, 88%→ **3**; **2** —180°, 82%→ **4**

A typical reaction is formulated in equation (I).

(I) $(CH_2)_2COCHN_2$ —$Rh_2(OAc)_4$, CH_2Cl_2, 25°→ [ ] —86%→ CHO

β-Lactams.[2] Decomposition of N-benzyl-N-*t*-butyldiazoacetoacetamide (**1**) with $Rh_2(OAc)_4$ in refluxing benzene results in the β-lactam **2** in 98% yield. The acetyl group on the diazo carbon is necessary for this insertion into the β-C—H bond rather than reaction of the carbene with the aromatic ring. The *t*-butyl group

$(CH_3)_3C$, N, Bzl, O, $COCH_3$, N_2 **1** —$Rh_2(OAc)_4$, C_6H_6, 98%→ CH_3C(=O), O, N, C_6H_5, $C(CH_3)_3$ **2**

is also essential for high yields. Insertion into a β-C—H bond is generally favored over insertion into a γ-C—H bond and can be further enhanced by use of a rhodium(II) benzoate as catalyst.

$(i\text{-Pr})_2N$–CO–$C(=N_2)$H ⟶ CH$_3$, CH$_3$, N–Pr-i (β-lactam) + CH$_3$, N–Pr-i (γ-lactam)

$Rh_2(OAc)_4$	95%	4.2 : 1
$Rh_2(OCOAr)_4$		6.2 : 1

α-Silyl esters or ketones.[3] Decomposition of α-diazo esters or ketones catalyzed by $Rh_2(OAc)_4$ in the presence of a trialkylsilane afford α-silyl esters or ketones in 85–95% yield.

$$N_2CHCOOC_2H_5 + (C_2H_5)_3SiH \xrightarrow[94\%]{Rh_2(OAc)_4} (C_2H_5)_3SiCH_2COOC_2H_5$$

Diazo ketone cyclization to hydroazulenones.[4] This reaction can be accomplished in high yield under high dilution and efficient stirring to minimize bimolecular reactions.

O=C, CHN$_2$ $\xrightarrow{Rh_2(OAc)_4,\ CH_2Cl_2}$ [O] $\xrightarrow[87\text{–}95\%]{Al_2O_3}$ O

[1] A. Padwa, T. J. Wisnieff, and E. J. Walsh, *J. Org.*, **54**, 299 (1989).
[2] M. P. Doyle, M. S. Shanklin, S.-M. Oon, H. Q. Pho, F. R. van der Heide, and W. R. Veal, *ibid.*, **53**, 3384 (1988).
[3] V. Bagheri, M. P. Doyle, J. Taunton, and E. E. Claxton, *ibid.*, **53**, 6158 (1988).
[4] L. T. Scott and C. A. Sumpter, *Org. Syn.*, submitted (1988).

Ruthenium(III) chloride–Tributylphosphite.

N-Methylation of anilines.[1] This system activates methanol for N-methylation of anilines. $RuCl_2[P(C_6H_5)_3]_3$ is more effective for N-methylation of N-alkylanilines.

$$C_6H_5NH_2 + CH_3OH \xrightarrow[180^\circ]{RuCl_3,\ P(OBu)_3} C_6H_5N(CH_3)_2 + C_6H_5NHCH_3$$

80% 6%

$$C_6H_5NHCH_3 + CH_3OH \longrightarrow C_6H_5N(CH_3)_2$$

$RuCl_3/P(OBu)_3$	60%
$RuCl_2[P(C_6H_5)_3]_3$	82%

[1] K.-T. Huh, Y. Tsuji, M. Kobayashi, F. Okuda, and Y. Watanabe, *Chem. Letters*, 449 (1988).

Ruthenium tetroxide.

$RCH_2NH_2 \rightarrow RCONH_2$.[1] The transformation of primary amines to amides can be effected by oxidation of N-Boc alkylamines with ruthenium tetroxide, generated *in situ* from RuO_2 (catalytic) and $NaIO_4$ (excess) in aqueous ethyl acetate. Deprotection is effected with TFA in CH_2Cl_2.

$$C_2H_5CH_2NHBoc \xrightarrow[90\%]{NaIO_4,\ RuO_4} C_2H_5\overset{O}{\overset{\|}{C}}NHBoc \xrightarrow[95\%]{TFA} C_2H_5CONH_2$$

$$BocNH(CH_2)_2CH_2NHBoc \xrightarrow[94\%]{} BocNH(CH_2)_2\overset{O}{\overset{\|}{C}}NHBoc \xrightarrow[98\%]{} H_2N(CH_2)_2\overset{O}{\overset{\|}{C}}NH_2$$

[1] K. Tanaka, S. Yoshifuji, and Y. Nitta, *Chem. Pharm. Bull.*, **36**, 3125 (1988).

S

Samarium(III) chloride.

Pinacol condensation.[1] This reaction can be effected with stoichiometric amounts of SmX_2, which is not available commercially. The same reaction can be carried out with $SmCl_3$, a stable and relatively inexpensive lanthanide, in catalytic amounts (5–10%) if conducted as an electrochemical reaction. Evidently, electrosynthesis can regenerate the divalent active species.

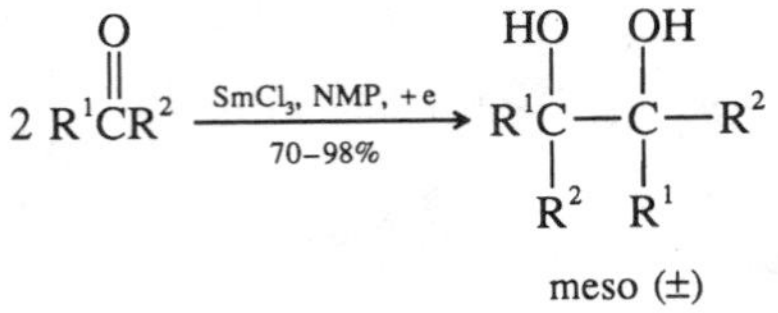

$$2\ R^1\overset{O}{\overset{\|}{C}}R^2 \xrightarrow[70-98\%]{SmCl_3,\ NMP,\ +e} R^1\underset{R^2}{\overset{HO}{C}}-\underset{R^1}{\overset{OH}{C}}-R^2$$

meso (±)

[1] E. Léonard, E. Duñach, and J. Périchon, *J.C.S. Chem. Comm.*, 276 (1989).

Samarium(II) iodide.

Cleavage of β-chlorotetrahydrofurans.[1] SmI_2 cleaves either *cis*- or *trans*-β-chloro-α-methyltetrahydrofurans (**1**) at reflux to give the (E)-unsaturated alcohol **2**. This cleavage can provide (E)-dienols (**4**) and (E)-enynols (**6**).

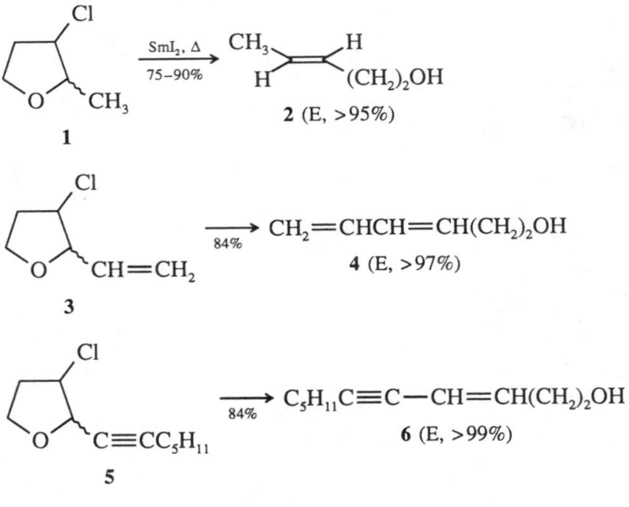

$$\mathbf{1} \xrightarrow[75-90\%]{SmI_2,\ \Delta} CH_3CH=CH(CH_2)_2OH$$

2 (E, >95%)

$$\mathbf{3}\ (\text{Cl},\ CH=CH_2) \xrightarrow[84\%]{} CH_2=CHCH=CH(CH_2)_2OH$$

4 (E, >97%)

$$\mathbf{5}\ (\text{Cl},\ C\equiv CC_5H_{11}) \xrightarrow[84\%]{} C_5H_{11}C\equiv C-CH=CH(CH_2)_2OH$$

6 (E, >99%)

Note that similar cleavage of α-alkyl-β-chlorotetrahydropyrans is best effected with sodium, which gives almost pure (E)-unsaturated alcohols from either the *cis*- or the *trans*-isomer.

Reductive cyclization.[2] Reduction of the unsaturated aldehyde **1** with SmI_2 in THF/HMPT (20:1) at 0° effects a tandem radical cyclization of the *trans*-3,5-disubstituted cyclopentene system to a linear triquinane unit (**2**) with surprisingly high *cis-anti-cis*-stereoselectivity.

SmI_2, THF/HMPT, 58%: **1** → **2**

Cyclopropanols.[3] Reaction of SmI_2 (2 equiv.) at 25° with 2-allyloxybenzoic acid chlorides (**1**) results in double cyclization to tricyclic products containing a cyclopropane ring.

SmI_2, THF, 25°, 51%: **1** ($COCl$, $OCH_2CH{=}CH_2$) → **2** (HO)

Decarbonylation of α-alkoxy acid chlorides. The reaction of an α-alkoxy acid chloride with a ketone mediated by SmI_2 results in a 1,2-glycol monoether and liberation of CO. The necessary molar ratios are 2:1:4, acid chloride/ketone/SmI_2.[4]

$$\text{cyclohexanone} + CH_3OCH_2\overset{O}{\overset{\|}{C}}Cl + SmI_2 \xrightarrow[53\%]{\text{THF, }25^\circ} \text{1-(}CH_2OCH_3\text{)cyclohexanol (OH, }CH_2OCH_3\text{)} + CO$$

α,β-Enones.[5] α-Haloketones react with aldehydes and SmI_2 (1 equiv.) to form α,β-enones (equation I). Practically no enones form if 2 equiv. of SmI_2 are used.

$$C_6H_5\overset{O}{\overset{\|}{C}}CH_2Br \xrightarrow[62-93\%]{1)\ SmI_2,\ THF;\ 2)\ RCHO} C_6H_5\overset{O}{\overset{\|}{C}}CH{=}CHR$$

E, 100%

[1] L. Crombie and L. J. Rainbow, *Tetrahedron Letters*, **29**, 6517 (1988); L. Crombie and R. D. Wyvill, *J.C.S. Perkin I*, 1983 (1985).
[2] T. L. Fevig, R. L. Elliott, and D. P. Curran, *Am. Soc.*, **110**, 5064 (1988).
[3] M. Sasaki, J. Collin, and H. B. Kagan, *Tetrahedron Letters*, **29**, 6105 (1988).
[4] *Idem, ibid.*, **29**, 4847 (1988).
[5] Y. Zhang, T. Liu, and R. Lin, *Syn. Comm.*, **18**, 2003 (1988).

Samarium(II) iodide–Tetrahydrofuran–Hexamethylphosphoric triamide.

Deoxygenation of oxides.[1] Sulfoxides, sulfones, N-oxides, and phosphine oxides can be deoxygenated by SmI_2 in THF/HMPT at 20–65°. HMPT is essential for high yields.

[1] Y. Handa, J. Inanaga, and M. Yamaguchi, *J.C.S. Chem. Comm.*, 298 (1989).

Selenocyanogen, $(SeCN)_2$. Preparation.[1]

Selenoketones.[2] The synthesis of selenoaldehydes via selenocyanates (**13**, 286) has been extended to a synthesis of selenoketones. Electron-deficient selenoketones are simply prepared by reaction of halides with KSeCN followed by base-induced elimination of HCN (equation I). However, this route fails in the case of precursors

$$\text{(I)}\quad R^1CH(X)R^2 \xrightarrow[75-95\%]{\text{KSeCN, THF, }CH_3CN} R^1CH(SeCN)R^2 \xrightarrow[55-95\%]{N(C_2H_5)_3} R^1C(=Se)R^2$$

$$\text{(II)}\quad CH_3CH_2SO_2C_6H_5 \xrightarrow[81\%]{\text{1) BuLi, }-78^\circ\text{; 2) CuCN; 3) }(SeCN)_2} CH_3CH(SeCN)SO_2C_6H_5 \xrightarrow[85\%]{N(C_2H_5)_3} CH_3C(=Se)SO_2C_6H_5$$

containing electron-accepting groups. In this case, the selenocyanate is obtained by reaction of selenocyanogen with cyanocuprates, as shown for the preparation of phenylsulfonyl methyl selenoketone (equation II).

These ketones undergo Diels–Alder reactions with dienes to give selenopyrans in good to excellent yields.

[1] L. Birckenbach and K. Kellermann, *Ber.*, **58**, 786 (1925).
[2] P. T. Meinke and G. A. Krafft, *Am. Soc.*, **110**, 8679 (1988).

Semicarbazide.

Cleavage of disulfides.[1] Diaryl disulfides are cleaved to thiols by semicarbazide in CH_2Cl_2-CH_3OH at room temperature within 2 hours. Cleavage of diheteroaryl disulfides proceeds more readily than that of diaryl disulfides.

$$\text{bis(5-methyl-1,3,4-thiadiazol-2-yl) disulfide} \xrightarrow[\text{88\%}]{H_2NCONHNH_2\,(\mathbf{1}),\ 25^\circ} \text{HS-(5-methyl-1,3,4-thiadiazol-2-yl)}$$

$$Cl-C_6H_4-S-S-C_6H_4-Cl \xrightarrow[\text{87\%}]{\mathbf{1},\ THF-CH_3OH,\ \Delta} 2Cl-C_6H_4-SH$$

[1] S. N. Maiti, M. P. Singh, P. Spevak, R. G. Micetich, and A. V. N. Reddy, *J. Chem. Res.* (*S*), 256 (1988).

Silica

Hydroxyisoxazolidines.[1] The reaction of phenylhydroxamic acid with crotonaldehyde in methanol results in a mixture of 3- and 5-hydroxyisoxazolidines (**1** and **2**) in which the former isomer predominates. The rate of this reaction is

$$CH_3CH{=}CHCHO + C_6H_5CONHOH \longrightarrow \mathbf{1} + \mathbf{2}$$

(**1**: 3-OH, 5-H_3C, N-COC_6H_5 isoxazolidine; **2**: 5-HO, 3-CH_3, N-COC_6H_5 isoxazolidine)

Conditions	1 : 2
CH_3OH, 5–7 hr. 60%	95–98 : 2–5
SiO_2, 30 min. 80–90%	100 : 0
DEAE-cellulose, 1 hr.	60 : 40
" 6 hr. 95–98%	4 : 96
Cellulose, 43%	75 : 25

markedly increased when carried out on an absorbent without a solvent. A reaction conducted on SiO_2 is complete within 30–40 minutes and results in formation of **1** only. Reactions conducted on diethylaminoethyl-cellulose (DEAE-cellulose) result in a mixture of **1** and **2** in the early stages and eventually of **2** almost exclusively. Apparently this adsorbent can convert **1** into the more stable product **2**. Cellulose itself has slight effect on the rate and provides a mixture of **1** and **2** in only modest yield and in the ratio 75:25.

[1] I. A. Motorina, L. A. Sviridova, G. A. Golebeva, and Y. G. Bundel, *Tetrahedron Letters*, **30**, 117 (1989).

Silicon(IV) fluoride, SiF_4. The gas is available from Matheson Co.

Fluorohydrins.[1] These have usually been obtained by reaction of epoxides with hydrogen fluoride at high temperatures. This reaction can be carried at 0° with SiF_4 in combination with a base such as diisopropylethylamine or even Bu_4NF. In all cases, the fluorine atom is introduced at the more substituted carbon. Highest

SiF_4, i-$Pr_2NC_2H_5$, $(CH_2Cl)_2$, 0–20°; 70%

SiF_4, H_2O, i-$Pr_2NC_2H_5$, ether, 0°; 80%

1 → **2**

68% $KN[Si(CH_3)_3]_2$

3

yields are obtained with 1,1-disubstituted oxiranes such as **1**. Reaction of the β-fluoro-β-silyl alcohol **2** with potassium hexamethyldisilazide affords a fluoroalkene. Desilylation of **2** with Bu_4NF affords the corresponding β-fluoro alcohol with retention of configuration of the C—F bond in about 75% yield.

2-Oxazolines.[2] Reaction of 1,1-disubstituted epoxides with nitriles and SiF_4 (1 equiv.) provides fluorohydrins, which on further exposure to SiF_4 are converted into 2-oxazolines, in which the nitrogen is attached to the substituted carbon. In the example cited, the yield is increased to 81% from reactions conducted at 80°.

SiF_4, 0° → $[(C_5H_{11})_2CF—CH_2OH]$ → CH_3CN, 65%

[1] M. Shimizu and H. Yoshioka, *Tetrahedron Letters*, **29**, 4101 (1988); *idem, ibid.*, **30**, 967 (1989).

[2] *Idem, Heterocycles*, **27**, 2527 (1988).

Silver fluoride/Calcium fluoride. AgF on CaF_2 can be prepared by evaporation of a mixture of Ag_2CO_3, HF, and CaF_2 in water to a dry granular powder.

Fluorination.[1] Alkyl bromides are converted to alkyl fluorides by reaction with this reagent in CH_3CN (35–92% yield). Bromofluorination of alkenes can be effected with AgF/CaF_2 and 1,3-dibromo-5,5-dimethylhydantoin (DBH). Iodo-

$$C_4H_9CH{=}CH_2 \xrightarrow[\text{76\% (isolated)}]{\text{DBH, AgF/CaF}_2,\ \text{CH}_2\text{Cl}_2} C_4H_9CH(F)CH_2Br$$

fluorination of alkenes is possible with AgF/CaF_2 in combination with I_2 or NIS (~60% yield).

[1] T. Ando, D. G. Cork, M. Fujita, T. Kimura, and T. Tatsuno, *Chem. Letters*, 1877 (1988).

Simmons–Smith reagent.

Diastereoselective cyclopropanation.[1] The optically active derivative **1** of cyclohexanone obtained with (R,R)-2,4-pentanediol (**12**, 375–378) undergoes diastereoselective cyclopropanation with the reagent obtained from Zn/Cu and CH_2I_2 or $(C_2H_5)_2Zn$ and CH_2I_2, but the highest optical yield (95% de) is obtained with the latter reagent in DME. The chiral auxiliary can be removed to provide the bicyclic alcohol **3**. The cyclopropane ring of **2** is cleaved on oxymercuration (98% yield).

$(C_2H_5)_2Zn$, CH_2I_2, DME, 20° (56%): **1** → **2** (95% de)

2 → 1) PCC (67%) 2) K_2CO_3 → **3**

[1] T. Sugimura, T. Futagawa, and A. Tai, *Tetrahedron Letters*, **29**, 5775 (1988).

Sodium–Ethanol.

Alkyl fluorides.[1] α-Fluoro sulfides, available by reaction of mercury(II) fluoride with thioacetals, are convertible into alkyl fluorides by reductive desulfurization with sodium in ethanol.

$$C_6H_5(CH_3)CHCHO \xrightarrow[2)\ HgF_2,\ CH_3CN]{1)\ C_6H_5SH,\ H^+} [C_6H_5(CH_3)CHCHF(SC_6H_5)] \xrightarrow[48\%]{Na,\ C_2H_5OH} C_6H_5(CH_3)CHCH_2F$$

[1] S. T. Purrington and J. H. Pittman, *Tetrahedron Letters*, **29**, 6851 (1988).

Sodium(potassium) amide.

Haller–Bauer[1] cleavage of α-silyl ketones. Paquette's group[2] has shown that reaction of $Na(K)NH_2$ with optically active α-silyl phenyl ketones such as **1** in refluxing benzene provides the corresponding silanes (**2**) with 88–92% retention of configuration. The same stereochemical control is observed in the cleavage of chiral α-silyl benzoylcycloalkanes such as **3**.

$$C_6H_5C(=O)C(Si(CH_3)_3)(CH_3)CH_2CH_2C_6H_5 \xrightarrow[C_6H_6,\ \Delta]{NaNH_2} HC(Si(CH_3)_3)(CH_3)CH_2CH_2C_6H_5$$

1 (89% ee) → **2** (81% ee)

3 → **4** (98:2)

[1] Review: E. M. Kaiser and C. D. Warner, *Synthesis*, 395 (1975).
[2] J. P. Gilday, J. C. Gallucci, and L. A. Paquette, *J. Org.*, **54**, 1399 (1989); L. A. Paquette, G. D. Maynard, C. S. Ra, and M. Hoppe, *ibid.*, **54**, 1408 (1989).

Sodium anthracenide (**1**). The reagent is obtained by sonication of sodium and anthracene in DME.

Sulfoximines, $R^1S(O)(=NH)R^2$.[1] Review.[2] The most attractive route to these imines is reaction of a sulfoxide with tosyl azide in the presence of CuX_2 or with O-(mesitylsulfonyl)hydroxylamine (MSH) to give an N-tosylsulfoximine intermediate. Cleavage of the tosyl group is then effected with a sodium arenide, in particular with sodium anthracenide.

$$CH_3\overset{O}{\overset{\|}{S}}CH_3 \xrightarrow{TsN_3} CH_3\overset{O}{\overset{\|}{S}}CH_3(=NTs) \xrightarrow[93\%]{1,\ DME} CH_3\overset{O}{\overset{\|}{S}}CH_3(=NH)$$

[1] C. R. Johnson and O. Lavergne, *J. Org.*, **54**, 986 (1989).
[2] C. R. Johnson, *Aldrichim. Acta*, **18**, 3 (1985).

Sodium azide-Dimethylformamide.

Selective cleavage of $ROSi(CH_3)(C_6H_5)_2$.[1] These ethers are cleaved to the alcohol by NaN_3 in DMF at 40° without effect on *t*-butyldimethyl- or *t*-butyldiphenylsilyl ethers.

[1] S. J. Monger, D. M. Parry, and S. M. Roberts, *J.C.S. Chem. Comm.*, 381 (1989).

Sodium benzeneselenolate, $NaSeC_6H_5$ (**1**).

Cleavage of oxetanes.[1] Oxetanes are cleaved by this salt to the corresponding γ-phenylselenenylpropanol at room temperature, whereas tetrahydrofurans (oxolanes) are cleaved only at 60–100°. In each case the less-substituted carbon is attacked by the reagent. Cleavage of oxiranes with this reagent is not regioselective.

$$\text{2-methyloxetane} \xrightarrow[60\%]{NaSeC_6H_5,\ 25°} HOCH(CH_3)(CH_2)_2SeC_6H_5$$

$$\text{2-(hydroxymethyl)tetrahydrofuran} \xrightarrow[84\%]{100°} HOCH(CH_2OH)(CH_2)_3SeC_6H_5$$

$$\text{(cyclic ether nucleoside, base)} \xrightarrow[86\%]{} \text{(5'-}C_6H_5Se\text{, 3'-OH nucleoside, base)}$$

[1] K. Haraguchi, H. Tanaka, H. Hayakawa, and T. Miyasaka, *Chem. Letters*, 931 (1988).

Sodium benzyloxide, $C_6H_5CH_2ONa$ (**1**). This alkoxide can be generated *in situ* from benzyl alcohol and NaH in THF at 25°.

Conjugate addition to nitroalkenes;* anti-*β-amino alcohols.[1] Sodium benzyloxide undergoes *anti*-selective conjugate addition to 2-nitro-2-butenes to give 2-

benzyloxy-3-nitrobutanes (**2**). These products are converted into β-amino alcohols (**3**) by catalytic hydrogenation in acidic ethanol.

1) **1**, THF
2) HOAc, −78°
50–70%

2 (85–94:15–6)

50–70% H_2, Pd/C, C_2H_5OH, HCl

3 (*anti/syn* = 80–90:20–10)

[1] A. Kamimura and N. Ono, *Tetrahedron Letters*, **30**, 731 (1989).

Sodium bis(2-methoxyethoxy)aluminum hydride (SMEAH) (**1**).

Reduction of 2,3-epoxycinnamyl alcohol (**2**).[1] This hydride can effect selective reduction of 2,3-epoxy alcohols to 1,3-diols, probably by initial attack of the free hydroxyl group followed by intramolecular delivery of the hydride. The regioselectivity is decreased by use of excess reductant or increased by the presence of a free alcohol (methanol).

1, DME

2 → **3** (22:1)

excess **1**, 4–9:1

[1] Y. Gao and K. B. Sharpless, *J. Org.*, **53**, 4081 (1988).

Sodium borohydride.

Chemoselective reductions. The reactivity of $NaBH_4$ can be decreased by use of a lower temperature or by a mixed solvent such as methanol or ethanolic methylene chloride. This simple strategy can be used to effect selective reduction of ketones in the presence of enones,[1] and of aldehydes in the presence of ketones.[2]

[1] D. E. Ward, C. K. Rhee, and W. M. Zoghaib, *Tetrahedron Letters*, **29**, 517 (1988).
[2] D. E. Ward and C. K. Rhee, *Syn. Comm.*, **18**, 1927 (1988).

Sodium borohydride–Cerium(III) chloride.

Reduction of diphenylphosphinoyl ketones and enones.[1] 1,2-Reduction of these substrates with $NaBH_4$ alone normally reduces these substrates to *threo* alcohols, whereas Luche conditions ($NaBH_4$ and $CeCl_3$) can result in *erythro* alcohols.

$(C_6H_5)_2P(O)$ Pr-*i* / C_6H_5 → *erythro* + *threo*

		erythro : threo
$NaBH_4$		~5 : 95
$NaBH_4 + CeCl_3$	90%	95 : 5

$(C_6H_5)_2P(O)$ CH_3 / Pr → (71%) + *threo*-isomer

$NaBH_4 + CeCl_3$ 50 : 50

Selective reduction of ketoses. In the presence of a large excess of $CeCl_3$, sodium borohydride reduces D-fructose (**1**) in the presence of D-glucose selectively to a mixture of D-mannitol and D-glucitol (**2**).[2]

(**1**) → $NaBH_4$, $CeCl_3$ → **2**

[1] J. Elliott, D. Hall, and S. Warren, *Tetrahedron Letters*, **30**, 601 (1989).
[2] R. Roy, S. Gervais, A. Gamiean, and J. Boratynski, *Carbohydrate Res.*, **177**, C5 (1988).

Sodium borohydride–Trifluoromethanesulfonic acid, $NaBH_4$–HOTf (**1**).

Ring expansion. Alicyclic groups attached to a hydroxymethyl (-CH_2OH) or carboxyl group undergo ring homologation on exposure to $NaOSO_2CF_3$ in an ethereal solvent at −78° to 25° via an intermediate carbenium ion.[1]

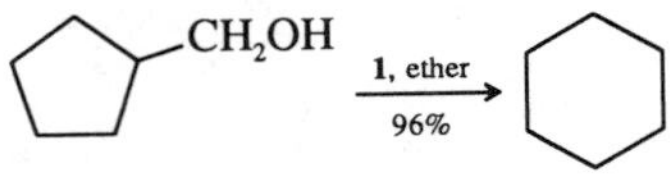

CH_2OH **1**, ether 96%

The same reagent converts unsaturated polycyclic compounds to diamondoid cage hydrocarbons.[2]

[1] G. A. Olah, A. Wu, and O. Farooq, *J. Org.*, **54**, 1452 (1989).
[2] G. A. Olah, A. Wu, and O. Farooq, and G. K. S. Prakash, *ibid.*, **54**, 1450 (1989).

Sodium borohydride–Zirconium(IV) chloride.

Reduction of C=O, C=N, and C≡N.[1] The combination of $NaBH_4$ and $ZrCl_4$ (4:1) reduces ketones, oximes, and nitriles in THF at 25° in 85–96% yields. Reduction of C=N and C≡N is slower than that of C=O. Reduction of oximes affords amines rather than hydroxylamines (obtained with B_2H_6 as reductant).

[1] S. Itsuno, Y. Sakurai, and K. Ito, *Synthesis*, 995 (1988).

Sodium bromite, $NaBrO_2$, **12**, 445.

Oxidation of ROH.[1] This oxidant can be used to oxidize alcohols in CH_2Cl_2 when it is supported on alumina; yields are generally >90%. It is also useful for oxidation of sulfides to sulfoxides.

[1] T. Morimoto, M. Hirano, Y. Aikawa, and X. Zhang, *J.C.S. Perkin I*, 2423 (1988).

Sodium dichloroisocyanurate, 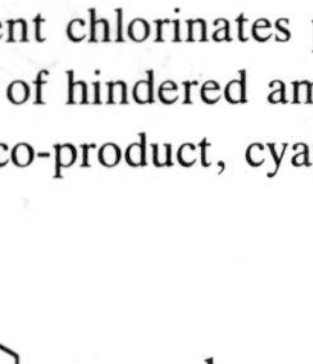 (**1**).

O, ClN, N^-Na^+, O, N, O, Cl

Supplier: Aldrich.

Chloramines.[1] This reagent chlorinates primary and secondary amines, but is most useful for chlorination of hindered amines. The reaction is carried out in water or benzene–water. The co-product, cyanuric acid, is removed by filtration.

O, CH_3, CH_3, N, H, CH_3, CH_3 → **1**, 69% → O, CH_3, CH_3, N, Cl, CH_3, CH_3

[1] J. Zakrzewski, *Syn. Comm.*, **18**, 2135 (1988).

Sodium O,O-diethyl phosphorotelluroate, 8, 455; **10**, 362.

Debromination.[1] The reagent converts *vic*-dibromoalkanes into alkenes in 74–89% yield.

$$\text{Br(H)(C}_6\text{H}_5\text{)C–C(C}_6\text{H}_5\text{)(H)Br} \xrightarrow[94\%]{(C_2H_5O)_2P(=O)TeNa} \text{C}_6\text{H}_5\text{(H)C=C(H)C}_6\text{H}_5 + Te$$

The debromination can also be effected with sodium O,O-diethyl phosphite, $(C_2H_5O)_2P(=O)Na$, and a catalytic amount of tellurium.

[1] X. Huang and Y. Q. Hou, *Syn. Comm.*, **18**, 2201 (1988).

Sodium hydrogen selenide.

Dealkylation of esters.[1] NaHSe, generated *in situ* from Se and $NaBH_4$ in DMF, is effective for cleavage of esters. C_6H_5SeNa also effects this reaction (**11**, 475), but this reagent is more difficult to prepare and handle.

[1] F. Kong, J. Chen, and X. Zhou, *Syn. Comm.*, **18**, 801 (1988).

Sodium hypochlorite, NaOCl.

Oxidative decarboxylation.[1] Sodium hypochlorite is known to effect oxidative decarboxylation of α-hydroxy carboxylic acids to form ketones and CO_2, but a hydroxyl group is not essential since trisubstituted acetic acids are also subject to this oxidation. Thus triphenylacetic acid is oxidized by NaOCl to triphenylmethanol and benzophenone. In the presence of a phase-transfer catalyst, the rate is enhanced

$$(C_6H_5)_3CCOOH \xrightarrow{NaOCl} \underset{(13-20\%)}{(C_6H_5)_3COH} + \underset{(24-42\%)}{(C_6H_5)_2C{=}O} + CO_2$$

and the ketone becomes the major product (80–84% yield). A similar, catalyzed oxidation of 2,2-diphenylpropanoic acid provides a mixture of benzophenone and acetophenone in the approximate ratio of 1:2:5.

[1] P. R. Elmore, R. T. Reed, T. Terkle-Huslig, J. S. Welch, S. M. Young, and R. G. Landolt, *J. Org.*, **54**, 970 (1989).

Sodium molybdate(VI)–Hydrogen peroxide, Na_2MoO_4–H_2O_2.

Singlet oxygen.[1] This combination results in sodium peroxomolybdenate, Na_2MoO_6. This reagent can be generated on a macroreticular resin such as Amberlyst and used as a chemical source of singlet oxygen, ^1O.

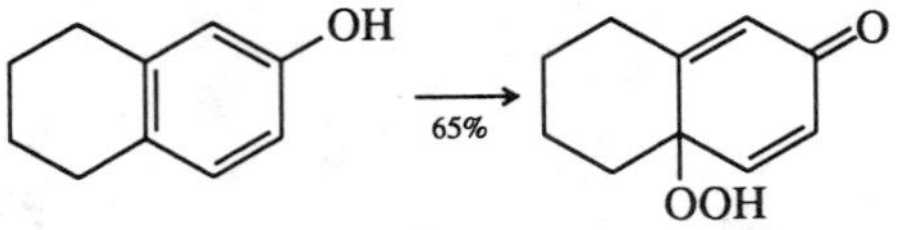

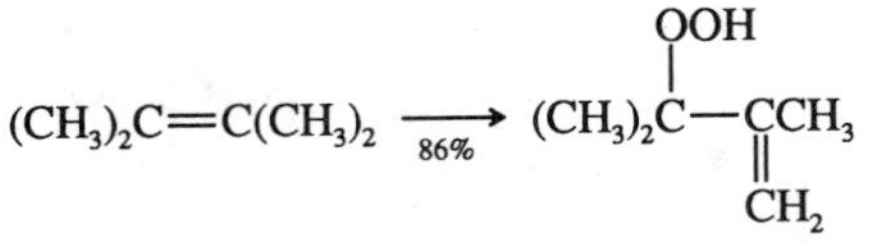

$$(CH_3)_2C{=}C(CH_3)_2 \xrightarrow{86\%} (CH_3)_2C(OOH){-}C(CH_3){=}CH_2$$

[1] E. C. McGoran and M. Wyborney, *Tetrahedron Letters*, **30**, 783 (1989).

Sodium periodate.

Glycol cleavage; quinones.[1] Glycol cleavage and oxidation of hydroquinones to quinones can be conveniently carried out in CH_2Cl_2 with sodium periodate supported on wet silica gel. Water is essential and the amount is the most critical factor. The procedure overcomes the problem of insolubility of $NaIO_4$ in organic solvents and shortens the reaction times to generally less than 15 minutes.

[1] M. Daumas, Y. Vo-Quang, L. Vo-Quang, and F. LeGaffe, *Synthesis*, 64 (1989).

Sodium phenylseleno(triethoxy)borate, $Na^+[C_6H_5SeB(OC_2H_5)_3]^-$ (**1**), **14**, 284.

Reduction of α,β-epoxy esters.[1] The reagent reduces these substrates to β-hydroxy esters in 80–90% yield.

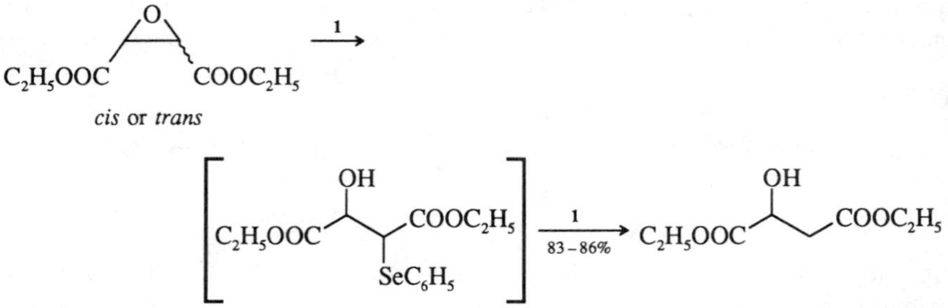

cis or *trans*

$$[C_2H_5OOC{-}CH(OH){-}CH(SeC_6H_5){-}COOC_2H_5] \xrightarrow[83-86\%]{1} C_2H_5OOC{-}CH(OH){-}CH_2{-}COOC_2H_5$$

[1] M. Miyashita, M. Hoshino, T. Suzuki, and A. Yoshikoshi, *Chem. Letters*, 507 (1988).

Sodium tetraborate (borax), $Na_2B_4O_7$; **4**, 461.

Hydration of cyanohydrins.[1] The cyanohydrins derived from aldehydes undergo hydration when treated with borax in water at 80° to give α-hydroxy amides in 65–85% yield. In some cases addition of a trace of KCN can increase the yield.

$$\underset{\displaystyle OH}{RCHCN} \xrightarrow[65-85\%]{Na_2B_4O_7,\ 80^\circ} \underset{\displaystyle OH}{RCHCONH_2}$$

[1] J. Jammot, R. Pascal, and A. Commeyras, *Tetrahedron Letters*, **30**, 563 (1989).

Sodium tetracarbonylhydridoferrate, $Na^+[HFe(CO)_4]^-$ (**1**).

$ArNO_2 \rightarrow ArNH_2$.[1] $Na^+[HFe(CO)_4]^-$ is a known reducing agent in a basic medium. However, the bis(triphenylphosphine)imine (PPN^+) salt of $HFe(CO)_4^-$[1] in combination with a strong Brønsted acid (TFA) is also a reducing agent, particularly for nitroarenes, even in the presence of aldehydes or acid halides.

$$C_6H_5NO_2 + PPN^+[HFe(CO)_4]^- \xrightarrow[98\%]{TFA,\ 25^\circ} C_6H_5NH_2$$

[1] P. L. Gaus, S. W. Gerritz, and P. M. Jeffries, *Tetrahedron Letters*, **29**, 5083 (1988).
[2] M. Y. Darensbourg, D. J. Darensbourg, and H. L. C. Barros, *Inorg. Chem.*, **17** (297) (1978).

Sodium tungstate, Na_2WO_4.

Oxidation of sec-amines to nitrones. In the presence of this tungstate, hydrogen peroxide oxidizes secondary amines to nitrones.[1]

$$CH_3(CH_2)_3\underset{\displaystyle H}{N}(CH_2)_3CH_3 \xrightarrow[89\%]{H_2O_2,\ CH_3OH,\ Na_2WO_4} CH_3(CH_2)_3\underset{\displaystyle O^-}{\overset{+}{N}}{=}CH(CH_2)_2CH_3$$

$$\text{2-methylpiperidine (N–H)} \xrightarrow[75-77\%]{} \text{2-methyl-3,4,5,6-tetrahydropyridine N-oxide } (\overset{+}{N}{=}C,\ N{-}O^-)$$

[1] H. Mitsui, S. Zenki, T. Shiota, and S.-I. Murahashi, *J.C.S. Chem. Comm.*, 874 (1984); *Org. Syn.*, submitted (1988).

Sorbic acid (2,4-Hexadienoic acid), $CH_3CH{=}CHCH{=}CHCOOH$ (**1**).

Pentadienyl hydroxy acids; 7-hydroxy-2,4-dienoic acids.[1] The dianion (**2**) of **1** reacts with carbonyl compounds to afford either pentadienyl hydroxy acids (**3**) or 7-hydroxy-2,4-dienoic acids (**4**), depending on the reaction conditions. Reaction

1 —$LiN(C_2H_5)_2$→ [$H_2C{=}CH{-}CH{=}CH{-}CH{=}C(OLi)(OLi)$] (**2**) —$C_2H_5CHO$, −70°; 75%→ $C_2H_5CH(OH){-}CH(COOH){-}CH{=}CH{-}CH{=}CH_2$ (**3**)

2 —$(C_6H_5)_2C{=}O$, 30°; 68%→ $(C_6H_5)_2C(OH){-}CH_2{-}CH{=}CH{-}CH{=}CH{-}COOH$ (**4**)

2 + $CH_3COCH{=}CHC_6H_5$ —−70°→ [$CH_3COCH_2CH(C_6H_5){-}CH(CH{=}CH_2){-}CH{=}CH{-}COOH$] (**5**) —20°→ $CH_3COCH_2CH(C_6H_5){-}CH(COOH){-}CH{=}CH{-}CH{=}CH_2$ (**6**)

of **2** with a substituted enone affords initially a 1,4-γ-adduct, which gradually rearranges to a 1,4-α-adduct (**6**).

[1] P. Ballester, A. Costa, A. Garcia-Raso, and R. Mestres, *J.C.S. Perkin I*, 2797 (1988).

Squaric acid.

1,4-Benzoquinones (**13**, 284). Full details of the use of a dialkyl squarate for a regioselective route to highly substituted benzoquinones are now available. The report includes a note of caution: These esters show allerginicity comparable to that of the active principal of poison ivy.[1]

[1] L. D. Foland, J. O. Karlsson, S. T. Perri, R. Schwabe, S. L. Xu, S. Patil, and H. W. Moore, *Am. Soc.*, **111**, 975 (1989).

N-Sulfinyl-*p*-toluenesulfonamide, $TsN{=}\overset{+}{S}\text{-}\overset{-}{O}$ (**1**).

Intramolecular cyclization of N-tosylimines.[1] The reagent converts unsaturated aldehydes into N-tosyliminium complexes, which undergo intramolecular cyclization to homoallylic amines.

TsNSO, $BF_3 \cdot O(C_2H_5)_2$; 52% ; 3 : 1

[1] M. J. Melnick, A. J. Freyer, and S. M. Weinreb, *Tetrahedron Letters*, **29**, 3891 (1988).

Sulfur.

O-Carbonylation of alcohols.[1] Reaction of an alcohol with CO and S in the presence of a tertiary amine (DBU) in THF at 80° results in a product (**1**) that on esterification with $C_6H_5CH_2Br$ affords S-benzyl carbonothioates (**2**). The reaction

$$ROH + CO + S \xrightarrow{\text{DBU, 80\%}} \underset{\mathbf{1}}{ROC(=O)S^-(DBU \cdot H)^+} \xrightarrow[72-85\%]{C_6H_5CH_2Br} \underset{\mathbf{2}}{ROC(=O)SCH_2C_6H_5}$$

can hardly involve carbonyl sulfide, which can be made from CO and S but only at high temperatures (>250°). The yield of **2** is highly dependent on the amine, being about 5% with Bu_3N and zero with pyridine.

[1] T. Mizuno, I. Nishiguchi, T. Hirashima, A. Ogawa, N. Kambe, and N. Sonoda, *Tetrahedron Letters*, **29**, 4767 (1988).

T

Tellurium.

Tellurobenzaldehyde. This aldehyde (**a**) is formed *in situ* by reaction of Te with $(C_6H_5)_3P{=}CHC_6H_5$ and trapped with 2,3-dimethylbutadiene to form the Diels–Alder adduct **1**. In the absence of a diene, **a** reacts with the Wittig reagent to form stilbene (61% yield).

$$(C_6H_5)_3P{=}CHC_6H_5 + Te \xrightarrow[105^\circ]{C_6H_5CH_3} \underset{\textbf{a}}{[Te{=}CHC_6H_5]} \xrightarrow[11\%]{CH_2{=}C(CH_3){-}C(CH_3){=}CH_2} \textbf{1}$$

(**1**: 3,4-dimethyl-6-phenyl-3,6-dihydro-2H-tellurin ring drawn with CH_3, CH_3, Te, C_6H_5)

[1] G. Erker and R. Hock, *Angew. Chem. Int. Ed.*, **28**, 179 (1989).

Telluronium ylides, $Bu_2Te{=}CHY$ ($Y{=}COOCH_3$, COC_6H_5, or CN) (1).

Preparation (cf., **12**, 159–160). Cyanomethyldibutyltelluronium chloride ($Y{=}CN$) and phenacyldibutyltelluronium bromide ($Y{=}COC_6H_5$) are prepared by reaction of Bu_2Te with $ClCH_2CN$ and $C_6H_5COCH_2Br$, respectively, followed by reaction of the salts with $KOC(CH_3)_3$.

Wittig reactions.[1] These ylides undergo Wittig reaction with aldehydes to give α,β-unsaturated esters, ketones, or nitriles. In the case of the latter two reactions, the E/Z ratio is 12–124:1. It is more convenient to carry out this reaction directly with Bu_2Te and the aromatic aldehyde plus XCH_2COOCH_3, $XCH_2COC_6H_5$, or XCH_2CN. This one-pot reaction gives higher yields (usually about 90%) than the two-step reaction through the telluronium salts.

$$RC_6H_4CHO + Bu_2Te \text{ and } XCH_2Y \xrightarrow[\sim 90\%]{THF,\ \Delta} RC_6H_4CH{=}CHY$$

[1] X. Huang, L. Xie, and H. Wu, *J. Org.*, **53**, 4862 (1988).

Tetrabutylammonium fluoride–Hexamethylphosphoric triamide, Bu_4NF–HMPT.

Cleavage of SEM ethers.[1] (Trimethylsilyl)ethoxymethyl (SEM) ethers are cleaved by Bu_4NF in combination with HMPT and 4-Å molecular sieves in 1 hour at 100° in 90–98% yield. The reaction requires 15 hours in the absence of the molecular sieve and yields are lower. SEM ethers can be cleaved selectively in the presence of a MOM ether by this method.

[1] T. Kan, M. Hashimoto, M. Yanagiya, and H. Shirahama, *Tetrahedron Letters*, **29**, 5417 (1988).

Tetrabutylammonium (tricarbonylnitrosyl)ferrate, $Bu_4N[Fe(CO_3)NO]$ (**1**), yellow crystals, m.p. 56°. The salt is prepared by reaction of $Fe(CO)_5$ in benzene with $NaNO_2$ and with Bu_4NBr in H_2O at 25°.

Alkylation of allylic carbonates.[1] In an atmosphere of CO, this salt catalyzes the alkylation of allylic carbonates with the sodium anion of diethyl malonate. The reaction occurs selectively at the carbon substituted by the leaving group with retention of configuration of the double bond. The reaction involves net retention of configuration, possibly by two consecutive S_N2 reactions.

$$CH_2{=}CHC(CH_3)_2{-}OCO_2C_2H_5 \xrightarrow[78\%]{\mathbf{1},\ CO,\ THF,\ NaCH(COOC_2H_5)_2} CH_2{=}CHC(CH_3)_2{-}CH(COOC_2H_5)_2 + (CH_3)_2C{=}CHCH_2(COOC_2H_5)_2 \quad (92:8)$$

COOCH$_3$ / OCO$_2$CH$_3$ cyclohexene $\xrightarrow{75\%}$ COOCH$_3$ / CH(COOC$_2$H$_5$)$_2$ cyclohexene

$$C_6H_5CH{=}CHCH(CH_3)OCO_2CH_3\ (73\%\ ee) \xrightarrow[78\%]{\mathbf{1},\ NaCH(COOCH_3)_2} C_6H_5CH{=}CHCH(CH_3)CH(COOCH_3)_2\ (73\%\ ee) + C_6H_5CH(CH(COOCH_3)_2)CH{=}CHCH_3\ (73\%\ ee) \quad (93:7)$$

[1] Y. Xu and B. Zhou, *J. Org.*, **52**, 974 (1987); *ibid.*, **53**, 4419 (1988).

Tetrachlorophthalic anhydride.

$R^1CONHR^2 \rightarrow R^1COOH$.[1] This conversion can be effected in 70–90% yield by tetrachloro- or tetrafluorophthalic anhydride in the absence of water or of a solvent in a melt at 135–170°. This reaction does not affect esters. Conversion of nitriles to carboxylic acids can be effected by a similar reaction with tetrafluoro- or tetrachlorophthalic acid at 135–180° for 4–6 days in 60–85% yield.

[1] J. T. Eaton, W. D. Rounds, J. H. Urbanowicz, and G. W. Gribble, *Tetrahedron Letters*, **29**, 6553, 6557 (1988).

2,6,9,10-Tetracyanoanthracene (TCA).

Photosensitized Diels–Alder reactions.[1] Diels–Alder reactions between electron-rich dienes and electron-rich dienophiles are generally unproductive. But in the presence of TCA or dicyanonaphthalene this cycloaddition can occur in reasonable yield. Thus 1,3-cyclohexadiene undergoes [4 + 2]cycloaddition with β-methylstyrene to form *endo*-6-methyl-5-phenylbicyclo[2.2.2]oct-2-ene with retention of the styrene configuration.

$C_6H_5CH{=}CHCH_3$ + 1,3-cyclohexadiene $\xrightarrow[h\nu\ >\ 400\ nm.]{TCA}$ *trans* C_6H_5 (CH_3) + *cis* C_6H_5 (CH_3) + [2 + 2]

	trans	*cis*	[2 + 2]
trans	81%		5%
cis	1%	61%	10%

[1] D. Hartsough and G. B. Schuster, *J. Org.*, **54**, 3 (1989).

Tetrakis(triphenylphosphine)palladium(0).

Intramolecular arylation. Aromatic substrates such as **1** on deprotonation (NaH) furnish stabilized enolates that undergo intramolecular arylation when heated

1 (Z,Z' = CN, COOR, COR), n = 1,2 $\xrightarrow[40-65\%]{NaH,\ Pd(0),\ DMF,\ 140°}$ **2**

at 140° in the presence of a Pd(0) catalyst to give benzo-fused, five- or six-membered rings. Yields are poor when Z (or Z') is H or CH_3.[1]

Example:

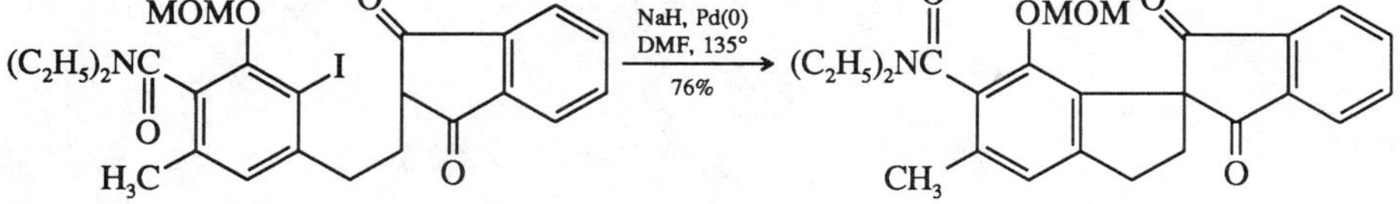

MOMO
$(C_2H_5)_2NC$
O
I
O
H_3C
O
NaH, Pd(0)
DMF, 135°
76%
O
OMOM
O
$(C_2H_5)_2NC$
CH_3
O

Cyclizative Heck coupling. This reaction is highly regioselective in the case of α,β-unsaturated carbonyl compounds such as **1** and **3**.[2]

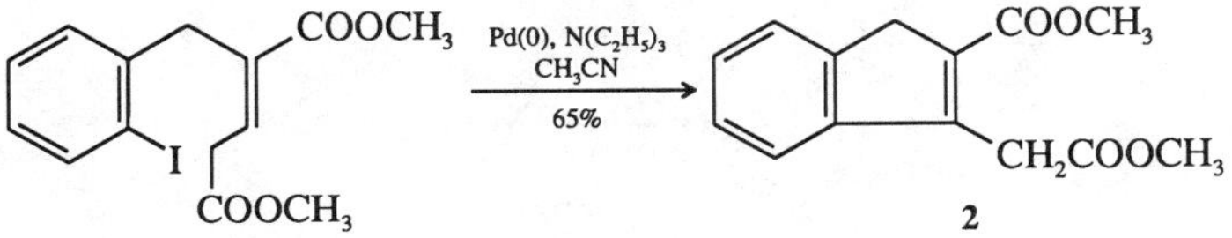

$COOCH_3$
I
$COOCH_3$
1
Pd(0), $N(C_2H_5)_3$
CH_3CN
65%
$COOCH_3$
CH_2COOCH_3
2

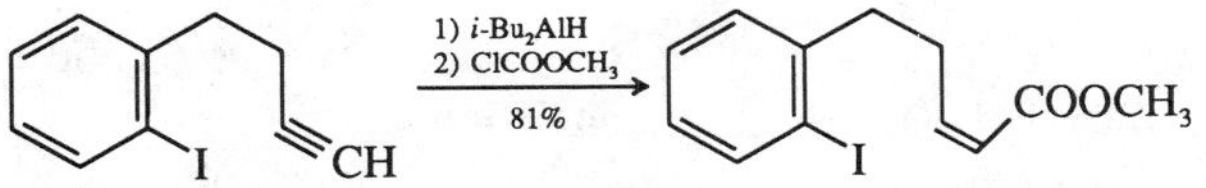

I
CH
1) i-Bu_2AlH
2) $ClCOOCH_3$
81%
I
$COOCH_3$
3
71% Pd(0)
$CHCOOCH_3$
4 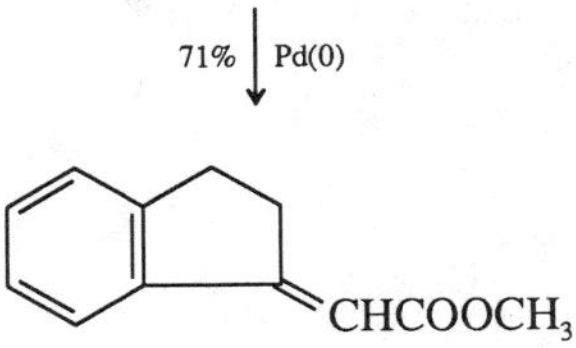

Various alkenyl ethers undergo Heck coupling in a basic medium in the presence of a Pd(0) catalyst to form cyclic ethers such as benzofurans and tetrahydrobenzopyrans.[3]

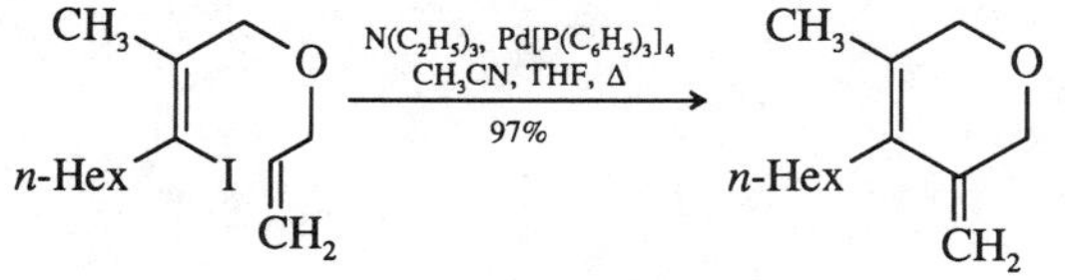

CH_3
O
n-Hex
I
CH_2
$N(C_2H_5)_3$, $Pd[P(C_6H_5)_3]_4$
CH_3CN, THF, Δ
97%
CH_3
O
n-Hex
CH_2

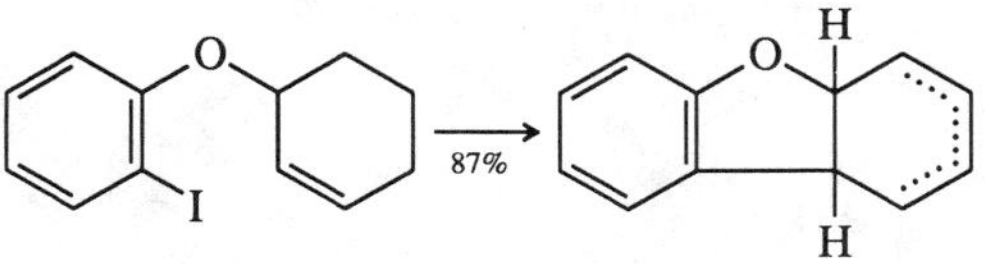

O
I
87%
O
H
H

[3 + 2]Annelation. The Pd(0) complex in combination with $N(C_2H_5)_3$ (1.5–2 equiv.) effects cyclic carbopalladation of substrates such as **1**, a cyclohexene substituted by a γ-iodoallyl electrophile group and activated by a carbonyl group,

COOCH₃ I Bu **1** —Pd(0), NR_3, THF, CH_3CN, 73%→ COOCH₃ H Bu **2**

COOCH₃ I + CO —Pd(0), NR_3, THF, CH_3CN, 67%→ COOCH₃ O

to undergo [3 + 2]annelation to a [6,5]bicyclic product (**2**). The second example is an example of carbonylative [3 + 2 + 1]annelation.[4]

Pd-catalyzed intramolecular ene reactions.[5] Both this complex and bis(dibenzylideneacetone)palladium, $Pd(dba)_2$, have been used to effect this intramolecular cyclization of dienes capable of forming allylpalladium complexes.

Y Y H AcO CH_2 (Y = $COOCH_3$) —$Pd(dba)_2$, $P(C_6H_5)_3$, THF, 80%→ Y Y CH_2 CH_2

Ts Ts AcO CH_2 —$Pd[P(C_6H_5)_3]_4$, HOAc, 63%→ Ts Ts CH_2 CH_2

Ts N OAc CH_2 CH_2 —63%→ Ts N CH_2 CH_2

Allylidenation of aldehydes.[6] This palladium complex in combination with a trialkylphosphine promotes a reaction between phenyl isocyanate and allylic alcohols to provide 1,3-dienes.

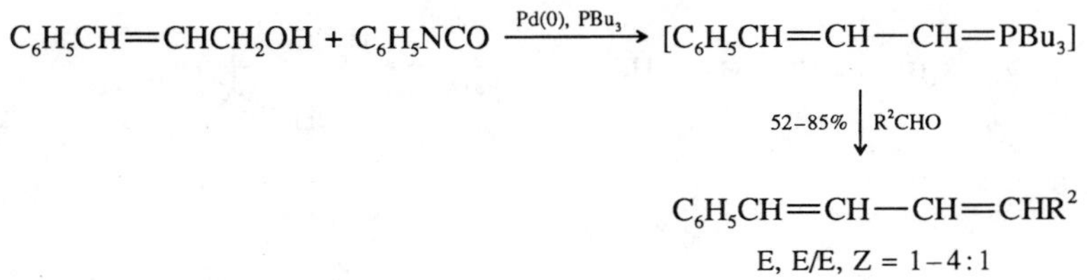

$$C_6H_5CH{=}CHCH_2OH + C_6H_5NCO \xrightarrow{Pd(0),\ PBu_3} [C_6H_5CH{=}CH{-}CH{=}PBu_3]$$

$$\xrightarrow[52-85\%]{R^2CHO} C_6H_5CH{=}CH{-}CH{=}CHR^2$$

E, E/E, Z = 1–4:1

This Wittig-like reaction provides a route to all-*trans* β-carotene (**1**).

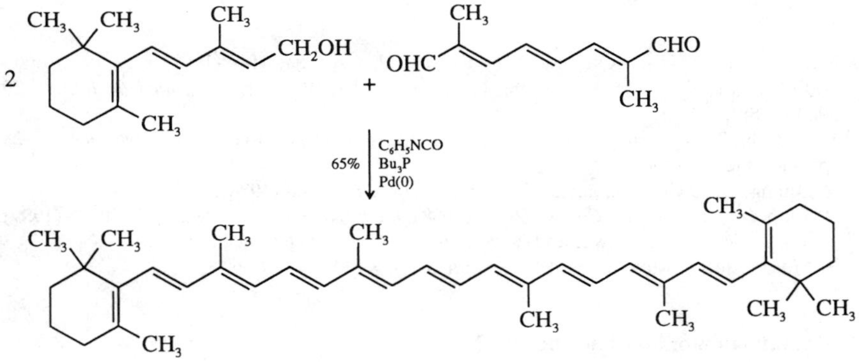

2 [retinol-type allylic alcohol, CH_2OH] + OHC–[dialdehyde, CHO] $\xrightarrow[65\%]{C_6H_5NCO,\ Bu_3P,\ Pd(0)}$ **1**

1

3-Aza-Cope rearrangement.[7] This reaction proceeds readily when catalyzed by Pd(0) and a strong protic acid, particularly trifluoroacetic acid or methanesulfonic acid. This isomerization is involved in a direct synthesis of δ,ε-unsaturated imines from allylamines with carbonyl compounds (equation I). The unsaturated

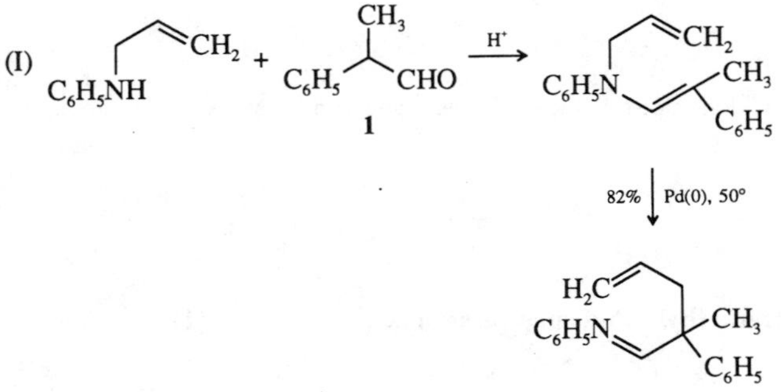

(I) $C_6H_5NHCH_2CH{=}CH_2$ + $C_6H_5CH(CH_3)CHO$ (**1**) $\xrightarrow{H^+}$ $C_6H_5N(CH_2CH{=}CH_2)CH{=}C(CH_3)C_6H_5$

$\xrightarrow[82\%]{Pd(0),\ 50^\circ}$ $C_6H_5N{=}CHC(CH_3)(C_6H_5)CH_2CH{=}CH_2$

2

imines can be reduced ($NaBH_4$) to unsaturated amines, or hydrolyzed to unsaturated carbonyl compounds (equation II). This rearrangement shows rather low asymmetric induction (one example).

$$\text{(II)}\ HN(CH_2CH{=}CH_2)_2 + \mathbf{1} \xrightarrow[84\%]{1)\ Pd(0),\ H^+;\ 2)\ H_3O^+} H_2C{=}CH{-}CH_2{-}C(CH_3)(C_6H_5){-}CHO$$

80% ↓ several steps

(2-methyl-4-phenyl-4-methyltetrahydrofuran: CH_3, C_6H_5, O, CH_3)

[1] M. A. Ciufolini, H.-B. Qi, and M. E. Browne, *J. Org.*, **53**, 4149 (1988).
[2] B. O'Connor, Y. Zhang, E. Negishi, F.-T. Luo, and J.-W. Cheng, *Tetrahedron Letters*, **29**, 3903 (1988).
[3] E. Negishi, T. Nguyen, B. O'Connor, J. M. Evans, and A. Silveira, Jr., *Heterocycles*, **28**, special issue, 55 (1989).
[4] Y. Zhang, B. O'Connor, and E. Negishi, *J. Org.*, **53**, 5588 (1988).
[5] W. Oppolzer, *Pure Appl. Chem.*, **60**, 39 (1988); *idem, Angew. Chem. Int. Ed.*, **28**, 38 (1989).
[6] N. Okukado, O. Uchikawa, and Y. Nakamura, *Chem. Letters*, 1449 (1988).
[7] S. Murahashi, Y. Makabe, and K. Kunita, *J. Org.*, **53**, 4489 (1988).

Tetrabutylammonium fluoride, Bu_4NF.

2-Aza-1,3-dienes.[1] The reaction of imines (**1**) of aldehydes obtained from bis(trimethylsilyl)methylamine with aldehydes catalyzed by Bu_4NF (Peterson alkenylation) provides 2-aza-1,3-dienes in 65–78% yield.

$$\underset{\mathbf{1}}{R^1CH{=}N{-}CH[Si(CH_3)_3]_2} \xrightarrow[65-78\%]{R^2CHO,\ Bu_4NF} \underset{\mathbf{2}\ (E/Z \sim 1:1)}{R^1CH{=}N{-}CH{=}CHR^2}$$

[1] J. Lasarte, C. Palomo, J. P. Picard, J. Dunogues, and J. M. Aizpurua, *J.C.S. Chem. Comm.*, 72 (1989).

2,2,6,6-Tetramethyl-2,6-disilapiperidine, (ring: $(CH_3)_2Si{-}NH{-}Si(CH_3)_2$ with $(CH_2)_3$) (**1**).

Preparation:[1]

$$CH_2{=}CHCH_2Si(CH_3)_3 \xrightarrow{71\%} CH_2CH_2CH_2Si(CH_3)_3(Si(CH_3)_3) \xrightarrow[87\%]{ClSi(CH_3)_3,\ AlCl_3}$$

$$1 \xleftarrow[67\%]{NH_3} (CH_3)_2SiCH_2CH_2CH_2Si(CH_3)_2 \text{ (each Si bearing Cl)}$$

Primary amines.[2] A version of the Gabriel synthesis involves alkylation of the potassium amide prepared by reaction of **1** with KH. Yields are generally in the range of 60 to 90%.

$$\mathbf{1} \xrightarrow[89\%]{1)\ KH,\ THF;\ 2)\ CH_3(CH_2)_7Br} (CH_2)_3[Si(CH_3)_2]_2N(CH_2)_7CH_3 \xrightarrow[84\%]{HCl\text{–}H_2O,\ \Delta} CH_3(CH_2)_7NH_2$$

[1] M. Ishikawa, M. Kumada, and H. Sakurai, *J. Organomet. Chem.*, **23**, 63 (1970).
[2] A. Hosomi, S. Kohra, Y. Tominaga, M. Inaba, and H. Sakurai, *Chem. Pharm. Bull.*, **36**, 2342 (1988).

Tetramethyl orthocarbonate, $C(OCH_3)_4$, b.p. 114°. Supplier: Aldrich.

Thiiranes.[1] In the presence of an acid catalyst, TsOH, this reagent converts 2-mercaptoalkanols into thiiranes in about 10 minutes at 25°. This reaction can

$$CH_3CH(OH)CH_2SH + C(OCH_3)_4 \xrightarrow[77\%]{TsOH} \text{2-methylthiirane} + O{=}C(OCH_3)_2 + 2CH_3OH$$

hardly involve a simple acid-catalyzed dehydration because an orthocarbonate is essential; possibly the actual reagent is $(CH_3O)_3C^+$.

[1] T. Takata and T. Endo, *Bull Chem. Soc. Japan*, **61**, 1818 (1988).

1,1,3,3-Tetramethylurea (TMU), **1**, 1146.

Glucosidation.[1] Glucosides can be prepared in 50–80% yield by reaction of 3,4,6-tri-O-acetyl-β-D-glucopyranosyl chloride with a primary or secondary alcohol in TMU as solvent and in the presence of 4-Å molecular sieves at a temperature of 110–120° with continuous removal of water from the reaction. Under these conditions, α- and β-glucosides are formed in about equal amounts.

[1] M. Nishizawa, Y. Kan, and H. Yamada, *Tetrahedron Letters*, **29**, 4597 (1988).

Tetraphenylantimony iodide, $(C_6H_5)_4SbI$.

γ- and δ-Lactones. In the presence of this reagent, diphenylketene (or ethylphenylketene) undergoes cycloaddition with oxiranes and even oxetanes to give lactones. Benzene is the solvent of choice for reactions with oxiranes.[1]

$$C_6H_5\underset{O}{CH-CH_2} + (C_6H_5)_2C{=}C{=}O \xrightarrow[91\%]{(C_6H_5)_4SbI,\ C_6H_6} \text{3-}C_6H_5\text{-2,2-}(C_6H_5)_2\text{-}\gamma\text{-butyrolactone}$$

$$CH_2{=}CH-\underset{O}{CH-CH_2} + (C_6H_5)_2C{=}C{=}O \xrightarrow[95\%]{(C_6H_5)_4SbI} \text{3-}(CH_2{=}CH)\text{-2,2-}(C_6H_5)_2\text{-}\gamma\text{-butyrolactone}$$

$$\text{2-}C_6H_5\text{-oxetane} + (C_6H_5)_2C{=}C{=}O \xrightarrow[91\%]{(C_6H_5)_4SbI} \text{4-}C_6H_5\text{-2,2-}(C_6H_5)_2\text{-}\delta\text{-valerolactone}$$

[1]M. Fujiwara, M. Imada, A. Baba, and H. Matsuda, *J. Org.*, **53**, 5974 (1988).

Tetraphenylantimony trifluoromethanesulfonate, $(C_6H_5)_4SbOTf$ (**1**). Preparation.[1]

β-Amino alcohols.[2] In the presence of this salt (**1**), amines (primary and secondary) react with oxiranes to give β-amino alcohols. The reaction with terminal oxiranes is highly chemoselective, resulting in β-cleavage almost exclusively. No rearrangement of oxiranes to allylic alcohols is observed. $(C_6H_5)_4SbI$ or $(C_6H_5)_4SbCl$ is less efficient than the triflate, which has the added advantage of solubility in CH_2Cl_2.

$$\text{cyclohexene oxide} + HN(C_2H_5)_2 \xrightarrow[70\text{–}75\%]{CH_2Cl_2,\ 40°} \text{2-}N(C_2H_5)_2\text{-cyclohexanol}$$

$$AcOCH_2\underset{O}{CH-CH_2} + HN(C_2H_5)_2 \xrightarrow[89\%]{} AcOCH_2CH(OH)CH_2N(C_2H_5)_2$$

[1] R. Rüther, F. Huber, and H. Preut, *J. Organomet. Chem.*, **295**, 21 (1985).
[2] M. Fujiwara, M. Imada, A. Baba, and H. Matsuda, *Tetrahedron Letters*, **30**, 739 (1989).

2,3,4,6-Tetra-O-pivaloyl-β-D-galactopyranosylamine (1).

(Structure 1: PivO, OPiv, O, PivO, NH_2, PivO)

***(R)*-α-Amino acids.**[1] This carbohydrate has been used as the chiral auxiliary as well as the amine in the Ugi four-component condensation for synthesis of α-amino acids (**7**, 366).[2]

$$\mathbf{1} + R^1CHO + (CH_3)_3CNC \xrightarrow{HCOOH,\ ZnCl_2} \mathbf{2}$$

(Structure 2: PivO, OPiv, O, HCO, N, $CONHC(CH_3)_3$, PivO, PivO, H, R^1)

$$\mathbf{2} \xrightarrow[82-90\%]{1)\ HCl,\ CH_3OH} H_2N\text{–}CH(R^1)\text{–}COOH$$

(R/S = 94–96:6–4)

[1] H. Kunz and W. Pfrengle, *Am. Soc.*, **110**, 651 (1988).
[2] R. Urban and I. Ugi, *Angew. Chem. Int. Ed.*, **14**, 61 (1975).

Thiolacetic acid, CH_3COSH.

Chiral dithioacetals.[1] In the presence of TsOH or ZnI_2, an aldehyde, a thiol, and thiolacetic acid react to form a mixed acyldithioacetal (**1**). These products can be deacylated (K_2CO_3, $NaOCH_3$, or CH_3Li) and then alkylated to provide chiral dithioacetals (**2**).

$$R^1CHO + HSR^2 + CH_3COSH \xrightarrow[65-90\%]{ZnI_2,\ CH_2Cl_2} R^1\text{—}CH(SCOCH_3)(SR^2)\ (\mathbf{1}) \xrightarrow[50-95\%]{K_2CO_3\ or\ CH_3Li,\ R^3X} R^1CH(SR^3)(SR^2)\ (\mathbf{2})$$

Optically active dithioacetals can be prepared by use of the thiolacid prepared from (R)-(−)-α-methoxyphenylacetic acid by reaction with oxalyl chloride and then with NaSH. This thiolacid reacts with an aldehyde and a thiol to form the mixed thioacetals corresponding to **1** as a 1:1 mixture of diastereomers, separable by chromatography (~75% yield). These are convertible into optically active dithioacetals corresponding to **2** with no loss of diastereomeric purity.

[1] J. Y. Gauthier, T. Henien, L. Lo, M. Thérien, and R. N. Young, *Tetrahedron Letters*, **29**, 6729 (1988); M. Thérien, J. Y. Gauthier, and R. N. Young, *ibid*, **29**, 6733 (1988).

Thiophosphoryl chloride, $S=PCl_3$.

Cyclothiophosphorylation.[1] The diasteromers of *myo*-inositol-1,2-cyclothiophosphate (**2**) can be prepared by reaction of 1,4,5,6-tetraacetyl-*myo*-inositol (**1**) with thiophosphoryl chloride followed by deprotection and addition of KCl.

OH, OH, RO, RO, OR, OR

1, R = Ac

1) $S=PCl_3$, CH_3CN, Py
2) TEA/CH_3OH
3) KCl
20%

KO, S, P, O, O, HO, HO, OH, OH

endo-**2**

\+ 1:1

S, P, KO, O, O, HO, HO, OH, OH

exo-**2**

[1] C. Schultz, T. Metschies, and B. Jastorff, *Tetrahedron Letters*, **29**, 3919 (1988).

Thiourea–Titanium(IV) isopropoxide.

2,3-Epithio alcohols.[1] Thiourea is known to convert epoxides into episulfides. In the presence of Ti(O-*i*-Pr)$_4$, *trans* 2,3-epoxy alcohols (**1**) also react with thiourea to provide epithio alcohols (**2**) without loss of enantioselectivity (equation I).

(I) n-C_8H_{17} (epoxide O) OH

1 (2S, 3S, 98% ee)

1) NH_2CSNH_2, Ti(O-*i*-Pr)$_4$
2) $NaHCO_3$
70–90%

n-C_8H_{17} (episulfide S) OH

2 (2R, 3R, 98% ee)

[1] Y. Gao and K. B. Sharpless, *J. Org.*, **53**, 4114 (1988).

Tin–Aluminum, Sn-Al.

Reaction of RCHO with $C_6H_5CH{=}CHCH_2Cl$.[1] This reaction can be regioselective and *threo*-selective when mediated by Sn and Al. The reaction of benzaldehyde with cinnamyl chloride provides the *anti*-homoallylic alcohol as the only detectable product. This diastereoselectivity obtains generally with aliphatic or aromatic aldehydes. The reaction with enals under the same conditions provides only the 1,2-adducts, again with *anti*-selectivity (equation I). The reaction of 2-phenylpropanal with **1** provides a mixture of homoallylic alcohols in the ratio 3:1, with *anti*-selectivity at the C—C bond formed in the reaction (equation II).

$$\text{(I)}\quad RCH{=}CHCHO + \underset{\mathbf{1}}{C_6H_5CH{=}CHCH_2Cl} \xrightarrow{\text{Sn, Al}} R\text{–}CH{=}CH\text{–}CH(OH)\text{–}CH(C_6H_5)\text{–}CH{=}CH_2$$

$$\text{(II)}\quad C_6H_5CH(CH_3)CHO + \mathbf{1} \longrightarrow C_6H_5CH(CH_3)\text{–}CH(OH)\text{–}CH(C_6H_5)\text{–}CH{=}CH_2 + C_6H_5CH(CH_3)\text{–}CH(OH)\text{–}CH(C_6H_5)\text{–}CH{=}CH_2 \quad (3:1)$$

[1] J. M. Coxon, S. J. van Eyk, and P. J. Steel, *Tetrahedron*, **45**, 1029 (1989).

Tin(II) chloride, $SnCl_2$.

Allylation with allylic alcohols.[1] In the presence of at least 1 equiv. of $SnCl_2$ and a trace of a palladium catalyst, $Cl_2Pd(C_6H_5CN)_2$, allyl alcohols and allyl acetates are converted into an allylic metal, $CH_2{=}CHCH_2Sn(IV)$, that can react with aldehydes to give homoallylic alcohols in reactions conducted at 25° in 1,3-dimethylimidazolidinone (DMI, **11**, 202), (equation I). Similar allylation of ketones is possible, but requires a higher temperature. SnF_2 and $Sn(OAc)_2$ cannot replace $SnCl_2$ in this allylation reaction.

$$\text{(I)}\quad CH_2{=}CHCH_2OH \xrightarrow{\text{Pd/SnCl}_2} [CH_2{=}CHCH_2Sn(IV)] \xrightarrow[\text{40–90\%}]{\text{RCHO, DMI}} CH_2{=}CHCH_2CH(OH)R$$

Example:

$$C_6H_5CHO + CH_3CH{=}CHCH_2OH \xrightarrow[\text{65\%}]{\text{Pd/SnCl}_2} H_2C{=}CH\text{–}CH(CH_3)\text{–}CH(OH)\text{–}C_6H_5 + H_2C{=}CH\text{–}CH(CH_3)\text{–}CH(OH)\text{–}C_6H_5 \quad (49:51)$$

Deoxygenation of 1,4-endoperoxides of 1,3-dienes.[2] These 1,4-endoperoxides are reconverted to 1,3-dienes by $SnCl_2$.

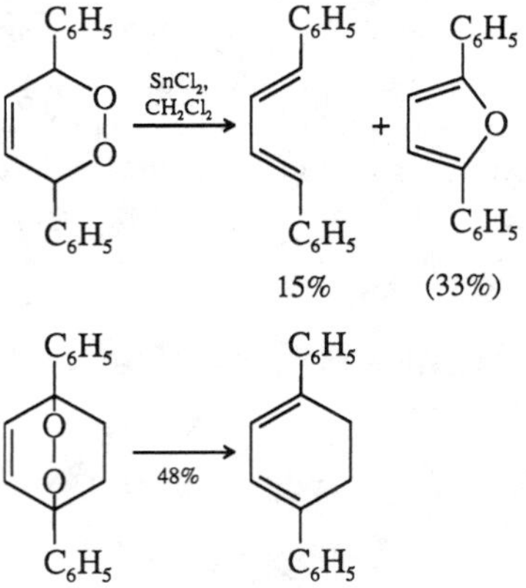

C_6H_5 C_6H_5 C_6H_5
$SnCl_2$, CH_2Cl_2 → +
C_6H_5 C_6H_5 C_6H_5
15% (33%)

C_6H_5 C_6H_5
→ 48%
C_6H_5 C_6H_5

[1] Y. Masuyama, J. P. Takahara, and Y. Kurusu, *Am. Soc.*, **110**, 4473 (1988).
[2] S. Kohmoto, S. Kasai, M. Yamamoto, and K. Yamada, *Chem. Letters*, 1477 (1988).

Tin(II) chloride–Chlorotrimethylsilane.

Aldol-type reaction of acetals with alkenes. Mildly activated alkenes can undergo addition reactions with acetals in the presence of $ClSi(CH_3)_3$ and $SnCl_2$, as in the reaction with dihydropyran. However, the corresponding reaction with styrene requires a full equivalent of $ClSi(CH_3)_3$ and a catalytic amount of $SnCl_2$ for a satisfactory yield.

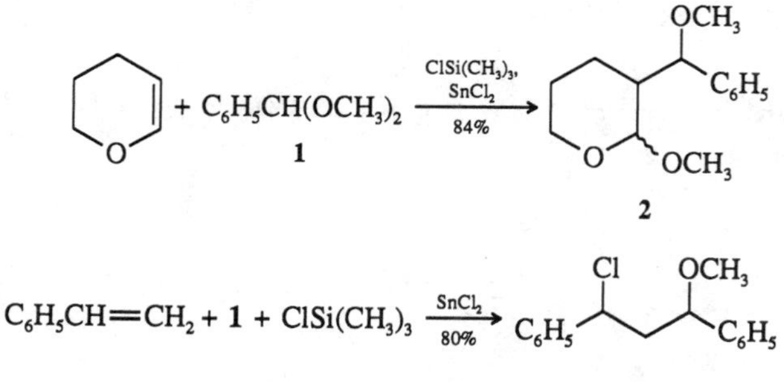

$$+ C_6H_5CH(OCH_3)_2 \ (\mathbf{1}) \xrightarrow[84\%]{ClSi(CH_3)_3,\ SnCl_2} \mathbf{2}$$

OCH_3, C_6H_5, O, OCH_3

$$C_6H_5CH{=}CH_2 + \mathbf{1} + ClSi(CH_3)_3 \xrightarrow[80\%]{SnCl_2} C_6H_5CH(Cl)CH_2CH(OCH_3)C_6H_5$$

These reactions may involve the active species **a**.

$$C_6H_5CH(OCH_3)_2 \rightleftharpoons \underset{\mathbf{a}}{C_6H_5CH(OCH_3)Cl} + (CH_3)_3SiOCH_3$$

$$\downarrow \text{dihydropyran}$$

$$\left[\text{2-chloro-3-}(CH(OCH_3)C_6H_5)\text{tetrahydropyran}\right] \xrightarrow{(CH_3)_3SiOCH_3} \mathbf{2}$$

[1] T. Mukaiyama, K. Wariishi, Y. Saito, M. Hayashi, and S. Kobayashi, *Chem. Letters*, 1101 (1988).

Tin(IV) chloride.

Regioselective O-debenzylation. Of a number of Lewis acids, $SnCl_4$ or $TiCl_4$ is the most effective for this reaction with polybenzylated sugars. The substrate

BzlO, BzlO, BzlO, OBzl, OBzl, OBzl $\xrightarrow{CH_2Cl_2,\ 25°}$ BzlO, BzlO, HO, OBzl, OBzl, OBzl + BzlO, HO, BzlO, OBzl, OBzl, OBzl; BzlO, BzlO, HO, OH, OBzl, OBzl

	BzlO/BzlO/HO/OBzl/OBzl/OBzl	BzlO/HO/BzlO/OBzl/OBzl/OBzl	BzlO/BzlO/HO/OH/OBzl/OBzl
1 $TiCl_4$ (0.75 hr.)	65%	19%	
1 $SnCl_4$ (7 hr.)	57%	23%	
2 $SnCl_4$	—	—	35%

OBzl, OBzl, OBzl ⟶ OBzl, OH, OBzl

$SnCl_4$	86%
$TiCl_4$	68%

requires at least three groups capable of coordination with $SnCl_4$ or $TiCl_4$. The order of complexation is primary > secondary > tertiary alkoxy group.[1]

2-O-Organosilyl glycosides; C-furanosides.[2] $SnCl_4$ (1–2 equiv.) effects condensation of furanosides and organosilanes bearing a leaving group to give C-furanosides. Thus the reaction of the furanosides **1** with allyltrimethylsilane gives

C-glycosides or C-furanosides (**2**), but without stereoselectivity. However, the reaction with a chlorodimethylsilane shows considerable selectivity. The difference

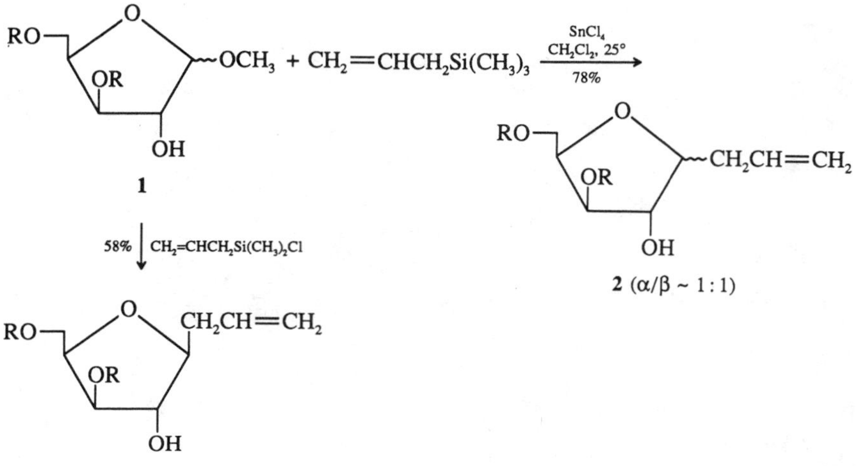

RO— [furanoside ring: O, OR, OH] ~OCH$_3$ + $CH_2{=}CHCH_2Si(CH_3)_3$ $\xrightarrow[78\%]{SnCl_4,\ CH_2Cl_2,\ 25°}$ RO— [furanoside ring: O, OR, OH] ~$CH_2CH{=}CH_2$

1 **2** (α/β ~ 1:1)

58% ↓ $CH_2{=}CHCH_2Si(CH_3)_2Cl$

RO— [furanoside ring: O, OR, OH] $CH_2CH{=}CH_2$

3 (β/α ~ 65:1)

is ascribed to formation of a 2-O-allyldimethylsilane in the latter reaction, which then rearranges to the 1,2-*trans*-C-glycoside **3**.

Another example:

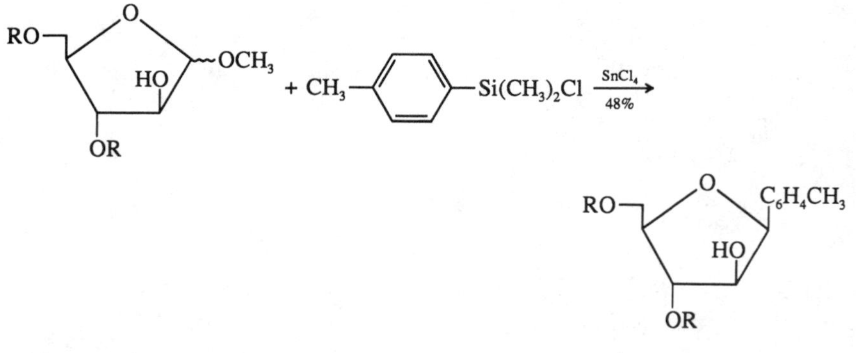

RO— [furanoside ring: O, HO, OR] ~OCH$_3$ + CH_3—C_6H_4—$Si(CH_3)_2Cl$ $\xrightarrow[48\%]{SnCl_4}$ RO— [furanoside ring: O, HO, OR] $C_6H_4CH_3$

Cyclization of acyclic acetals.[3] Reaction of the γ-hydroxy-γ-vinyl acetal (**1**) with $SnCl_4$ (1 equiv.) in CH_2Cl_2 at −78° provides *cis*-hydroindenes **2** and **3**, which are convenient precursors to the dione **4**.

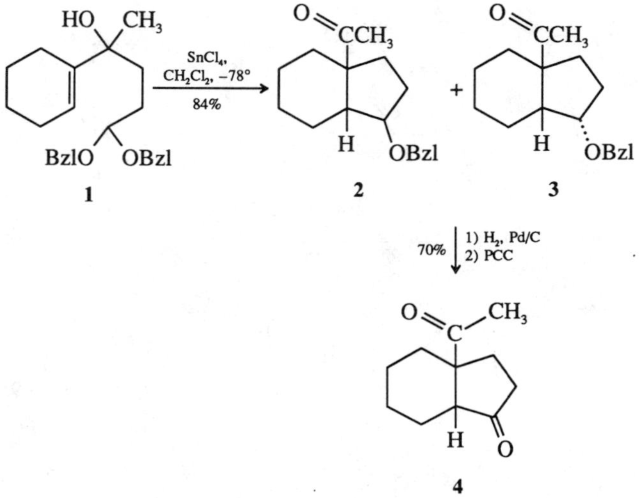

Another example:

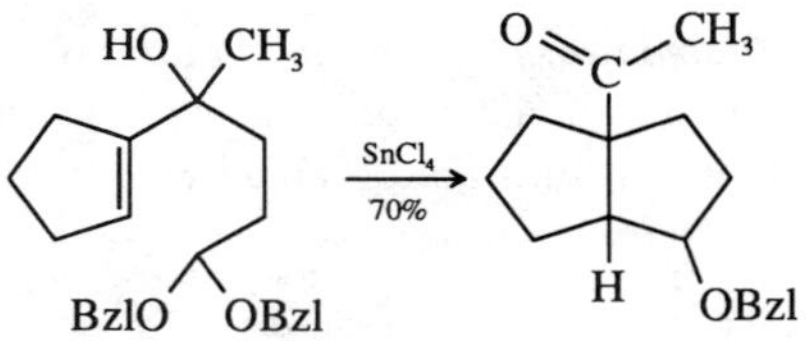

Acyl isocyanates.[4] $SnCl_4$ is the best catalyst for preparation of these products by reaction of ArCOCl and NaOCN, with $ZnCl_2$ being a close second. 1,2-Dichlorobenzene is the preferred solvent. Yields are lower in reactions of aliphatic acyl chlorides.

[1] H. Hori, Y. Nishida, H. Ohrui, and H. Meguro, *J. Org.*, **54**, 1346 (1989).
[2] O. R. Martin, S. P. Rao, K. G. Kurz, and H. A. El-Shenawy, *Am. Soc.*, **110**, 8698 (1988).
[3] M. Sworin and W. L. Neumann, *J. Org.*, **53**, 4894 (1988).
[4] M.-Z. Deng, P. Caubere, J. P. Senet, and S. Lecolier, *Tetrahedron*, **44**, 6079 (1988).

Tin(II) trifluoromethanesulfonate.

Enantioselective Michael additions.[1] The addition of tin(II) enolates to enones can proceed enantioselectively when activated by trimethylsilyl triflate and in the presence of a chiral diamine such as **1**.

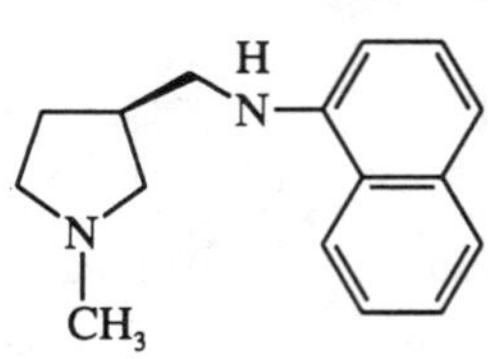

(1)

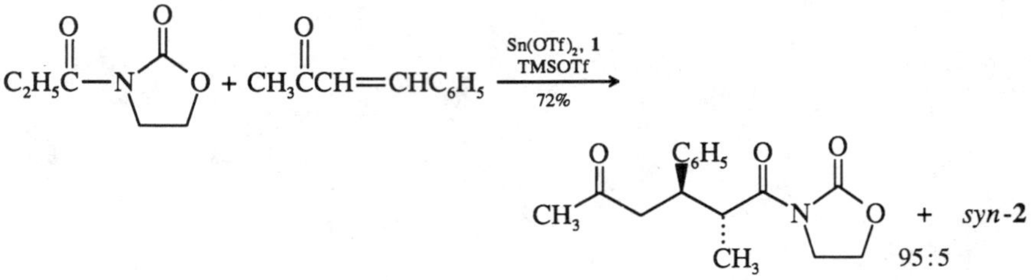

$$C_2H_5COX + CH_3COCH{=}CHC_6H_5 \xrightarrow[72\%]{Sn(OTf)_2,\ 1,\ TMSOTf} \text{anti-2} + \text{syn-2} \quad 95:5$$

anti-**2** (93% ee)

$$CH_3CSSCH_3 + CH_3COCH{=}CHC_6H_5 \xrightarrow[82\%]{Sn(OTf)_2,\ 1,\ TMSOTf} CH_3COCH_2CH(C_6H_5)CH_2CSSCH_3$$

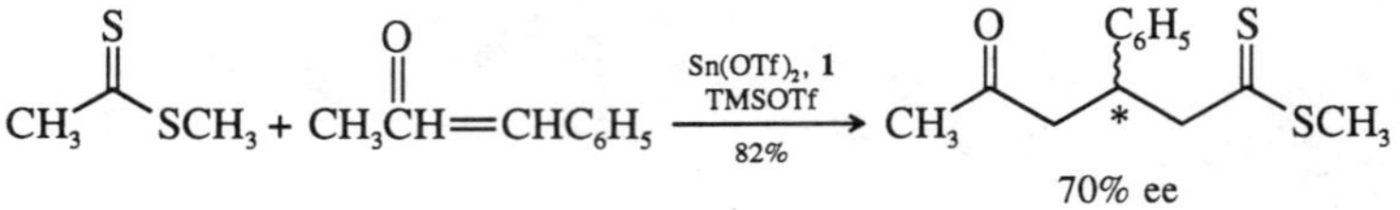

70% ee

2-(1′-Hydroxyalkyl)butenolides.[2] The tin(II) furanolates (**2**) generated from 2-(5*H*)-furanone (**1**) react with an aldehyde to give a 2-substituted butenolide (**3**) as the major product. Similar results are obtained with boron furanolates.

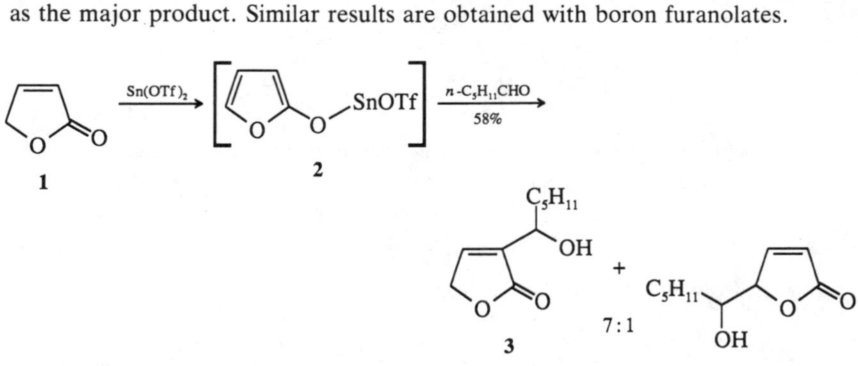

$$\mathbf{1} \xrightarrow{Sn(OTf)_2} [\mathbf{2}\ (OSnOTf)] \xrightarrow[58\%]{n\text{-}C_5H_{11}CHO} \mathbf{3} + \mathbf{4} \quad 7:1$$

[1] T. Yura, N. Iwasawa, K. Narasaka, and T. Mukaiyama, *Chem Letters*, 1021, 1025 (1988).
[2] C. W. Jefford, D. Jaggi, and J. Boukouvalas, *J.C.S. Chem. Comm.*, 1595 (1988).

Tin(II) trifluoromethanesulfonate–Tributyltin fluoride.

Aldol reactions. In the presence of these two promoters and in combination with the chiral diamine (S)-1-methyl-2-[(piperidinyl)methyl]pyrrolidine (**13**, 302),

1, aldehydes undergo asymmetric aldol reactions with silyl enol ethers in 78–95% ee. The active species (**2**) is presumed to be a complex of the amine, Bu_3SnF, and $Sn(OTf)_2$, which unlike the individual components, is soluble in CH_2Cl_2.[1]

$$C_6H_5CHO + CH_2{=}C(OSi(CH_3)_3)(SC_2H_5)\ (\mathbf{3}) \xrightarrow[73\%]{\mathbf{2},\ CH_2Cl_2,\ -78^\circ} C_2H_5SC(O)CH_2CH(OH)C_6H_5$$

(S), 82% ee

$$(CH_3)_3CHO + \mathbf{3} \xrightarrow[90\%]{} C_2H_5SC(O)CH_2C^*H(OH)C(CH_3)_3$$

(>95% ee)

This catalytic aldol reaction has been extended to α,β-enals.[2] Thus these reactions catalyzed by the chiral acyloxyboranes **3a** and **3b** in addition to $Sn(OTf)_2$ and Bu_3SnF lead to optically active products in as high as 97% ee even when **3** is

3a (2R, 3R)
3b (2S, 3S)

present in only 10 mole %. Reactions with acrolein are slightly less enantioselective (84% ee), but β-substitution of the enal decreases the enantioselectivity to 2% ee.

$$CH_2{=}C(CH_3)CHO + \text{cyclopentadiene} \xrightarrow[85\%]{\mathbf{3a},\ CH_2Cl_2} \text{norbornene-CHO, CH}_3$$

(96% ee)

[1] S. Kobayashi and T. Mukaiyama, *Chem. Letters*, 297 (1989).
[2] K. Furuta, S. Shimizu, Y. Miwa, and H. Yamamoto, *J. Org.*, **54**, 1481 (1989).

Titanium(0).

Low-valent complexes. Pons and Santelli[1] have reviewed reductions with low-valent complexes of transition metals, particularly those of titanium and vanadium (301 references).

[1] J.-M. Pons and M. Santelli, *Tetrahedron*, **44**, 4295 (1988).

Titanium(IV) butoxide/Copper(I) chloride.

Rearrangement of propargylic alcohols to enals or enones.[1] In the presence of a carboxylic acid (generally ~7 equiv. of *p*-toluic acid or crotonic acid), propargylic alcohols are rearranged to enals or enones by $Ti(OBu)_4$ in combination with CuCl, both in about 0.02 mole %. Silver trifluoroacetate can replace CuCl, but yields are lower. The acid is not mandatory, but can increase the rate and yield.

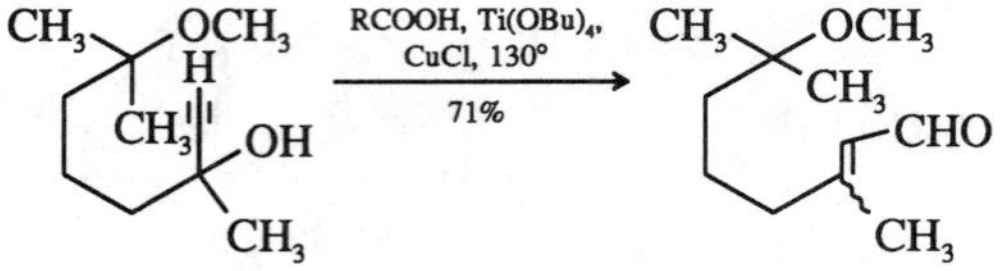

RCOOH, $Ti(OBu)_4$, CuCl, 130° — 71%

$CH_3C{\equiv}CCH_2OH \xrightarrow{47\%} CH_3\overset{O}{\overset{\|}{C}}CH{=}CH_2$

[1] P. Chabardes, *Tetrahedron Letters*, **29**, 6253 (1988).

Titanium(III) chloride–Potassium/Graphite, $TiCl_3$–C_8K.

Stereoselective McMurry coupling. Clive and co-workers[1] report that a modified McMurry reagent prepared in DME with C_8K and $TiCl_3$ (~2:1) can effect highly stereoselective coupling of keto aldehydes. Thus the key step in a synthesis

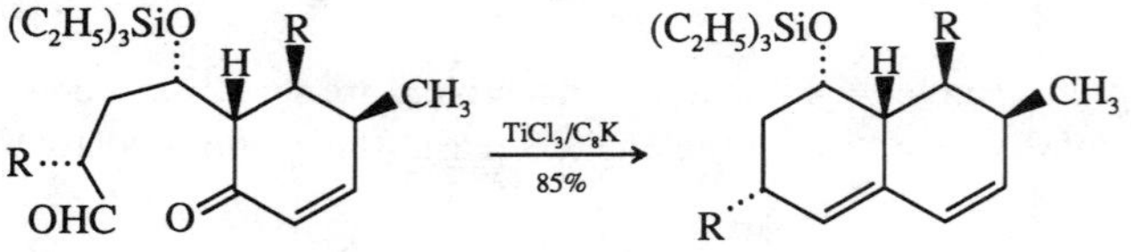

$TiCl_3/C_8K$, 85%: **1** → **2**

of (+)-compactin (**3**) is the coupling of the keto aldehyde (**1**) to give the A/B ring system (**2**). Using the modified Ti(0) reagent, the asymmetric center adjacent to the ketone is not epimerized.

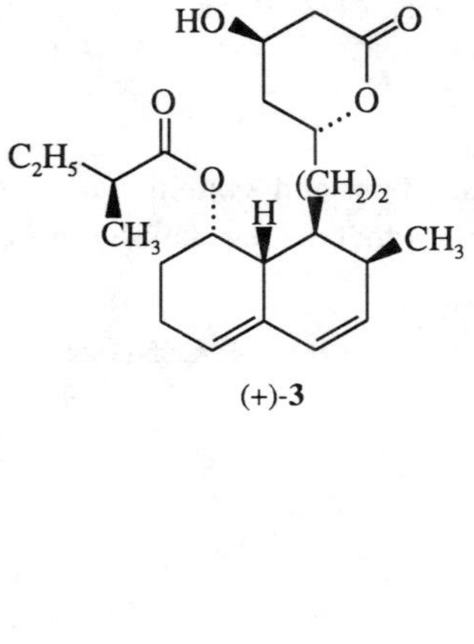

(+)-**3**

[1] D. L. J. Clive, K. S. Keshava Murthy, A. G. H. Wee, J. Siva Prasad, G. V. J. da Sielva, M. Majewski, P. C. Anderson, R. D. Haugen, and L. D. Heerze, *Am. Soc.*, **110**, 6914 (1988).

Titanium(III) chloride–Sodium cyanoborohydride.

>C = NOH → >CNH₂. $TiCl_3$ is known to reduce oximes to imines (**4**,506). If this reduction is conducted in the presence of $NaBH_3CN$ (buffered), the imine is reduced to an amine before it is hydrolyzed to a ketone. This mild method is particularly useful for complex substrates with acid- and base-sensitive groups.[1]

[1] J. P. Leeds and H. A. Kirst, *Syn. Comm.*, **18**, 777 (1988).

Titanium(III) chloride–Zinc/copper couple.

1,2-Cycloalkanediols. Six- to fourteen-membered cyclic pinacols can be prepared by reaction of α,ω-dials with $TiCl_3$ and Zn/Cu couple (1:9) in DME at 25°. *cis*-Pinacols predominate in six- to eight-membered rings, but *trans*-products predominate in rings of 10 or more members.[1]

CHO CHO —($TiCl_3$, Zn/Cu, 85%)→ OH OH

cis, 100%

This cyclization was used to obtain the 14-membered sesquiterpene sarcophytol B (**1**) (last step, 46% yield).[2]

CH_3 CH_3 $CH(CH_3)_2$ OH OH CH_3

1

[1] J. E. McMurry and J. G. Rico, *Tetrahedron Letters*, **30**, 1169 (1989).
[2] J. E. McMurry, J. G. Rico, and Y. Shih, *ibid*, **30**, 1173 (1989).

Titanium(IV) chloride.

Michael additions of ketene alkyl silylacetals.[1] In the presence of $TiCl_4$ (1 equiv.) these acetals undergo Michael addition to ethyl propiolate via a titanate intermediate.

$$(CH_3)_2C{=}C(OCH_3)(OSi(CH_3)_3) + HC{\equiv}CCOOC_2H_5 \xrightarrow[81\%]{TiCl_4,\ CH_2Cl_2,\ -78°} (CH_3)_2C(COOCH_3)CH{=}CHCOOC_2H_5$$

(E), 100%

$\downarrow TiCl_4$

$$\left[(CH_3)_2C(COOCH_3)CH{=}C(TiCl_3)COOC_2H_5\right] \xrightarrow[86\%]{NBS} (CH_3)_2C(COOCH_3)CH{=}C(Br)COOC_2H_5$$

(E/Z) = 99:1

The same reaction, but in the presence of a catalytic amount of zirconium(IV) chloride, results in [2+2] cycloaddition.

$$(CH_3)_2C{=}C(OCH_3)(OSi(CH_3)_3) + HC{\equiv}CCOOC_2H_5 \xrightarrow[70\%]{ZrCl_4,\ CH_2Cl_2} \text{cyclobutene: } (H_3C)_2C\text{–}C(OCH_3)(OSi(CH_3)_3)\text{–}C(COOC_2H_5){=}CH$$

Conjugate crotylation. The Hosomi–Sakurai reaction[2] has been extended to conjugate crotylation of cyclohexenone with crotylsilanes.[3] In this case the stereoselectivity is highly dependent on the geometry of the crotyl group and to a less extent on the groups on silicon. (E)-Crotyltrialkylsilanes show high *erythro*-selectivity, being highest with (E)-(crotyl)methyldiphenylsilane (equation I).

(I) cyclohexenone + $CH_3CH{=}CHCH_2SiMe(C_6H_5)_2$ $\xrightarrow[78\%]{TiCl_4}$ 3-[$CH(CH_3)CH{=}CH_2$]cyclohexanone

(15.6:1)

(Z)-Crotylsilanes are inherently mildly *threo*-selective. The highest *threo*-selectivity can be obtained when silicon bears an alkoxy substituent, but 2 equiv. of $TiCl_4$ is then required, and yields are only moderate because of a secondary reaction (equation II).

(II) cyclohexenone + (Z)-$CH_3CH{=}CHCH_2Si(CH_3)_2OC_2H_5$ $\xrightarrow[65\%]{TiCl_4}$ 3-[$CH(CH_3)CH{=}CH_2$]cyclohexanone

(11.2:1)

Fries rearrangement.[4] Fries rearrangement of the phenyl ester **1** with $AlCl_3$ at 120° results in a mixture of eight products, only one (**2**) of which is the expected product of *ortho* rearrangement. The others are formed by migration or elimination of the isopropyl group. However, rearrangement catalyzed by $TiCl_4$ at 120° gives **2** in 79% yield.

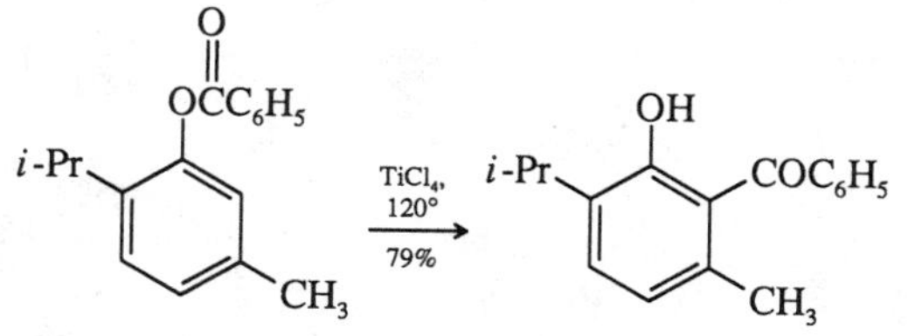

Intramolecular Mukaiyama aldol condensation.[5] The silyl ketene acetal **1** cyclizes to the tetrahydrofuran **2** in 32% yield on exposure to $TiCl_4$ (1 equiv.) in CH_2Cl_2 at 0°. The product is convertible into **3**, an analog of cycloleucine.

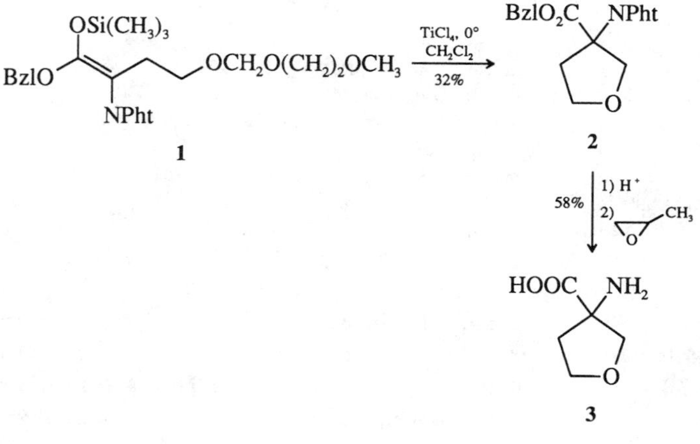

Carboxamides from C_6H_5COOH.[6] Carboxamides can be prepared from carboxylic acids directly rather than via the acid chlorides if the reaction with an amine and $N(C_2H_5)_3$ is carried out in the presence of a Lewis acid. In reactions with a primary amine, BF_3 etherate is completely ineffective. In contrast, $TiCl_4$ is about equal to BF_3 etherate in reactions with a secondary amine and has the further advantage that the rate of amidation is faster with $TiCl_4$ than with BF_3 etherate.

Polyarenes.[7] $TiCl_4$ effects condensation of the enol silyl ether of a cycloalkanone with 2-arylacetaldehydes to provide a product that can be dehydrogenated to an angular polyarene.

CH_3 CHO + $(CH_3)_3SiO$ $\xrightarrow[62\%]{TiCl_4}$ CH_3

Acylalkylation of arenes.[8] In the presence of $TiCl_4$, arenes add to nitroalkenes to form a nitronate adduct, which undergoes a Nef reaction to provide substituted arenes.

NO_2 + $C_6H_5CH_3$ $\xrightarrow[CH_2Cl_2]{TiCl_4}$ [$\overset{+}{N}$ $OTiCl_3$ O^- $C_6H_4CH_3$-*p*] $\xrightarrow[94\%]{H_3O^+}$ O $C_6H_4CH_3$-*p*

[1] A. Quendo and G. Rousseau, *Tetrahedron Letters*, 29 6443 (1988).
[2] A. Hosomi and H. Sakurai, *Am. Soc.*, **99**, 1673 (1977) (**11**, 529–530).
[3] T. Tokoroyama and L.-R. Pan, *Tetrahedron Letters*, **30**, 197 (1989).
[4] R. Martin and P. Demerseman, *Synthesis*, 25 (1989).
[5] D. M. Walker and E. W. Logusch, *Tetrahedron Letters*, **30**, 1181 (1989).
[6] A. Nordahl and R. Carlson, *Acta Chem. Scand.*, **B42**, 28 (1988).
[7] P. Di Raddo and R. G. Harvey, *Tetrahedron Letters*, **29**, 3885 (1988).
[8] K. Lee and D. Y. Oh, *ibid.*, **29**, 2977 (1988).

Titanium(IV) chloride–Lithium aluminum hydride, 6, 596.

Chlorofluorocarbene.[1] This carbene can be obtained by reaction of $CFCl_3$ (b.p. 23.7°) with the titanium species obtained from $TiCl_4$ and $LiAlH_4$ (1:1). It converts alkenes to chlorofluorocyclopropanes at 0° in generally good yields.

$(CH_3)_2C{=}CHCH_3$ $\xrightarrow[89\%]{CFCl_3,\ TiCl_4-LiAlH_4}$ H, CH_3, CH_3, Cl, CH_3, F + H, CH_3, CH_3, F, CH_3, Cl

2.2:1

$\xrightarrow[52\%]{}$ Cl, F + F, Cl

2.2:1

[1] W. R. Dolbier, Jr. and C. R. Burkholder, *Tetrahedron Letters*, **29**, 6749 (1988).

Titanium(IV) chloride–Magnesium, 7, 373.

Coupling of aldimines.[1] The Ti(II) reagent obtained from $TiCl_4$ and Mg effects reductive coupling of aryl aldimines to 1,2-diarylethylenediamines as a 1:1 mixture of *meso*- and *dl*-isomers.

$$C_6H_5CH{=}\overset{+}{N}(CH_3)_2 \xrightarrow[92\%]{Ti(II)} C_6H_5CH[N(CH_3)_2]CH[N(CH_3)_2]C_6H_5$$

meso/dl = 1:1

[1] C. Betschart, B. Schmidt, and D. Seebach, *Helv.*, **71**, 1999 (1988).

Titanium(IV) chloride–Titanium(IV) isopropoxide.

[2+2]Cycloaddition or [3+2]cycloaddition of alkenes to quinones.[1] In the presence of 1 equiv. of a Ti(IV) catalyst composed of $TiCl_4$ and $Ti(O\text{-}i\text{-}Pr)_4$, β-methyl styrenes react with 2-alkoxy-1,4-benzoquinones to form *cis*-cyclobutanes (**1**) or dihydrobenzofurans (**2**), which can be formed by an acid-catalyzed rearrangement of **1**. The composition of the Ti(IV) catalyst largely controls the type of cycloadduct. $TiCl_4$ favors [3+2]cycloaddition, whereas [2+2]cycloaddition requires a mixed $TiCl_4$–$Ti(O\text{-}i\text{-}Pr)_4$ catalyst; the rate of reaction determines to a large extent the stereochemistry of the cyclobutanes, which are invariably *cis* at the ring juncture.

4-$CH_3OC_6H_4$CH=CHCH$_3$ (E/Z = 14:1) + 2-methoxy-1,4-benzoquinone $\xrightarrow{TiCl_4:Ti(O\text{-}i\text{-}Pr)_4\ 1.6:1}$

2 (72%) + **1** (12%, >19:1)

1 $\xrightarrow[>85\%]{H_2SO_4}$ **2**

CH_3 ... 3,4-$(CH_3O)_2C_6H_3$ (Z/E = 22:1) + CH_3 ... OBzl quinone; $TiCl_4$-Ti(O-*i*-Pr)$_4$ 1.8:1, 49% → 34:1

[1] T. A. Engler, K. D. Combrink, and J. E. Ray, *Am. Soc.*, **110**, 7931 (1988).

Titanium(IV) isopropoxide, Ti(O-*i*-Pr)$_4$ (**1**).

Epoxy alcohols. A few years ago Mihelich[1] was granted a patent for preparation of epoxy alcohols by photooxygenation of alkenes in the presence of titanium or vanadium catalysts. Adam *et al.*[2] have investigated this reaction in detail and find that Ti(IV) isopropoxide is the catalyst of choice for epoxidation of di-, tri-, and tetrasubstituted alkenes, acyclic and cyclic, to provide epoxy alcohols. When applied to allylic alcohols, the reaction can be diastereo- and enantioselective. The reaction actually proceeds in two steps: an ene reaction to provide an allylic hydroperoxide followed by intramolecular transfer of oxygen catalyzed by Ti(O-*i*-Pr)$_4$. The latter step is a form of Sharpless epoxidation and can be highly stereoselective.

$(CH_3)_2C{=}CH$-*t*-Bu → (O_2, Ti(O-*i*-Pr)$_4$, 84%) epoxy alcohol (95:5)

cyclohexene → 60% + (7%)

[1] E. D. Mihelich, U. S. Patent 4,345,984 [*C. A.*, **98**, 125739c (1983)].
[2] W. Adam, M. Braun, A. Griesbeck, V. Lucchini, E. Staab, and B. Will, *Am. Soc.*, **111**, 203 (1989).

p-Toluenesulfonyl azide, TsN_3.

α-Alkyl-β-keto amides. Tosyl azide reacts with 1,3-diketones to give 2-diazo-1,3-diketones in high yield.[1] These products on irradiation (254 nm) undergo Wolff rearrangement to form α-keto ketenes (**a**), which can be trapped by a primary or secondary amine to provide α-alkyl-β-keto amides.[2]

$$CH_3COCH_2COCH_3 \xrightarrow[\sim 60\%]{TsN_3} CH_3COC(=N_2)COCH_3 \xrightarrow{h\nu} [CH_3COC(CH_3)=C=O]\ (\mathbf{a}) + N_2$$

$$\mathbf{a} \xrightarrow[88\%]{H_2NCH_2C\equiv CH} CH_3COCH(CH_3)CONHCH_2C\equiv CH$$

[1] R. M. Moriarty, B. R. Bailey, III, O. Prakash, and I. Prakash, *Am. Soc.*, **107**, 1375 (1985).
[2] J. Cossy, D. Belotti, A. Thellend, and J. P. Pete, *Synthesis*, 720 (1988).

Trialkylchlorosilanes.

Conjugate additions of organocoppers. The beneficial effect of R_3SiCl on the rate and stereoselectivity of conjugate addition of organocopper reagents has been studied in detail by Nakamura *et al.*[1] In the absence of $(CH_3)_3SiCl$, the copper-catalyzed reaction of Grignard reagents to enones gives 1,2- and 1,4-adducts usually in about equal amounts. Addition of 1–2 equiv. of both $(CH_3)_3SiCl$ and HMPT to the reaction results in almost exclusive 1,4-addition at an accelerated rate. This effect is particularly marked in addition to enals to give enol silyl ethers, usually with high E-selectivity.

$(CH_3)_3SiCl$ also improves conjugate addition of stoichiometric alkylcopper reagents.

[1] S. Matsuzawa, Y. Horiguchi, E. Nakamura, and I. Kuwajima, *Tetrahedron*, **45**, 349 (1989).

Trialkylstibines. These organoantimony compounds are obtained by reaction of RMgX with $SbCl_3$.

β-Hydroxy ketones.[1] α-Bromo ketones undergo a Reformatsky-type reaction with aldehydes which is promoted by R_3Sb but which is only possible with iodine as catalyst. The function of I_2 is probably for a Br/I exchange, since I_2 is not required in reactions with α-iodo ketones.

$$i\text{-BuCHO} + CH_3CH(Br)CH_2COCH_3 + Bu_3Sb \xrightarrow[60\%]{I_2,\ 25°} i\text{-Bu}CH(OH)CH(CH_3)COCH_3 + (Bu_3SbBr)_2O$$

syn/*anti* = 57:43

$$\text{2-bromocyclohexanone} + p\text{-}BrC_6H_4CHO + Bu_3Sb \xrightarrow[84\%]{I_2} p\text{-}BrC_6H_4CH(OH)\text{-(2-oxocyclohexyl)}$$

(*syn*/*anti* = 71:29)

[1] Y.-Z. Huang, C. Chen, and Y. Shen, *J.C.S. Perkin I*, 2855 (1988).

Trialkyltin hydrides.

Cyclopropropylmethyl radicals.[1] Generation of a radical center adjacent to a cyclopropane can result in ring opening of the cyclopropylmethyl system. This reaction can be used to convert allylic cyclohexenols to alkylcyclohexenes, as shown for conversion of **1** into **3** via the phenyl selenide **2**. When the cyclopropyl ring is

$$\mathbf{1} \xrightarrow[82\%]{C_6H_5SeCN,\ PBu_3,\ THF} \mathbf{2}\ (SeC_6H_5) \xrightarrow[95\%]{(C_6H_5)_3SnH,\ h\nu,\ -10°} \mathbf{3}\ (CHCl_2, CH_3)$$

fused onto a cyclopentane, ring expansion is a competing, but minor, pathway (equation I). Photolysis of a benzoate or bromide rather than a selenide can also

$$\text{(I)}\quad (CH_3,\ SeC_6H_5,\ H) \xrightarrow[86\%]{Bu_3SnH,\ h\nu,\ 0°} (CH_3,\ H,\ CH_3) + \text{isomeric cyclohexene}\quad 84:16$$

be used to generate the radical. The Simmons–Smith reagent is suitable for cyclopropanation since the stereochemistry can be controlled by the hydroxyl group. Intramolecular cyclopropanation (**12**, 52) is also useful. The advantage of this system is the regio- and stereoselectivity.

[1] D. L. J. Clive and S. Daigneault, *J.C.S. Chem. Comm.*, 332 (1989).

Tributylarsine, Bu_3As (**1**).

RCHO → RCH=CHCO₂CH₃.[1] Conversion of aldehydes to (E)-α,β-unsaturated esters can be effected by reaction with methyl bromoacetate and Bu_3As (1 equiv.) catalyzed by zinc powder. This reaction proceeds with alkyl, aryl, and heteroaryl aldehydes.

$$RCHO + BrCH_2COOCH_3 \xrightarrow{Bu_3As,\ Zn} RCH\overset{(E)}{=}CHCO_2CH_3$$

$R = C_6H_5$	80%
$R = CH_3(CH_2)_5$	93%

[1] Y. Shen, B. Yang, and G. Yuan, *J.C.S. Chem. Comm.*, 144 (1989).

Tributylphosphine–Zinc.

Wittig-like reaction.[1] The reaction of an aldehyde with methyl bromoacetate promoted by Bu_3P (1 equiv.) and Zn (1 equiv.) provides α,β-unsaturated esters.

$$CH_3(CH_2)_{10}CHO + BrCH_2COOCH_3 \xrightarrow[71\%]{Bu_3P,\ Zn,\ 100°} CH_3(CH_2)_{10}CH\overset{(E)}{=}CHCOOCH_3$$

[1] Y. Shen, Y. Xin, and J. Zhao, *Tetrahedron Letters*, **29**, 6119 (1988).

Tributylstibine, Bu_3Sb.

α,β-Unsaturated nitriles.[1] The reaction of chloroacetonitrile with aldehydes and Bu_3Sb (1 equiv.) results in these unsaturated nitriles as a mixture of *cis*- and *trans*-isomers.

$$n\text{-}C_5H_{11}CHO + ClCH_2CN + Bu_3Sb \xrightarrow[92\%]{} n\text{-}C_5H_{11}CH{=}CHCN\ [+ Bu_3SbCl_2 + Bu_3Sb(OH)_2]$$

(*trans*/*cis* = 62:38)

[1] Y.-Z. Huang, Y. Shen, and C. Chen, *Syn. Comm.*, **19**, 83 (1989).

Tributyltin hydride.

Regioselective cleavage of* gem-*difluoropropanes. 1,1-Difluoro-2-(bromomethyl)cyclopropane is cleaved by Bu_3SnH/AIBN to 3,3-difluoro-1-butene as the

$$\text{1,1-difluoro-2-(}CH_2Br\text{)cyclopropane} \xrightarrow{Bu_3SnH,\ AIBN,\ C_6H_6} CH_3CF_2CH{=}CH_2$$

only product.[1] This radical reaction can be used to obtain (E)-difluoroallylic compounds because of preferential cleavage of the C_2—C_3 bond.[2]

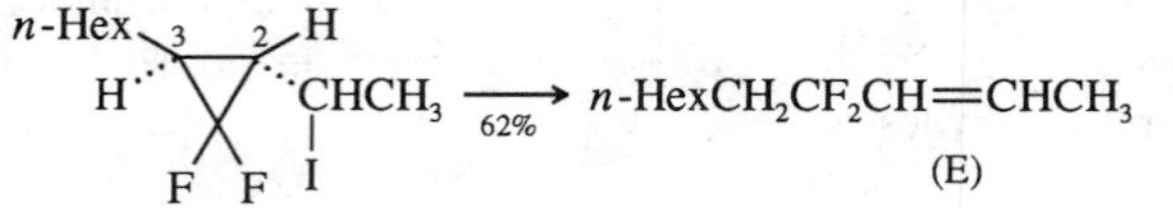

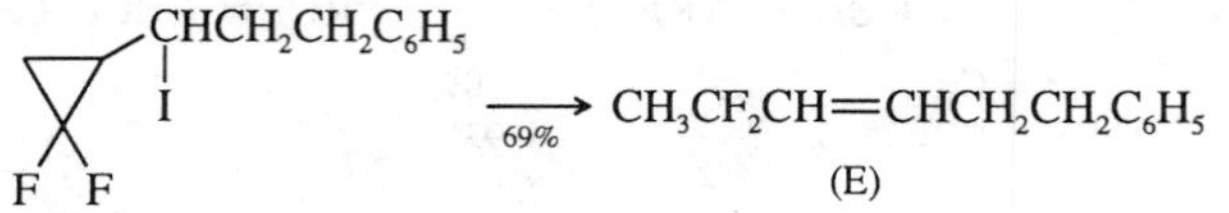

$$n\text{-HexCH}_2\text{CF}_2\text{CH=CHCH}_3 \text{ (E) } (62\%)$$

$$\text{CH}_3\text{CF}_2\text{CH=CHCH}_2\text{CH}_2\text{C}_6\text{H}_5 \text{ (E) } (69\%)$$

Intramolecular 1,5-hydrogen atom transfer cyclizations.[3] In reactions with low concentrations of the tin hydride (**14**, 313), an original vinyl radical can undergo a 1,5-hydrogen shift to produce a new radical from a C—H bond, which then undergoes cyclization (equation I).

(I) Bu_3SnCl, AIBN, $NaCNBH_3$, $(CH_3)_3COH$ → 88%

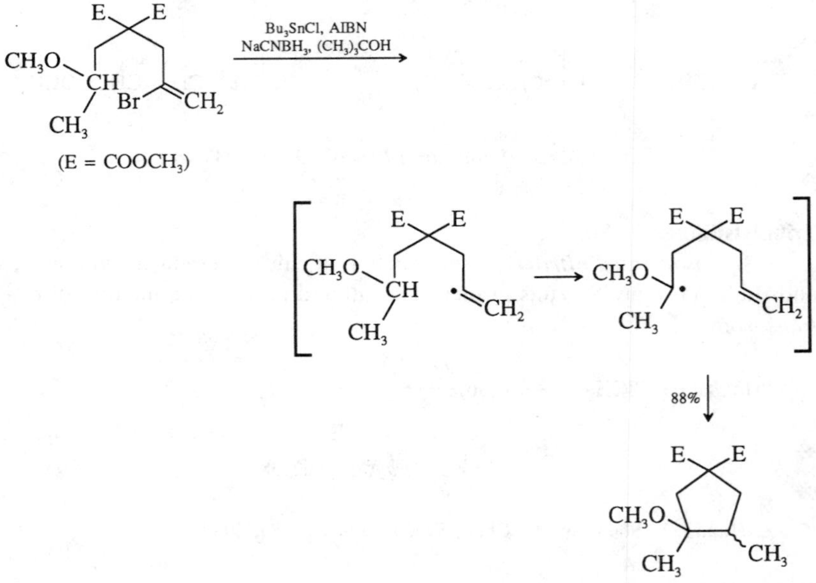

(E = $COOCH_3$)

(*cis*/*trans* = 18:82)

Example:

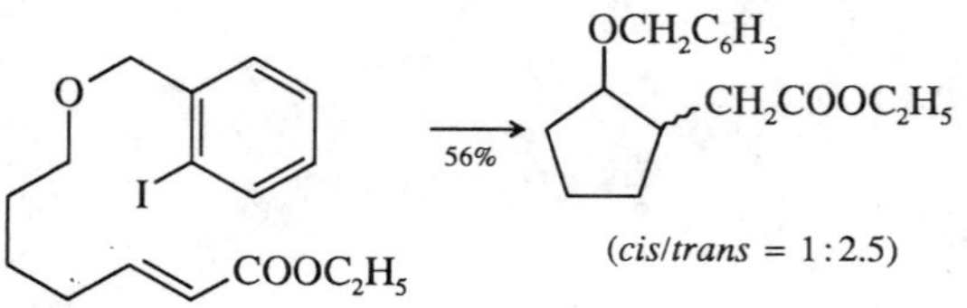

56%

$OCH_2C_6H_5$, $CH_2COOC_2H_5$

(*cis*/*trans* = 1:2.5)

2-Deoxy sugars.[4] Reduction of acetylated or benzoylated glycosyl halides with Bu_3SnH is accompanied by rearrangement of the ester group at C_2 to the anomeric center. The migration is *cis*-selective. Reduction of 6- or 4-iodoglucosyl iodide by Bu_3SnH is not accompanied by this rearrangement.

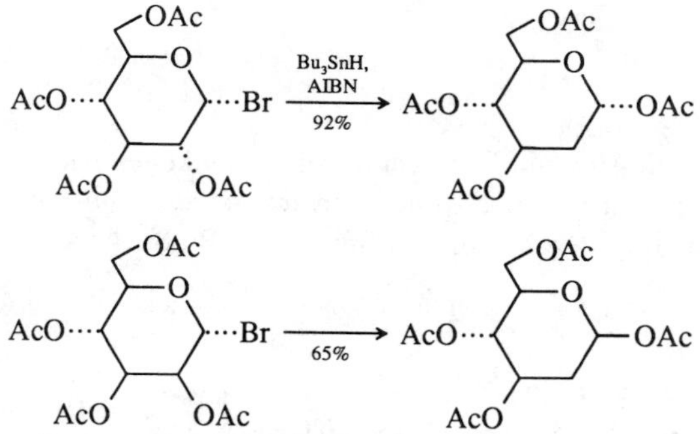

OAc, O, AcO, Br, AcO, OAc → (Bu_3SnH, AIBN, 92%) OAc, O, AcO, OAc, AcO

OAc, O, AcO, Br, AcO, OAc → (65%) OAc, O, AcO, OAc, AcO

Allyl radical cyclization.[5] Allylic radicals are less reactive than vinyl ones, but can undergo cyclization with Bu_3SnH/AIBN.

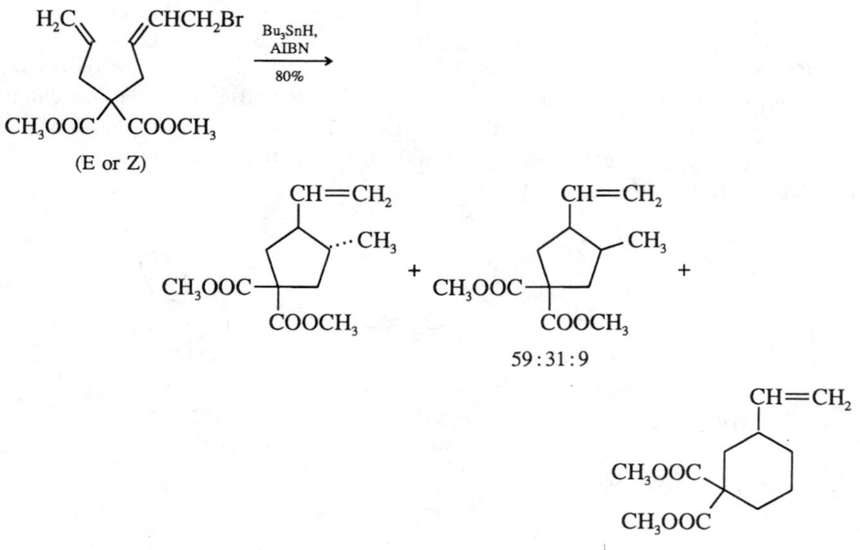

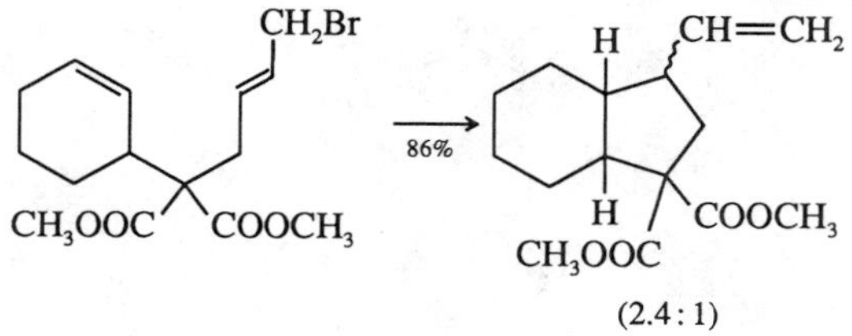

H_2C, $CHCH_2Br$, CH_3OOC, $COOCH_3$ (E or Z) → (Bu_3SnH, AIBN, 80%)

$CH{=}CH_2$, CH_3, CH_3OOC, $COOCH_3$ + $CH{=}CH_2$, CH_3, CH_3OOC, $COOCH_3$ + $CH{=}CH_2$, CH_3OOC, CH_3OOC

59 : 31 : 9

CH_2Br, CH_3OOC, $COOCH_3$ → (86%) H, $CH{=}CH_2$, H, CH_3OOC, $COOCH_3$

(2.4 : 1)

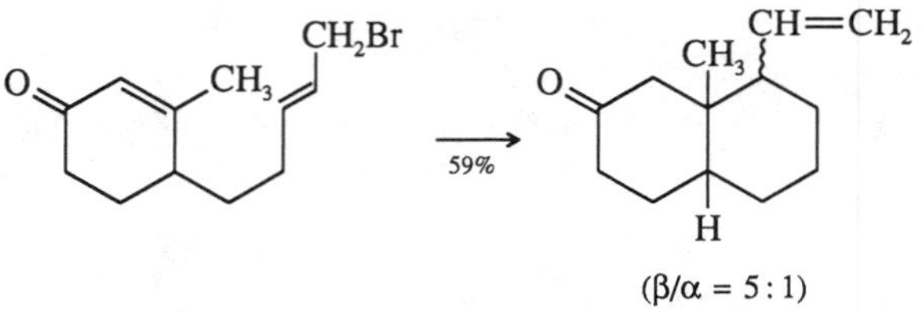

59%

($\beta/\alpha = 5:1$)

anti-*Radical elimination*.[6] Reaction of nitroalkenes with lithium benzenethiolate and then with aqueous formaldehyde provides *anti*-γ-phenylthio-β-nitro alcohols (**1**), which undergo radical elimination (Bu_3SnH, AIBN, 110°) to (E)-allylic alcohols.

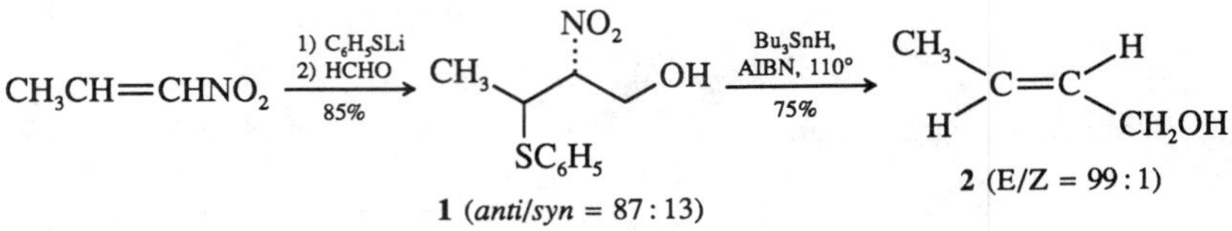

$CH_3CH{=}CHNO_2$ — 1) C_6H_5SLi, 2) HCHO, 85% → **1** (*anti/syn* = 87 : 13) — Bu_3SnH, AIBN, 110°, 75% → **2** (E/Z = 99 : 1)

***RI* → *RCN*.**[7] One step required in a recent synthesis of 9,11-dehydrodigitoxigenin (**3**) is an iodide/cyanide exchange with the steroid **1**. Direct exchange requires protection of the 14β-hydroxyl group and proceeds in only 40% yield. The exchange can be achieved by a free-radical reaction initiated by Bu_3SnCl, $NaCNBH_3$, and AIBN in 82% yield.

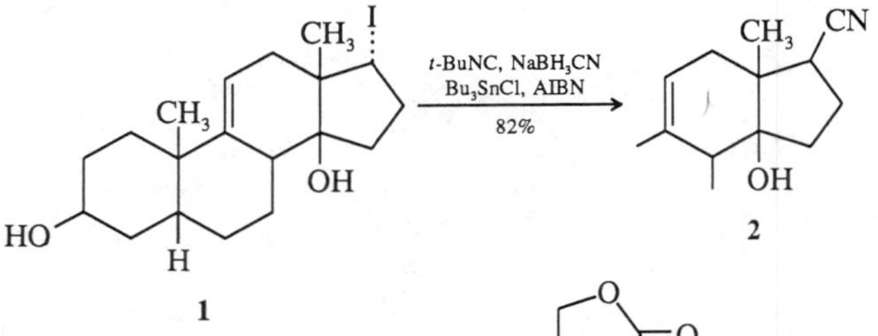

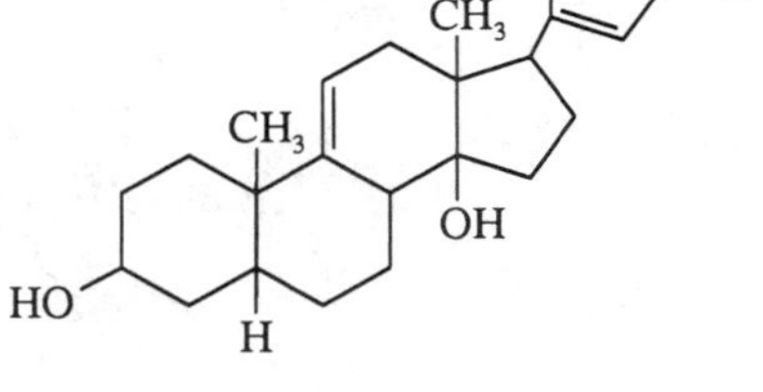

1 — *t*-BuNC, $NaBH_3CN$, Bu_3SnCl, AIBN, 82% → **2**

3

Acyl radicals.[8] Tributyltin hydride in combination with AIBN can generate acyl radicals from phenylselenol esters, which are a better source than acyl chlorides or phenylthiol esters for acyl radicals for intramolecular cyclization.

$COSeC_6H_5$ — Bu_3SnH, AIBN, C_6H_6, reflux, 82% → (*trans*/*cis* = 62:38)

$COSeC_6H_5$, $COOCH_3$ — 71% → CH_2COOCH_3

Acylation of isoquinolines.[9] A mixture of an isoquinoline and Bu_3SnH reacts at −78° with a 3-halopropionyl chloride in CH_2Cl_2 to give N-acyl-1,2-dihydroisoquinolines in 80–95% yield (*cf.*, **13**, 10). These can undergo radical cyclization with Bu_3SnH/AIBN to furnish benzoindolizidin-3-ones.

R + $ClCCH_2CH_2I$ (C=O) — Bu_3SnH, CH_2Cl_2, 85–95% → R

45–80% | Bu_3SnH, AIBN

Acylamino cyclization.[10] The indolizidine alkaloid (−)-swainsonine (**3**) has been synthesized by radical cyclization of the acylamine **1**, derived from D-tartaric acid, to provide the indolizidinones **2** in 71% yield. Conversion to the alkaloid (**3**) included removal of the keto group by conversion to the thiolactam by Lawesson's reagent (97%) followed by desulfurization with Raney nickel (96% yield).

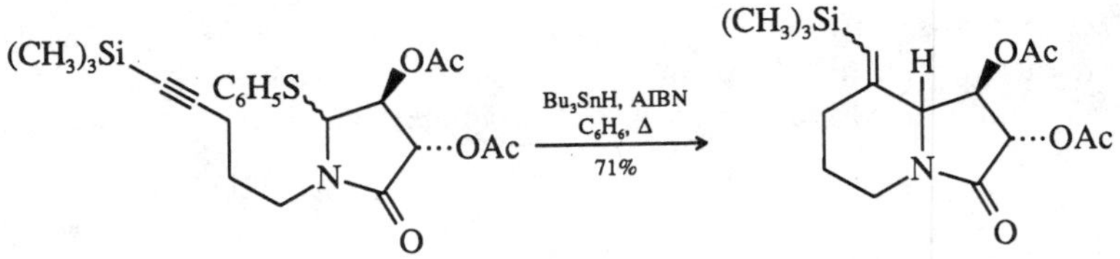

Bu$_3$SnH, AIBN
C$_6$H$_6$, Δ
71%

1

2 (~1 : 1)

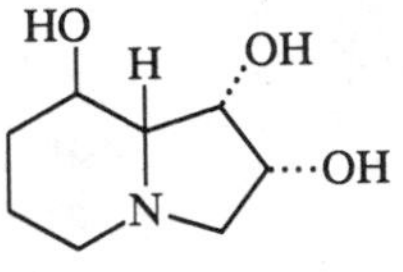

3

β-Glycosides.[11] The construction of β-glycosides has relied heavily on the anomeric effect (equatorial C$_2$-acetoxy substituent). A recent approach relies on generation of a radical at an alkoxy-substituted anomeric position. The precursors (**1**) can be obtained as shown in equation (I). The same strategy can be applied to construction of β-linked disaccharides.

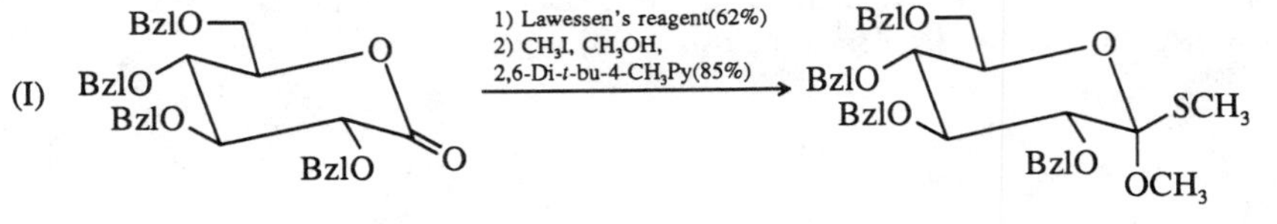

(I)

1) Lawessen's reagent(62%)
2) CH$_3$I, CH$_3$OH,
2,6-Di-*t*-bu-4-CH$_3$Py(85%)

1

87% Bu$_3$SnH, AIBN
C$_6$H$_5$CH$_3$, 30°

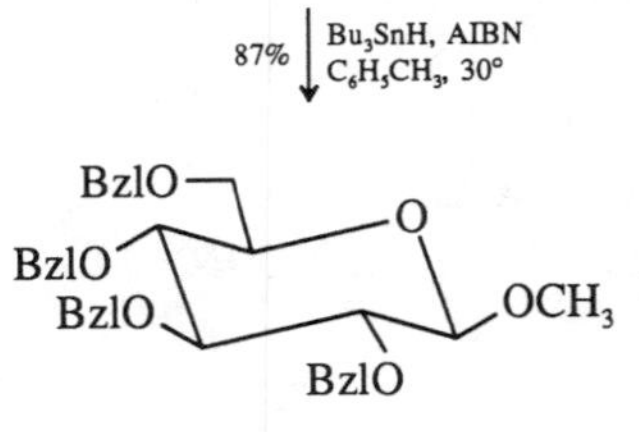

2 (β/α = 12 : 1)

Cyclization of selenoimidates.[12] Reaction of the selenoimidate **1** with Bu$_3$SnH (AIBN) generates an imidoyl radical that cyclizes to the nitrile **2** and the imine **3** of a chromanone. Only **3** is obtained if the benzyl substituent on nitrogen is changed to an alkyl group. The reaction follows a different course when an N-tolyl selenoimidate (**4**) is the starting imine (equation I).

$$\text{1 (o-}(\text{CH}_2\text{CH=CHC}_6\text{H}_5\text{O})\text{C}_6\text{H}_4\text{C(=N-Bzl)SeC}_6\text{H}_5) \xrightarrow{\text{Bu}_3\text{SnH, AIBN}} \mathbf{2}\ (50\%) + \mathbf{3}\ (50\%)$$

2 (50%): o-(CN)C$_6$H$_4$OCH$_2$CH=CHC$_6$H$_5$

3 (50%)

(II) 4 → (cyclized product bearing N, CH_3, C_6H_5, O)

Thioimidates can undergo similar radical cyclization.

Deamination of RNH_2.[13] A new method for reductive deamination of primary amines (**1**) involves conversion to an aromatic imidoyl chloride (**2**), which is converted to the hydrocarbon by Bu_3SnH and AIBN (**10**, 412–413). Yields are satisfactory in the deamination of benzylamine, but are only moderate in the reaction of simple alkylamines.

$$\underset{\mathbf{1}}{RNH_2} \xrightarrow[2)\ SOCl_2]{1)\ C_6H_5COCl} \underset{\mathbf{2}}{C_6H_5C(Cl){=}NR} \xrightarrow{Bu_3SnH,\ AIBN} RH + C_6H_5CN$$

R	RH	C_6H_5CN
$R = C_6H_5CH_2$	(82%)	(82%)
$= C_6H_{13}$-c	(63%)	(60%)
$= C_8H_{17}$-n	(38%)	(37%)

Stereoselective radical cyclization of 1,5-cyclooctadienes. Winkler and Sridar[14] have noted transannular effects on the stereochemistry of cyclization of these octadienes (equation I). Thus radical cyclization of *cis*-**1** and of the corresponding ketone (**3**) is highly *trans*-selective and is more *trans*-selective than that of the monosubstituted diene **5**. Cyclization of a 3,8-disubstituted cyclooctene is

(I) cis-1 → 2 (99 : 1 trans); trans-1 → 2 (5 : 1 trans) [Bu_3SnH, AIBN, $h\nu$; substrate bears OH and $(CH_2)_3I$]

3 [$(CH_2)_3I$] → trans-4 (95 : 5)

5 [$(CH_2)_3I$] → trans-6 (7 : 1)

completely *trans*-selective. The observed *trans*-selectivity can be correlated with transannular effects in unsaturated eight-membered rings.

Thionolactones.[15] The reaction of thionocarbonic acid derivatives of 4-phenyl-3-butenol with Bu_3SnH (AIBN) can provide thionolactones as the major product via a radical cyclization.

$C_6H_5CH{=}CHCH_2CH_2OC(S)SCH_3 \xrightarrow[77\%]{Bu_3SnH,\ AIBN,\ C_6H_6,\ \Delta}$ Bzl-thionolactone $+ Bu_3SnSCH_3$

[1] W. R. Dolbier, Jr., B. H. Al-Sader, S. F. Sellers, and H. Koroniak, *Am. Soc.*, **103**, 2138 (1981).
[2] T. Morikawa, M. Uejima, and Y. Kobayashi, *Chem. Letters*, 1407 (1988).
[3] D. P. Curran, D. Kim, H. T. Liu, and W. Shen, *Am. Soc.*, **110**, 5900 (1988).
[4] B. Giese, S. Gilges, K. S. Gröninger, C. Lamberth, and T. Witzel, *Ann.*, 615 (1988); B. Giese and K. S. Gröninger, *Org. Syn.*, submitted (1988).
[5] G. Stork and M. E. Reynolds, *Am. Soc.*, **110**, 6911 (1988).
[6] A. Kamimura and N. Ono, *J.C.S. Chem. Comm.*, 1278 (1988).
[7] A. R. Daniewski, M. M. Kabat, M. Masnyk, J. Wicha, W. Wojciechowska, and H. Duddeck, *J. Org.*, **53**, 4855 (1988).

[8] D. L. Boger and R. J. Mathvink, *ibid.*, **53**, 3377 (1988).
[9] R. Yamaguchi, T. Hamasaki, and K. Utimoto, *Chem. Letters*, 913 (1988).
[10] J. M. Dener, D. J. Hart, and S. Ramesh, *J. Org.*, **53**, 6022 (1988).
[11] D. Kahne, D. Yang, J. J. Lim, R. Miller, and E. Paguaga, *Am. Soc.*, **110**, 8716 (1988).
[12] M. D. Bachi and D. Denenmark, *ibid.*, **111**, 1886 (1989).
[13] T. Wirth and C. Rüchardt, *Chimia*, **42**, 230 (1988).
[14] J. D. Winkler and V. Sridar, *Tetrahedron Letters*, **29**, 6219 (1988); *Idem*, *Am. Soc.*, **108**, 1708 (1986).
[15] M. D. Bachi and E. Bosch, *J. Org.*, **54**, 1234 (1989).

Tributyltin hydride–Tributylphosphine oxide.

Reduction of carbonyl compounds.[1] α-Chloro aldehydes or ketones are reduced to chlorohydrins by Bu_3SnH in combination with 1 equiv. of Bu_3PO or HMPT.

[1] I. Shibata, T. Suzuki, A. Baba, and H. Matsuda, *J.C.S. Chem. Comm.*, 882 (1988).

Tributyltin hydride–Triethylborane.

Deoxygenation of alcohols.[1] Dithiocarbonates are converted to the corresponding alkane on treatment with Bu_3SnH (1 equiv.) and $(C_6H_5)_3B$ (1 equiv.) in benzene.

$$\mathrm{RO\overset{S}{\overset{\|}{C}}SCH_3} \xrightarrow[83-93\%]{Bu_3SnH,\ (C_2H_5)_3B} \mathrm{RH}$$

Butyrolactones.[2] The same reagent effects cyclization of dithiocarbonates of homopropargylic or homoallylic alcohols to γ-butyrolactones.

C_6H_5 SCH_3 C=S O $\xrightarrow[-Bu_3SnSCH_3]{Bu_3SnH,\ (C_2H_5)_3B}$ [C_6H_5 S O] $\xrightarrow[78\%]{H_3O^+}$ C_6H_5 O O

C_2H_5 SCH_3 S O $\xrightarrow[82\%]{}$ C_2H_5 CH_2 O O

[1] K. Nozaki, K. Oshima, and K. Utimoto, *Tetrahedron Letters*, **29**, 6125 (1988).
[2] *Idem*, *ibid.*, **29**, 6127 (1988).

Tri-μ-carbonylhexacarbonyldiiron, $Fe_2(CO)_9$.

Tricarbonyliron complexes of enones.[1] Reaction of $Fe_2(CO)_9$ with α,β-enones in toluene at 70–75° results in tricarbonyliron complexes, which on reaction with Grignard reagents, organolithium, or organocuprates are converted into 1,4-diketones.

$$C_6H_5CH{=}CHC(O)Bu \xrightarrow[35\%]{Fe_2(CO)_9} C_6H_5CH{=}CHC(O)Bu\cdot Fe(CO)_3 \xrightarrow[73\%]{1)\ CH_3Li\quad 2)\ CH_3Br} CH_3C(O)CH(C_6H_5)CH_2C(O)Bu$$

[1] T. N. Danks, D. Rakshit, and S. E. Thomas, *J.C.S. Perkin I*, 2091 (1988).

Tri-μ-carbonylnonacarbonyltetrarhodium [Tetrarhodium dodecacarbonyl, $Rh_4(CO)_{12}$]. Preparation.[1]

Silylformylation of alkynes.[2] β-Silyl-α,β-enals can be prepared by reaction of dimethylphenylsilane and carbon monoxide with alkynes catalyzed by $Rh_4(CO)_{12}$. Yields are improved if triethylamine is also present. Regioselectivity in the reaction with internal alkynes is controlled by steric factors.

$$CH_3C{\equiv}CH + CO + (CH_3)_2C_6H_5SiH \xrightarrow[99\%]{Rh_4(CO)_{12},\ N(C_2H_5)_3,\ C_6H_6,\ 100^\circ} (CH_3)(OHC)C{=}CHSi(CH_3)_2C_6H_5$$

(Z/E = 80:20)

$$C_6H_5C{\equiv}CCH_3 + CO + R_3SiH \xrightarrow[95\%]{} (C_6H_5)(OHC)C{=}C(CH_3)(SiR_3) + (C_6H_5)(R_3Si)C{=}C(CH_3)(CHO)$$

89:11

$$CH_3C{\equiv}CCOOCH_3 + CO + R_3SiH \xrightarrow[67\%]{} (CH_3)(OHC)C{=}C(COOCH_3)(SiR_3)$$

(Z, 100%)

[1] S. Martinengo, G. Giordano, and P. Chini, *Inorg. Syn.*, **20**, 209 (1980).
[2] I. Matsuda, A. Ogiso, S. Sato, and Y. Izumi, *Am. Soc.*, **111**, 2332 (1989).

Tricarbonyl(triphenylphosphine)nickel, $(CO)_3Ni[P(C_6H_5)_3]$ (**1**). Preparation.[1] This metal carbonyl is stable and less toxic than $Ni(CO)_4$.

Intramolecular allylation/methoxycarbonylation of alkenes and alkynes.[2] $Ni(COD)_2$ is an efficient catalyst for intramolecular allylation of alkenes (equation I).

(I) [AcO–CH₂CH=CH–CH₂–C(SO₂C₆H₅)₂–CH₂CH=CH₂] $\xrightarrow[83\%]{Ni(COD)_2,\ dppb}$ [3-vinyl-4-methylene-1,1-bis(phenylsulfonyl)cyclopentane] + HOAc

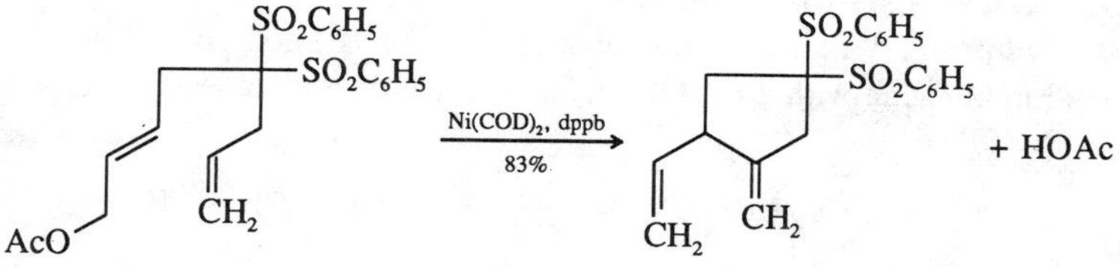

The intermediate cyclized allylnickel species can be trapped by methoxycarbonylation when CO and CH_3OH are present and tricarbonyl(triphenylphosphine)nickel is the catalyst.

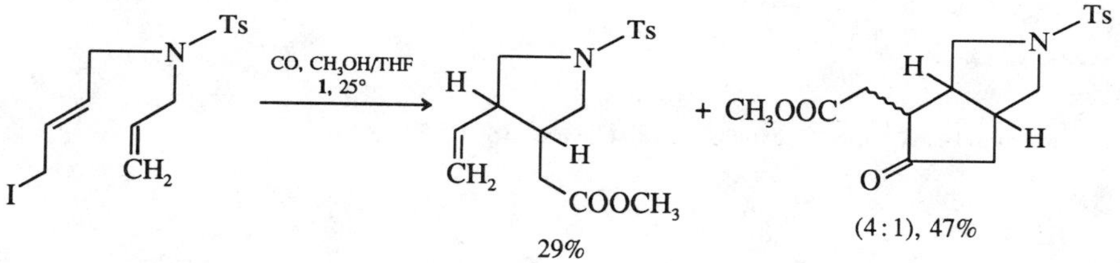

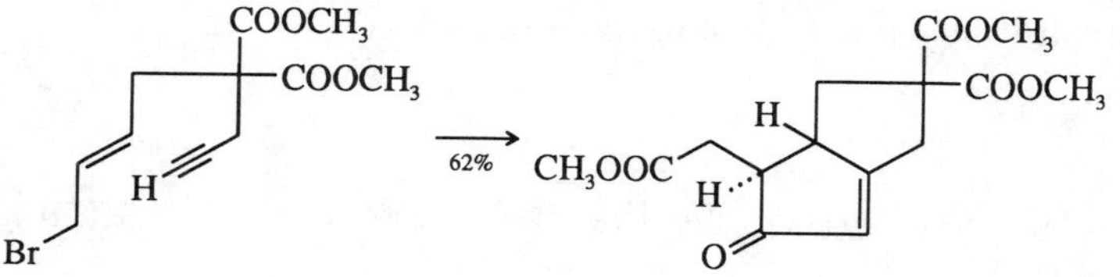

[I–CH₂CH=CH–CH₂–N(Ts)–CH₂CH=CH₂] $\xrightarrow[1,\ 25°]{CO,\ CH_3OH/THF}$ [N-Ts pyrrolidine with vinyl and CH₂COOCH₃ substituents] 29% + [CH_3OOC-substituted bicyclic N-Ts ketone] (4:1), 47%

[(Z) Br–CH₂CH=CH–CH₂–C(COOCH₃)₂–CH₂C≡CH] $\xrightarrow{62\%}$ [CH_3OOC-substituted bicyclic enone bearing C(COOCH₃)₂]

(Z)

[1] W. F. Edgell, M. P. Dunkle, *Inorg. Chem.*, **4**, 1629 (1965).
[2] W. Oppolzer, M. Bedoya-Zurita, and C. Y. Switzer, *Tetrahedron Letters*, 6433 (1988).

Trichloroisopropoxytitanium, Cl_3TiO-*i*-Pr (**1**).

Nitro aldol reaction (Henry reaction[1]). The reaction of the anion (BuLi) of primary nitroalkanes with aromatic aldehydes to form β-nitro alcohols is *anti*-selective if carried out in the presence of this titanium reagent (1 equiv.).[2]

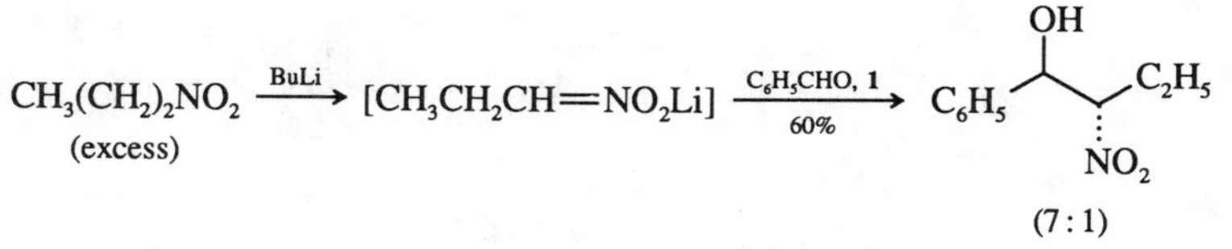

$$CH_3(CH_2)_2NO_2 \xrightarrow{BuLi} [CH_3CH_2CH{=}NO_2Li] \xrightarrow[60\%]{C_6H_5CHO,\ 1} C_6H_5CH(OH)CH(NO_2)C_2H_5$$

(excess) (7:1)

[1] H. B. Hass and E. F. Riley, *Chem. Revs.*, **32**, 406 (1943).
[2] A. G. M. Barrett, C. Robyr, and C. D. Spilling, *J. Org.*, **54**, 1233 (1989).

1,1,1-Trichloro-3,3,3-trifluoroacetone.

Dehydration.[1] Primary and secondary alcohols undergo dehydration when refluxed in benzene with CF_3COCCl_3 and a trace of *p*-toluenesulfonic acid.

$$p\text{-}CH_3OC_6H_4CH_2CH_2OH \xrightarrow[88\%]{} p\text{-}CH_3OC_6H_4CH{=}CH_2$$

$$HO\text{-}C_6H_{10}\text{-}OH \xrightarrow[92\%]{} \text{1,3-cyclohexadiene}$$

$$CH_3CH(OH)(CH_2)_5CH_3 \xrightarrow[75\%]{} CH_3CH{=}CH(CH_2)_4CH_3 \ (E)$$

[1] S. Abdel-Baky and A. Moussa, *Syn. Comm.*, **18**, 1795 (1988).

Triethylamine.

Seleno–Pummerer rearrangement. Diacetonylselenium dichlorides (**1**) when treated with triethylamine rearrange to α-chloro selenides (**2**).[1]

$$2\ CH_3COC(CH_3)_3 \xrightarrow{SeOCl_2} [(CH_3)_3CCOCH_2]_2SeCl_2\ (\mathbf{1}) \xrightarrow[90\%]{N(C_2H_5)_3} (CH_3)_2HCC(O)CH_2\text{-}Se\text{-}CHClC(O)CH(CH_3)_2\ (\mathbf{2})$$

[1] J. Nakayama and Y. Sugihara, *Chem. Letters*, 1317 (1988).

Triethylammonium bis(catecholato)alkenylsiliconates (1).

$$(C_2H_5)_3\overset{+}{N}H\left[RCH{=}CH\text{-}Si(O_2C_6H_4)_2\right]^-$$

1a, R = H
b, R = C_6H_5

The pentacoordinate organosilicon compounds are obtained by reaction of alkenyltrialkoxysilanes with catechol and $N(C_2H_5)_3$.

Coupling with halides and triflates.[1] In the presence of several palladium catalysts, these reagents couple with vinyl or aryl iodides and aryl triflates in moderate to high yield.

$$p\text{-}O_2NC_6H_4I + \mathbf{1a} \xrightarrow[84\%]{PdCl_2(C_6H_5CN)_2,\ \text{dioxan}} p\text{-}O_2NC_6H_4CH{=}CH_2$$

$$C_6H_5CH{=}CHBr + \mathbf{1b} \xrightarrow[48\%]{} C_6H_5CH{=}CHCH{=}CHC_6H_5$$

[1] A. Hosomi, S. Kohra, and Y. Tominaga, *Chem. Pharm. Bull.*, **36**, 4622 (1988).

Triethylborane.

Radical addition of R_3SnH to —C≡C—.[1] $B(C_2H_5)_3$ is as efficient as AIBN for initiation of this radical reaction resulting in vinyltins selectively. The reaction is sluggish in the absence of oxygen.

$$RC{\equiv}CH + (C_6H_5)_3SnH \xrightarrow[75\text{–}85\%]{B(C_2H_5)_3,\ 25°} (R)(H)C{=}C(H)(Sn(C_6H_5)_3)\ (E) + (Z)\text{-isomer} \quad 80:20$$

$$n\text{-}C_5H_{11}C{\equiv}CC_5H_{11}\text{-}n + (C_6H_5)_3SnH \xrightarrow[86\%]{} (n\text{-}C_5H_{11})(H)C{=}C(Sn(C_6H_5)_3)(C_5H_{11}\text{-}n)$$

Triethylborane can also initiate radical cyclization of unsaturated alkynes to vinylstannanes.

$$\text{O–CH}_2\text{C≡CH substrate (CH}_3\text{, CH}_3\text{, CH}_3) \xrightarrow[96\%]{(C_6H_5)_3SnH,\ B(C_2H_5)_3} \text{tetrahydrofuran with }{=}CHSn(C_6H_5)_3\text{, CH}_3\text{, CH(CH}_3)_2$$

[1] K. Nozaki, K. Oshima, and K. Utimoto, *Tetrahedron*, **45**, 923 (1989).

Triethyloxonium tetrafluoroborate.

ω-Amino esters.[1] These compounds can be prepared from lactams (**1**) by conversion to lactim ethers (**2**) with $(C_2H_5)_3O^+BF_4^-$ to provide the salts (**3**), which are hydrolyzed by water to the salts (**4**) of ω-amino esters.

$(CH_2)_n$ lactam (1) $\xrightarrow[CHCl_3]{(C_2H_5)_3O^+BF_4^-}$ 2 (OC_2H_5, BF_4^-) $\xrightarrow{H_2O}$ $(CH_2)_n$ $COOC_2H_5$ / $\overset{+}{N}H_3BF_4^-$

1, n = 1	98%	2	87%
n = 2	80%		90%

[1] R. Menezes and M. B. Smith, *Syn. Comm.*, **18**, 1625 (1988).

Triethylsilane–Titanium(IV) chloride.

Reduction of bicyclic ketals.[1] Reductive cleavage of the bicyclic ketal **1** in the presence of a Lewis acid favors *cis*-products, regardless of the reductant. The highest *cis*-selectivity is obtained with $(C_2H_5)_3SiH$ and $TiCl_4$. DIBAL shows high

1 (H, $(CH_2)_2OBzl$) $\xrightarrow{CH_2Cl_2}$ *cis*-**2** (HO, H, H, $(CH_2)_2OBzl$) + *trans*-**2**

$TiCl_4/(C_2H_5)_3SiH$, −78°	100%	99.93:0.07
DIBAL, 0°	100%	4:96

trans-selectivity. This reaction provides one step in the conversion of L-malic acid into the optically active form of a civet cat acid (**3**).

H, $(CH_2)_2OBzl$ $\xrightarrow[\sim 100\%]{(C_2H_5)_3SiH,\ TiCl_4}$ HO, H, H, $(CH_2)_2OBzl$ $\longrightarrow$ H, CH_3, H, COOH

(−)-**3**

[1] H. Kotsuki, Y. Ushio, I. Kadota, and M. Ochi, *Chem. Letters*, 927 (1988).

Trifluoroacetic acid.

(E)-α-Methyl-α,β-enals.[1] The anion of the *t*-butyl- or cyclohexylimine of α-trialkylsilylpropionaldehyde (**1**) is known (**13**, 323) to react with aldehydes to provide α-methyl-α,β-enals as a mixture of (E)- and (Z)-isomers in the ratio 1.3–11:1. However, the crude imine intermediates isomerize mainly to the (E)-isomers when treated at room temperature with trifluoroacetic acid.

$$R_3{}^1SiCH(CH_3)CH{=}NC(CH_3)_3 \ (\mathbf{1}) + RCHO \xrightarrow{sec\text{-BuLi}} [RCH{=}C(CH_3)CH{=}NC(CH_3)_3] \xrightarrow[2)\ H_2O]{1)\ CF_3COOH,\ 20°} (H)(R)C{=}C(CHO)(CH_3)$$

(E/Z > 100:1)

[1] R. Desmond, S. G. Mills, R. P. Volante, and I. Shinkai, *Tetrahedron Letters*, **29**, 3895 (1988).

Trifluoromethanesulfonic acid, CF_3SO_3H.

Protonation of nitroalkenes. This strong acid diprotonates nitroalkenes, even nitroethylene, to give N,N-dihydroxyiminium carbenium ions, which react with arenes to give arylated oximes. This overall process provides a route to α-aryl methyl ketones from 2-nitropropene.[1]

$$H_2C{=}C(CH_3)NO_2 \xrightarrow{CF_3SO_3H} [H_2\overset{+}{C}{-}C(CH_3){=}\overset{+}{N}(OH)_2] \xrightarrow{C_6H_6} [C_6H_5CH_2C(CH_3){=}NOH] \xrightarrow[85\%]{H_2O} C_6H_5CH_2COCH_3$$

[1] K. Okabe, T. Ohwada, T. Ohta, and K. Shudo, *J. Org.*, **54**, 733 (1989).

Trifluoromethanesulfonic anhydride (1).

Ritter reaction. The original version for this conversion of a tertiary alcohol to an amide involved a reaction with nitriles in a strongly acidic medium.[1] In a new version, the alcohol is converted into a triflate, which need not be isolated, but is treated with a nitrile (2 equiv.) in CH_2Cl_2 and then with aqueous $NaHCO_3$. Overall yields are 50–98%. The advantage of this version is that it is applicable to primary and secondary alcohols as well as tertiary ones.[2]

$$ROH \longrightarrow \left[ROTf \xrightarrow{R'CN} R\overset{+}{N}{\equiv}CR'OTf^- \right] \xrightarrow[50-98\%]{NaHCO_3} RNHCOR'$$

[1] L. I. Krimen and D. J. Cota, *Org. React.*, **17**, 213 (1969).
[2] A. G. Martínez, R. M. Alvarez, E. T. Vilar, A. G. Fraile, M. Hanack, and L. R. Subramanian, *Tetrahedron Letters*, **30**, 581 (1989).

N-*p*-Trifluoromethylbenzylcinchoninum bromide, 12, 379–381; **13**, 325.

Asymmetric synthesis of α-amino acids.[1] This optically active phase-transfer catalyst (**1**) can effect enantioselective alkylation of the imine (**2**) derived from glycine. The highest enantioselectivity obtains with the *t*-butyl ester; the lowest with benzylic type esters. Alkylation of **2** results in a mixture of enantiomers, but crystallization of the mixture removes the racemic product and leaves the highly pure optically active amino ester in solution.

$$\underset{\mathbf{2}}{(C_6H_5)_2C{=}NCH_2COOC(CH_3)_3} \xrightarrow[75\%]{p\text{-}ClC_6H_4CH_2Br,\ \mathbf{1};\ CH_2Cl_2,\ H_2O,\ NaOH}$$

$$\underset{(R)\text{-}\mathbf{3}}{(C_6H_5)_2C{=}NCH(CH_2C_6H_4Cl\text{-}p)COOC(CH_3)_3} \quad + \quad \underset{(S)\text{-}\mathbf{3}}{(C_6H_5)_2C{=}NCH(CH_2C_6H_4Cl\text{-}p)COOC(CH_3)_3}$$

83 : 17

[1] M. J. O'Donnell, W. D. Bennett, and S. Wu, *Am. Soc.*, **111**, 2353 (1989).

Trifluoromethyl bromide.

Trifluoromethylation of arenes.[1] Arenes are trifluoromethylated by CF_3Br on irradiation (315–440 nm) in acetonitrile. This photochemical reaction with uracil provides 5-trifluoromethyluracil in 11% yield.

Naphthalene $\xrightarrow{CF_3Br,\ h\nu}$ 1-CF_3-naphthalene (78%) + 2-CF_3-naphthalene (22%)

[1] T. Akiyama, K. Kato, M. Kajitani, Y. Sakaguchi, J. Nakamura, H. Hayashi, and A. Sugimori, *Bull. Chem. Soc. Japan*, **61**, 3531 (1988).

Trifluoromethyltrimethylsilane, $(CH_3)_3SiCF_3$ (**1**). The reagent is prepared from $BrCF_3$, $ClSi(CH_3)_3$, and hexaethylphosphoric triamide in 65–70% yield.[1]

Trifluoromethylation of carbonyl compounds.[2] In the presence of Bu_4NF, this reagent adds to a carbonyl group to give a silyl ether that is hydrolyzed to the free trifluoromethylated alcohol in 60–92% yield.

1) **1**, Bu_4NF, THF; 2) H_3O^+; 92% → OH, CF_3

[1] I. Ruppert, K. Schlich, and W. Volbach, *Tetrahedron Letters*, **25**, 2195 (1984).
[2] G. K. S. Prakash, R. Krishnamurti, and G. A. Olah, *Am. Soc.*, **111**, 393 (1989).

Trimethoxysilane–Dilithium 2,3-butanediolate $(CH_3O)_3SiH–[LiOCH(CH_3)]_2$, (**1**).

S,N-Acetals. Reduction of alkylthiomethyleniminium salts with **1** in THF at 0° results in S,N-acetals (**2**) as the major product. Reduction of these salts with $LiAlH_4$ or $NaBH_4$ results in tertiary amines or a mixture of **2** and the corresponding tertiary amine.[1]

$$p\text{-}CH_3OC_6H_4C(SC_2H_5)\!=\!N^+(\text{morpholine})\ I^- \xrightarrow[91\%]{1,\ THF,\ 0°} p\text{-}CH_3OC_6H_4CH(SC_2H_5)N(\text{morpholine})$$

2

[1] Y. Tominaga, Y. Matsuoka, H. Hayashida, S. Kohra, and A. Hosomi, *Tetrahedron Letters*, **29**, 5771 (1988).

Trimethylaluminum.

Cleavage of (2S,3S)-3-phenylglycidol (**1**).[1] This epoxide (**1**) is cleaved by a number of nucleophiles regioselectively at the benzylic carbon to give 1,2-diols. The cleavage usually occurs with almost complete inversion at the benzylic center (>99:1). Trimethylaluminum, however, reacts with retention (32:1, equation I). Triethylaluminum reacts with **1** to give the corresponding adducts in a 1:1 ratio.

(I) C_6H_5–(epoxide)–CH_2OH (**1**) $\xrightarrow[87\%]{Al(CH_3)_3}$ $C_6H_5CH(CH_3)CH(OH)CH_2OH$ + $C_6H_5CH(CH_3)CH(OH)CH_2OH$ (diastereomers), 32:1

[1] S. Takano, M. Yanase, and K. Ogasawara, *Heterocycles*, **29**, 249 (1989).

Trimethylgermanium halides, $(CH_3)_3GeBr(Cl)$ (**1**).

γ-Reactivity of germanium dienolates of α,β-unsaturated esters.[1] The germanium dienolate **2**, obtained by reaction of the lithium dienolate of the enoate with **1**, reacts with electrophiles at the γ-position.

$$(CH_3)_2C{=}CHCOOC_2H_5 \xrightarrow[]{\text{1) LDA, THF; 2) } \mathbf{1}} CH_2{=}C(CH_3){-}CH(Ge(CH_3)_3)COOC_2H_5 \quad \mathbf{2}$$

$$\mathbf{2} + C_6H_5CH(OCH_3)_2 \xrightarrow[87\%]{TiCl_4,\ -78^\circ} C_6H_5CH(OCH_3)CH_2C(CH_3){=}CHCOOC_2H_5$$

(E/Z = 9.5 : 1)

[1] Y. Yamamoto, S. Hatsuya, and J. Yamada, *J.C.S. Chem. Comm.*, 1639 (1988).

Trimethyl orthoformate, $HC(OCH_3)_3$.

Benzylidenes. These derivatives of sugars are generally prepared by reaction with benzaldehyde catalyzed by an acid (generally TsOH). But a large excess of the aldehyde is generally required. One expedient is use of trimethyl orthoformate to remove the water generated during acetalization.[1]

$$HOCH(COOH)CH(OH)COOH + C_6H_5CHO + HC(OCH_3)_3 \xrightarrow[85-90\%]{CH_3OH,\ H^+} C_6H_5\text{-dioxolane}(COOCH_3)_2$$

[1] H.-S. Byun and R. Bittman, *Org. Syn.*, submitted (1989).

Trimethylsilyl azide.

Azidohydrins.[1] Epoxides are converted to *trans*-siloxy azides by reaction with $N_3Si(CH_3)_3$ (1.5 equiv.) and aluminum isopropoxide (1 equiv.) in CH_2Cl_2. The reaction involves selective attack of azide on the less-substituted carbon of the epoxide.

$$\text{cyclohexene oxide} \xrightarrow[74\%]{N_3Si(CH_3)_3,\ Al(O\text{-}i\text{-}Pr)_3} \text{trans-2-azido-1-}OSi(CH_3)_3\text{-cyclohexane}$$

$$CH_3CO_2CH_2\text{-oxirane} \xrightarrow{77\%} CH_3CO_2CH_2CH(OSi(CH_3)_3)CH_2N_3$$

Primary amines.[2] $(CH_3)_3SiN_3$ in the presence of methanol is in equilibrium with $(CH_3)_3SiOCH_3$ + HN_3. This *in situ* generation of hydrazoic acid can be used to convert trialkylboranes into primary amines.

$$R_3B + (CH_3)_3SiN_3 + CH_3OH \longrightarrow \left[R_2\overset{-}{B}(R)\text{—}NH\text{—}N^+\equiv N \right] \xrightarrow{45\text{–}70\%} RNH_2 + N_2$$

Example:

$$\text{cyclohexene} \xrightarrow[64\%]{1)\ BH_3\text{-THF};\ 2)\ (CH_3)_3SiN_3,\ CH_3OH} \text{cyclohexyl-}NH_2$$

[1] M. Emziane, P. Lhoste, and D. Sinou, *Synthesis*, 541 (1988).
[2] G. W. Kabalka, N. M. Goudgaon, and Y. Liang, *Syn. Comm.*, **18**, 1363 (1988).

Trimethylsilylmethylmagnesium chloride (1).

1-Trimethylsilylbutadienes.[1] These dienes can be obtained by reaction of this Grignard reagent (**1**) with allylic dithioacetals catalyzed by $Cl_2Ni[P(C_6H_5)_3]_2$.

$$C_6H_5CH{=}CH\text{-}(1,3\text{-dithiolan-2-yl}) \xrightarrow[91\%]{1,\ Ni(II),\ ether,\ C_6H_6} C_6H_5CH{=}CHCH{=}CHSi(CH_3)_3$$

$$CH_3C(C_6H_5){=}CH\text{-}(1,3\text{-dithiolan-2-yl}) \xrightarrow{88\%} C_6H_5C(CH_3){=}CHCH{=}CHSi(CH_3)_3$$

E/Z = 94:6

[1] Z.-J. Ni and T.-Y. Luh, *J. Org.*, **53**, 5582 (1988).

Trimethylsilyldiazomethane.

Silylcyclopropanes.[1] Photochemical or metal-catalyzed reactions of $(CH_3)_3SiCHN_2$ with aliphatic alkenes result in silylcyclopropanes.

$$C_6H_5CH{=}CHCOOC_2H_5\ (\mathbf{1}) + (CH_3)_3SiCHN_2 \xrightarrow[C_6H_6,\ 60^\circ]{PdCl_2}$$

trans-1-phenyl-2-(ethoxycarbonyl)-3-trimethylsilylcyclopropane (34%) + isomer with $Si(CH_3)_3$ on the other face (45%)

$$C_6H_5CH{=}CHCOCH_3\ (\mathbf{2}) + (CH_3)_3SiCHN_2 \xrightarrow{38\%} \mathbf{3}$$

(3: 4-phenyl-3-acetylpyrazole, C_6H_5, $COCH_3$, N–H)

In the absence of a catalyst, the reagent can undergo a 1,3-dipolar addition to furnish a pyrazole, as in the case of (E)-benzylideneacetone (**2**).

(E)-1-Trimethylsilyl-1-alkenes.[2] These alkenes can be prepared by reaction of the lithium anion (**1**) of trimethylsilyldiazomethane with primary alkyl halides followed by decomposition with CuCl (86–96% yield). (E)-2-Aryl-1-trimethylsilylethylenes are obtained directly by reaction of trimethylsilyldiazomethane with benzylsulfonyl chlorides and triethylamine.

$$CH_3(CH_2)_8CH_2X \xrightarrow{(CH_3)_3SiC(Li)N_2} CH_3(CH_2)_8CH_2C(=N_2)Si(CH_3)_3$$

$$\xrightarrow{CuCl,\ C_6H_6,\ \Delta} CH_3(CH_2)_8CH{=}CHSi(CH_3)_3 \quad (E/Z = 95:5)$$

[1] T. Aoyama, Y. Iwamoto, S. Nishigaki, and T. Shioiri, *Chem. Pharm. Bull.* **37**, 253 (1989).
[2] T. Aoyama and T. Shioiri, *Tetrahedron Letters*, **29**, 6295 (1988).

β-(Trimethylsilyl)ethoxymethyl chloride, 10, 431.

SEM ethers.[1] The synthesis of orelline (**4**), a 2,2-bipyridine derivative, from the SEM ether (**1**) of 2-bromo-3-hydroxypyridine involves metal–halogen exchange followed by coupling to provide the bipyridine derivative **2**. Thanks to the SEM groups, **2** is lithiated at the 4,4′-position to provide **3**, which is converted in several steps to the toadstool toxin **4**.

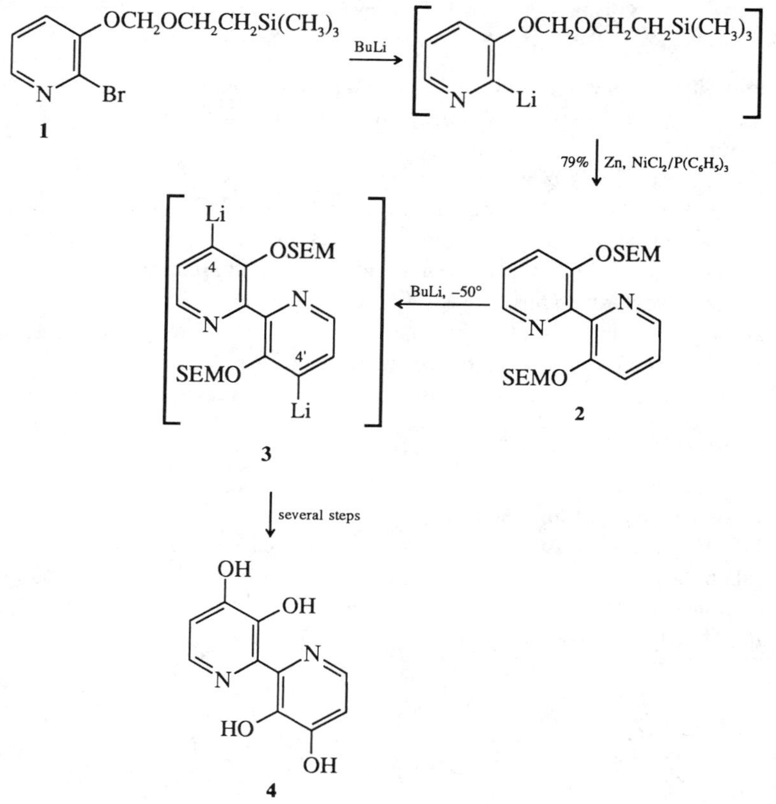

[1] H.-A. Hasseberg and H. Gerlach, *Helv.*, **71**, 957 (1988).

(2-Trimethylsilylethylidene)tris(2-methylphenyl)phosphorane,
$(2\text{-}CH_3C_6H_4)_3P{=}CHCH_2Si(CH_3)_3$ (**1**).

(Z)-Allyltrimethylsilanes.[1] Reaction of aldehydes with 2-trimethylsilylethylidinetriphenylphosphorane has been used to convert aldehydes into allyltrimethylsilanes as a mixture of (E)- and (Z)-isomers (**9**,492). Replacement of triphenylphosphine by tris(2-methylphenyl)phosphine enhances (Z)-selectivity, particularly in reactions with aliphatic saturated aldehydes. These aldehydes can be converted into the (Z)-allylsilanes (Z/E > 94%) with **1**, and with use of BuLi at 0° for generation of the phosphorane.

[1] H. Iio, M. Ishii, M. Tsukamoto, and T. Tokoroyama, *Tetrahedron Letters*, **29**, 5965 (1988).

Trimethylsilyl isoselenocyanate, $(CH_3)_3SiNCSe$, b.p. 75–76°/30 mm. The reagent is prepared by reaction of $ClSi(CH_3)_3$ with KSeCN in CH_3CN or DME at 25° (80–90% yield).

O-Trimethylsilyl cyanohydrins.[1] Substantially no reaction occurs between aldehydes or ketones with **1** in the absence of a catalyst. In the presence of $ZnCl_2$, aldehydes react with **1** in cyclohexane to give O-trimethylsilyl cyanohydrins and elemental selenium in high yields. The main advantage of this reagent over $(CH_3)_3SiCN$ is the chemoselectivity. Thus it reacts selectively with an aldehyde in the presence of a ketone or in the presence of an aryl aldehyde. However, there is only slight selectivity between a primary and a secondary aldehyde. Since the reagent is sensitive to moisture and air, a one-pot reaction with an aldehyde is usually desirable.

[1] K. Sukata, *J. Org.*, **54**, 2015 (1989).

2-Trimethylsilyloxy-1,3-butadiene.

[2+2]Photocyclization.[1] The diene **1** undergoes [2+2]photocyclization to 2-cycloalkenones to give bicyclic vinylcyclobutanols that can undergo Pd(II)-mediated ring expansion to bicyclic α-methylenecyclopentanones when treated with $Cl_2Pd(C_6H_5CN)_2$.

COOC₂H₅ / CH₂ / CH₃ / CH₃ / O + CH₂ / H₂C / OSi(CH₃)₃ / **1** → (hv, 65%) C₂H₅OOC / OSi(CH₃)₃ / CH=CH₂ / CH₃ / CH₃ / O / H (*cis*, 12 : 1) → (72%, $Cl_2Pd(C_6H_5CN)_2$) C₂H₅OOC / CH₂ / CH₃ / CH₃ / =O / O / H

[1] M. Demuth, B. Pandey, B. Wietfeld, H. Said, and J. Viader, *Helv.*, **71**, 1392 (1988).

Trimethylsilyl trifluoromethanesulfonate.

Acetals→ vinyl ethers. The classical method for this transformation is thermal elimination of an alcohol substituent. A new method[1] is the reaction of dialkyl acetals with $(CH_3)_3SiOTf$ (1.1–1.7 equiv.) and an equivalent amount of diisopropylethylamine in CH_2Cl_2 at −20 to 25°. Dioxanes and dioxolanes are also converted by this method to enol ethers, but longer reaction periods are necessary.[2]

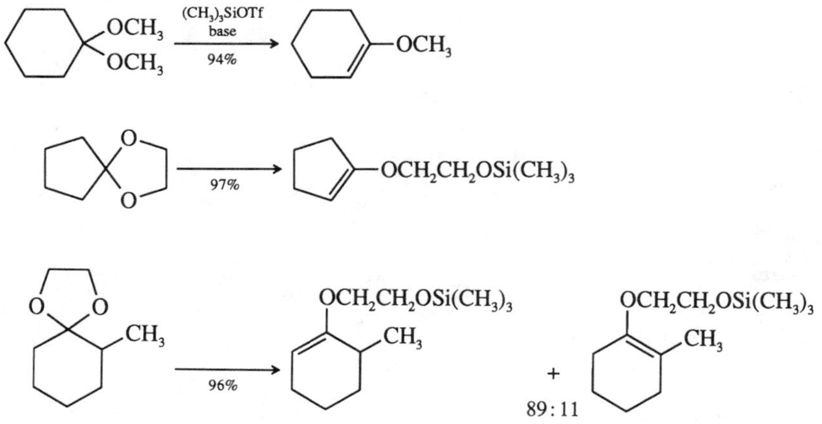

$(CH_3)_3SiOTf$, base; 94%

$OCH_2CH_2OSi(CH_3)_3$; 97%

$OCH_2CH_2OSi(CH_3)_3$, CH_3; 96%; + 89:11

This reaction can be used as the first step in a ring-expansion of acetals; remaining steps are cyclopropanation and photochemical cleavage of the cyclopropane in the presence of 4,4′-dicyanobiphenyl (4,4′-DCBi, **2**) as sensitizer.[3]

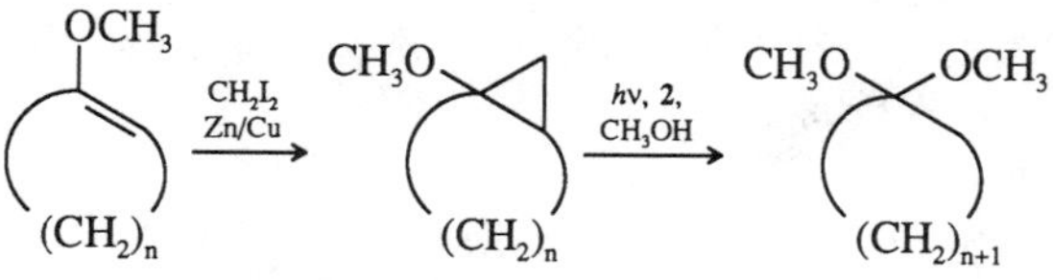

CH_2I_2, Zn/Cu; $h\nu$, **2**, CH_3OH; $(CH_2)_n$ → $(CH_2)_{n+1}$

1 n = 3	14%	45%
= 5	78%	31%
= 6	60%	31%

Cyclopentanones.[4] The adducts **2**, obtained by reaction of trimethylstannyllithium with cyclohexenone followed by trapping with an electrophile, when treated with a Lewis acid can give rise to cyclohexanones and/or cyclopentanones. $TiCl_4$ tends to give mixtures of both products, whereas trimethylsilyl triflate favors formation of cyclopentanones. Thus the adduct **5** on treatment with TMSOTf provides only the cyclopentanone **6**. Reaction of **5** with $TiCl_4$ gives **6** and **7** in the ratio 59:41.

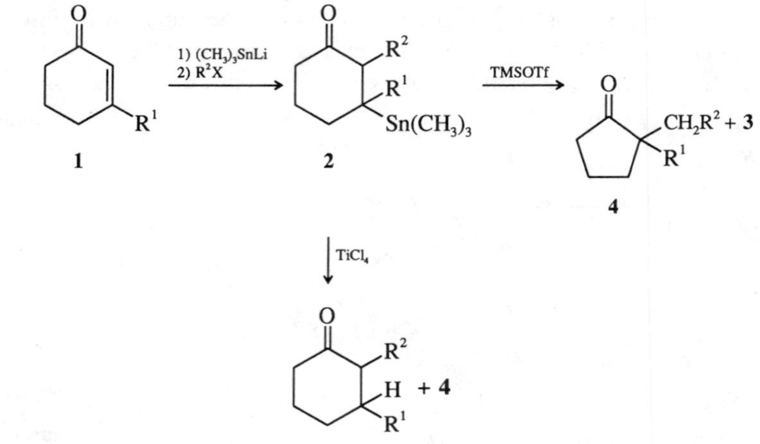

1) $(CH_3)_3SnLi$
2) R^2X

TMSOTf

R^1 1 — R^2, R^1, $Sn(CH_3)_3$ 2 — CH_2R^2 + 3, R^1 4

$TiCl_4$

R^2, H + 4, R^1 3

Example:

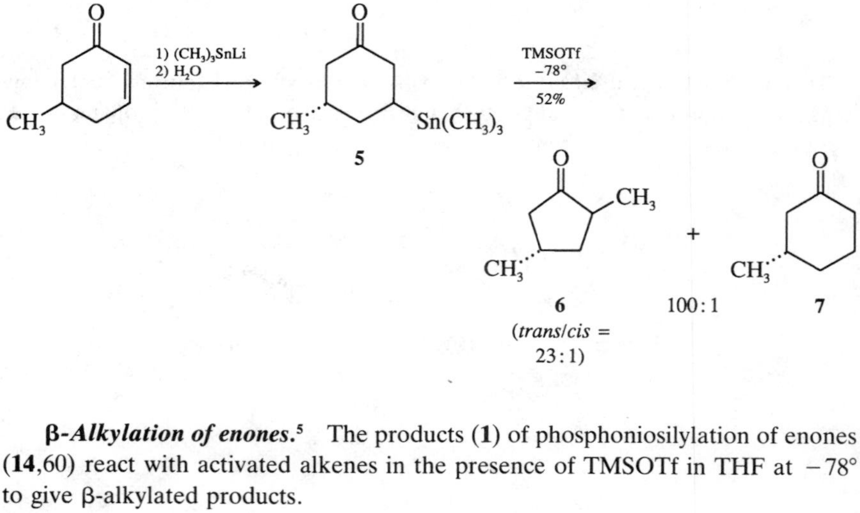

1) $(CH_3)_3SnLi$
2) H_2O

TMSOTf
−78°
52%

CH_3 — CH_3 5 $Sn(CH_3)_3$

CH_3, CH_3 6 (*trans*/*cis* = 23:1) + CH_3 7 — 100:1

β-Alkylation of enones.[5] The products (**1**) of phosphoniosilylation of enones (**14**,60) react with activated alkenes in the presence of TMSOTf in THF at −78° to give β-alkylated products.

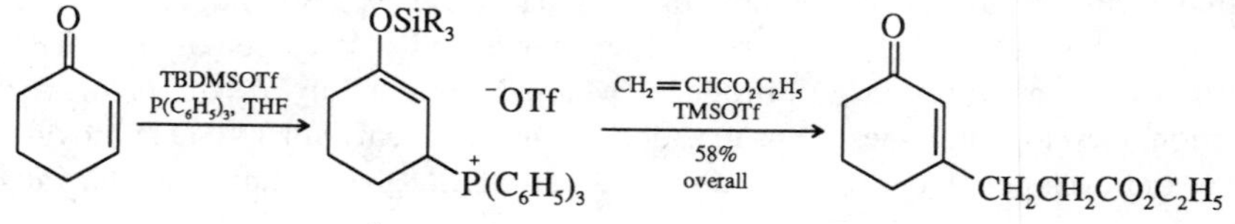

TBDMSOTf
$P(C_6H_5)_3$, THF

$OSiR_3$, ^-OTf, $\overset{+}{P}(C_6H_5)_3$ 1

$CH_2{=}CHCO_2C_2H_5$
TMSOTf
58% overall

$CH_2CH_2CO_2C_2H_5$

$$\text{cyclopentenone} \longrightarrow \text{3-}P^+(C_6H_5)_3\text{-1-}OSiR_3\text{-cyclopentene} \; {}^-OTf \xrightarrow{c\text{-}C_6H_{11}CHO}$$

$$\left[\text{1-}OSiR_3\text{-3-}(C_6H_5)_3P^+\text{-3-}CH(OSi(CH_3)_3)C_6H_{11}\text{-}c\text{-cyclopentene} \; {}^-OTf \right] \xrightarrow[\text{63\% overall}]{Bu_4NF} \text{3-}CH(OH)C_6H_{11}\text{-}c\text{-cyclopent-2-enone}$$

Aldol reaction.[6] This triflate is an effective catalyst for an aldol-type reaction between silyl enol ethers and acetals at −78°. The reaction shows moderate to high *syn*-selectivity regardless of the geometry of the enol ether.

$$C_6H_5C(OSiR_3)=CHCH_3 + C_6H_5CH(OCH_3)_2 \xrightarrow[83\%]{1,\ CH_2Cl_2,\ -78°} C_6H_5COCH(CH_3)CH(OCH_3)C_6H_5$$

(*syn*/*anti* = 71:29)

$$CH_3OC(OSiR_3)=CHCH_3 + C_6H_5CH(OCH_3)_2 \xrightarrow{74\%} CH_3OCOCH(CH_3)CH(OCH_3)C_6H_5$$

(*syn*/*anti* = 55:45)

Acetals of hexopyranosides.[7] The pyruvic acid acetal (**3**) of the methyl glucopyranoside **2** can be prepared by reaction with ethyl pyruvate in CH_2Cl_2 (−5 → 3°) in the presence of this triflate.

$$\mathbf{2} + CH_3COCOOC_2H_5 \xrightarrow[31\%]{1,\ CH_2Cl_2} \mathbf{3}$$

2, $R_3 = SiMe_2C(CH_3)_3$ (substituents: R_3SiO, $OSiR_3$, BzlO, BzlO, OCH_3)

3, (R/S) = 31:24 (substituents: $COOC_2H_5$, CH_3, BzlO, BzlO, OCH_3)

[1] P. G. Gassman and S. J. Burns, *J. Org.*, **53**, 5574 (1988).
[2] *Idem*, *ibid.*, **55**, 5576 (1988).

[3] L. Friedman and H. Schecter, *ibid.*, **26**, 2522 (1961).
[4] T. Sato, T. Watanabe, T. Hayata, and T. Tsukui, *J.C.S. Chem. Comm.*, 153 (1989).
[5] S. Kim and P. H. Lee, *Tetrahedron Letters*, **29**, 5413 (1988).
[6] S. Murata, M. Suzuki, and R. Noyori, *Tetrahedron*, **44**, 4259 (1988).
[7] H. Hashimoto, K. Hiruma, and J. Tamura, *Carbohydrate Res.*, **177**, C9 (1988).

(Trimethylstannyl)copper reagents.

Additions to α,β-acetylenic esters. Piers *et al.*[1] have prepared a number of these reagents, and conclude that (trimethylstannyl)copper(I)·dimethyl sulfide, $(CH_3)_3SnCu \cdot S(CH_3)_2$ (**1**), is the best reagent of this type for conversion of 1-alkynes to 2-trimethylstannyl-1-alkenes (equation I) or for addition to ethyl 2-butynoate to

$$\text{(I)}\quad HC{\equiv}C(CH_2)_nX \xrightarrow{\mathbf{1}} CH_2{=}C\begin{matrix} Sn(CH_3)_3 \\ (CH_2)_nX \end{matrix}$$

$(X = Cl, OH, OR)$

$$\text{(II)}\quad \underset{\mathbf{3}}{CH_3C{\equiv}CCOOC_2H_5} \xrightarrow[78\%]{\mathbf{1}}$$

$$\underset{(E)\text{-}\mathbf{4}}{\begin{matrix}(CH_3)_3Sn \\ CH_3\end{matrix} C{=}C \begin{matrix} H \\ COOCH_5\end{matrix}} \qquad >99:<1 \qquad \underset{(Z)\text{-}\mathbf{4}}{\begin{matrix}(CH_3)_3Sn \\ CH_3\end{matrix} C{=}C \begin{matrix} COOC_2H_5 \\ H\end{matrix}}$$

provide (E)-3-trimethylstannyl-2-alkenoates (equation II). This reaction is a general route to (E)-1,4-adducts of α,β-acetylenic esters. On the other hand, the (Z)-1,4-adducts can be obtained by use of the phenylthiocuprate $[(CH_3)_3SnCuSC_6H_5]Li$ (**2**).

$$\text{(III)}\quad CH_3C{\equiv}CCOOC_2H_5 \xrightarrow[76\%]{\mathbf{2}} \underset{98:2}{(Z)\text{-}\mathbf{4} \;+\; (E)\text{-}\mathbf{4}}$$

[1] E. Piers, J. M. Chong, and H. E. Morton, *Tetrahedron*, **45**, 363 (1989).

Triphenylarsonium ylides, 10, 445; 13,137.

Unsaturated amides.[1] The organoarsonium bromide **1**, prepared as shown in equation (I), reacts with aldehydes to form (E)-unsaturated or (2E,4E)-diunsa-

$$\text{(I)}\quad (C_6H_5)_3As + BrCH_2\overset{O}{\overset{\|}{C}}{-}N\langle\text{piperidine}\rangle \xrightarrow[96\%]{} \underset{\mathbf{1}}{(C_6H_5)_3\overset{+}{A}sCH_2\overset{O}{\overset{\|}{C}}{-}N\langle\text{piperidine}\rangle \; Br^-}$$

$$\text{(II)}\quad \text{RCHO} + \mathbf{1} \xrightarrow[98-99\%]{K_2CO_3,\ \text{THF},\ 25°} \text{RCH{=}CHCO{-}N(CH}_2)_5$$

E, 100%

turated amides. This reaction provides a route to pellitorine (**2**), a natural insecticide.

$$\underset{(E)}{\text{RCH{=}CHCHO}} + \mathbf{1} \xrightarrow[80-90\%]{} \underset{2E,\ 4E,\ 100\%}{\text{R(CH{=}CH)}_2\text{CO{-}N(CH}_2)_5}$$

$$CH_3(CH_2)_4CHO + (C_6H_5)_3As^+CH_2CHO\ Br^- \xrightarrow[81\%]{K_2CO_3} \underset{(E)\text{-}\mathbf{1}}{CH_3(CH_2)_4CH{=}CHCHO}$$

$$(E)\text{-}\mathbf{1} + (C_6H_5)_3As^+CH_2\overset{O}{\overset{\|}{C}}NHCH_2CH(CH_3)_2\ Br^- \xrightarrow[79\%]{K_2CO_3,\ 25°}$$

$$\underset{\mathbf{2}\ (E,E,\ 100\%)}{CH_3(CH_2)_4CH{=}CHCH{=}CH\overset{O}{\overset{\|}{C}}NHCH_2CH(CH_3)_2}$$

Triphenylarsonium ethylide. *trans*-Epoxides can be prepared stereoselectively via arsonium ylides under special conditions (**10**,445). A simpler method involves transylidation of a phosphonium ylide to an arsonium ylide (equation I).[2]

$$\text{(I)}\quad (C_6H_5)_3\overset{+}{P}CH(CH_3)_2\ I^- \xrightarrow{1)\ BuLi;\ 2)\ (CF_3CO)_2O} \left[CF_3CO_2^-\ (C_6H_5)_3\overset{+}{P}C(CH_3)_2C({=}O)CF_3 \xrightarrow[-(C_6H_5)_3P=O]{2(C_6H_5)_3As=CH_2} (C_6H_5)_3As^+\overline{C}H{-}C(CF_3){=}C(CH_3)_2 \right] \xrightarrow[35-65\%]{ArCHO} \text{Ar(H)C{-}O{-}C(H)C(CF}_3){=}C(CH_3)_2$$

[1] Y. Z. Huang, L. Shi, J. Yang, and J. Zhang, *Tetrahedron Letters*, **28**, 2159 (1987).
[2] Y. Shen, Q. Liao, and W. Qiu, *J.C.S. Chem. Comm.*, 1309 (1988).

Triphenylbismuth diacetate, 12,548.

N-Phenylation of α-amino acids. α-Amino esters can be monophenylated in high yield by reaction with $(C_6H_5)_3Bi(OAc)_2$ in CH_2Cl_2 at 25° when catalyzed by Cu(0) or $Cu(OAc)_2$. Diphenylation is more difficult, and dibasic amino esters (histidine or arginine esters) are not phenylated.[1]

[1] D. H. R. Barton, J.-P. Finet, and J. Khamsi, *Tetrahedron Letters*, **30**, 937 (1989).

Triphenylmethylphosphonium halide–Potassium *t*-butoxide.

Wittig-Smiles reaction.[1] The reaction of a triarylmethylphosphonium halide and a base with an α-keto dicarboxylic acid halide imide such as **1** results in a Wittig reaction followed by a Smiles rearrangement to provide a pyrrolidine dione (**2**).[2]

$$ClOCC(CH_3)_2COC(=NCH_3)Cl\ (\mathbf{1}) + (C_6H_5)_3PHCHCH_3 \xrightarrow[28\%]{KOC(CH_3)_3} \mathbf{2}$$

(**2**: 3,3,1-trimethyl-4-($=CHC_6H_5$)-pyrrolidine-2,5-dione; labels: CH_3, CHC_6H_5, CH_3, O, N, O, CH_3)

[1] J. F. Bunnett and R. E. Zahler, *Chem. Rev.*, **49**, 362 (1951).
[2] L. Capuano, A. Bettinger, M. Engel, M. Pfordt, and V. Huch, *Ann.*, 913 (1988).

Triphenylphosphine–Carbon tetrabromide.

ROTHP → RBr. A variety of primary and secondary tetrahydropyranyl ethers are converted to the corresponding bromide with inversion by reaction with $2P(C_6H_5)_3/CBr_4$ in CH_2Cl_2 in 75–85% yield.[1] The reaction with tertiary tetrahydropyranyl ethers is best carried out in CH_3CN with added LiBr (1.5 equiv.). The yield in the only example reported is 62%, as opposed to a 44% yield under the standard conditions.[1]

[1] A. Wagner, M.-P. Heitz, and C. Mioskowski, *Tetrahedron Letters*, **30**, 557 (1989).

Triphenylphosphine–Diisopropyl azodicarboxylate (1).

Oxidation of esters of α-amino acids. These substrates are oxidized by the Mitsunobu reagent (**1**) at the α-position to give either α-keto esters or adducts. For example, the N-benzyloxyphenylglycine methyl ester **2** is oxidized to methyl phenylglyoxalate **3** in 93% yield. In contrast, reaction of N-formylphenylglycine methyl ester (**4**) results in a triazoline (**5**) in 80% yield.[1]

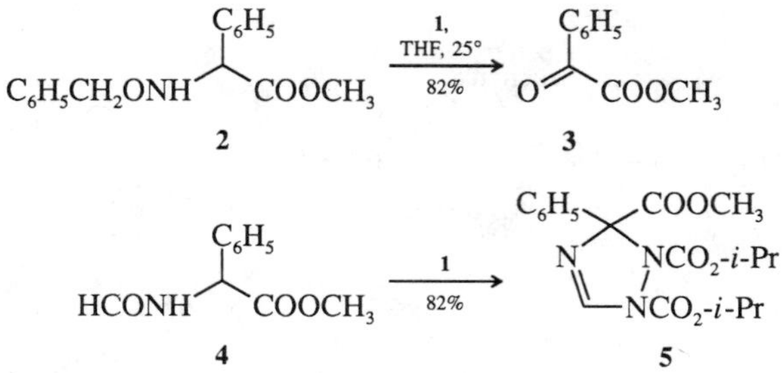

[1] T. Kolasa and M. J. Miller, *Tetrahedron Letters*, **29**, 4661 (1988).

Triphenylphosphine hydrobromide, m.p. 189°. Preparation from $P(C_6H_5)_3$ + HBr (gas).[1]

Tetrahydropyranylation of tert-alcohols.[2] The reaction of *tert*-alcohols with dihydropyran in CH_2Cl_2 at 0° catalyzed by this reagent provides tetrahydropyranyl ethers in 78–96% yield.

[1] J. D. Surmatis and A. Ofner, *J. Org.*, **28**, 2735 (1963).
[2] V. Bolitt, C. Mioskowski, D.-S. Shin, and J. R. Falck, *Tetrahedron Letters*, **29**, 4583 (1988).

2,4,6-Triphenylpyrylium tetrafluoroborate, 8,520. Supplier: Aldrich.

Photosensitized dehydrogenation.[1] Irradiation of flavanones with this sensitizer provides flavones in moderate (~60%) yield. This dehydrogenation has been carried out with CAN or DDQ.

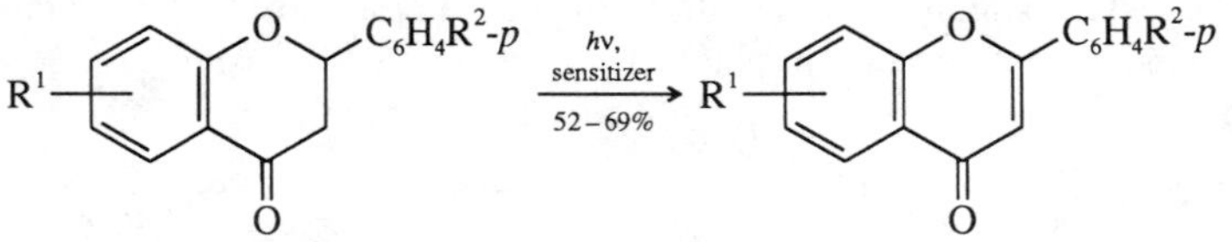

$R^1 = H, CH_3, OCH_3$
$R^2 = H, OCH_3$

[1] M. J. Climent, H. Garcia, S. Iborra, M. A. Miranda, and J. Primo, *Heterocycles*, **29**, 115 (1989).

Tris(dibenzylideneacetone)dipalladium (chloroform), $(dba)_3Pd_2\cdot CHCl_3$ (**1**).

Isomerization/cyclization of α-dienyl-ω-allyl acetates (carbonates).[1] This Pd catalyst effects isomerization and cyclization of these substrates to *vic*-disubstituted cyclopentanes. This reaction can also afford *cis*-fused bicyclic products with an *endo*-substituent (equation I).

Reductive cyclization of 1,6-diynes.[2] Cyclization of these substrates to 1,2-dialkylidenecyclopentanes can be effected with **1** and a phosphine as catalyst and triethylsilane (10-fold excess) in place of polymethylhydrosiloxane.

Examples:

Carbonylation of allylic acetates.[3] This reaction is effected by catalysis with this and a few other Pd(0) complexes, but requires bromide ion as a cocatalyst. It provides a route to (E)-β,γ-unsaturated esters in generally high yield from primary allylic acetates.

$$C_6H_5CH{=}CHCH_2OAc \xrightarrow[84\%]{Pd(0),\ NaBr,\ C_2H_5OH,\ CO,\ i\text{-}Pr_2NC_2H_5,\ 50^\circ} C_6H_5CH{=}CHCH_2COOC_2H_5$$

$$PrCH{=}CHCH(OAc)Pr \xrightarrow{51\%} PrCH{=}CHCH(COOC_2H_5)Pr$$

$$\text{1-vinylcyclohexyl acetate} \xrightarrow{80\%} \text{cyclohexylidene}{=}CHCH_2COOC_2H_5$$

[1] B. M. Trost and J. I. Luengo, *Am. Soc.*, **110**, 8239 (1988).
[2] B. M. Trost and D. C. Lee, *ibid.*, **110**, 7255 (1988).
[3] S.-J. Murahashi, Y. Imada, Y. Tankguchi, and S. Higashiura, *Tetrahedron Letters*, **29**, 4945 (1988).

Tris(diethylamino)sulfonium difluorotrimethylsiliconate (TASF), **10**, 452.

α-Halo carbanions.[1] TASF reacts with α-haloalkylsilanes to generate an α-halo carbanion as shown by addition to aldehydes (equation I). These carbanions

$$\text{(I)}\quad (CH_3)_3SiCHCl_2 \xrightarrow[THF,\ 25^\circ]{TASF} [^-CHCl_2] \xrightarrow[74\%]{C_6H_5CHO} C_6H_5CH(OH)CHCl_2$$

are stable at 25°, in contrast to dichloromethyllithium, which decomposes at temperatures above −110°. However, a similar reaction is not observed with (chloromethyl)trimethylsilane.

[1] M. Fujita, M. Obayashi, and T. Hiyama, *Tetrahedron*, **44**, 4135 (1988).

Tris(6,6,7,7,8,8,8-heptafluoro-2,2-dimethyl-3,5-octanedionato)europeum or -praseodymium, Eu(fod)$_3$ or Pr(fod)$_3$.

Diels–Alder catalysts.[1] These lanthanide complexes markedly improve the rate and *endo*-selectivity of Diels–Alder reactions of cyclopentadiene with allenic esters.

$$\text{cyclopentadiene} + CH_3CH{=}C{=}CHCO_2C_2H_5 \xrightarrow{C_6H_6,\ \Delta} \textit{endo} + \textit{exo}$$

		endo : exo
	38%	2.5 : 1
+ Eu(fod)$_3$	79%	5.4 : 1

[1] R. P. Gandhi, M. P. S. Ishar, and A. Wali, *J.C.S. Chem. Comm.*, 1074 (1988).

Tris(6,6,7,7,8,8,8-heptafluoro-2,2-dimethyl-3,5-octanedionato)ytterbium, Yb(fod)$_3$.

Diels–Alder reactions with cycloalkenones.[1] (E)-1-Methoxy-1,3-butadiene undergoes cycloaddition with cyclohexenone at 160° to afford the *endo*-adduct (**2**) in 47% yield. Catalysis with the usual Lewis acid results in resins, but Yb(fod)$_3$ catalysis results in cycloaddition at 110° and leads to *endo*- and *exo*-**3** in 55% yield. The adducts **3** arise from a Diels–Alder reaction with the dienone **4**, which is formed when **2** is heated at 100° with Yb(fod)$_3$.

	1	**2**	**3**
	160°, 47%		—
Yb(fod)$_3$	110°, 55%	—	(*endo*/*exo* = 1.5 : 1)

2 $\xrightarrow[88\%]{100°}$ **4** $\xrightarrow{\mathbf{1},\ Yb(fod)_3}$ **3** (*endo*/*exo* = 1.5 : 1)

[1] F. Fringuelli, L. Minuti, L. Radics, A. Taticchi, and E. Wenkert, *J. Org.*, **53**, 4607 (1988).

Tris[2-(2-methoxyethoxy)ethyl]amine (TDA-1), **13**,336–337; **14**,341–342.

Smiles rearrangement.[1] Attempted N-alkylation of 3-(2,4,6-trichlorophenoxyacetamido)pyridine (**1**) catalyzed by TDA-1 is accompanied by a Smiles rearrangement to give an N-alkyl-3-(2,4,6-trichlorophenylamino)pyridine (**2**) in 63–71% yield. Normal N-alkylation of **1** obtains when one or two of the chlorines of **1** are replaced by hydrogen.[2]

Wittig reactions.[3] Yields in the reaction of an aldehyde with cyclopropyltriphenylphosphonium bromide (**2**,95–96) under the usual conditions (NaH, DMF) are negligible, but can be improved to about 15% by addition of 18-crown-6. A dramatic increase in yield is obtained by use of NaH in THF with TDA-1 as a transfer catalyst. Best results are obtained by heating the phosphonium salt suspended in THF with TDA-1 before addition of the carbonyl compound. With this technique, alkylidenecyclopropanes can be obtained from carbonyl compounds in yields of 60–95%.

Glycosylation.[4] The β-D-ribonucleoside tubercidin (**4**) can be prepared by phase-transfer catalyzed glycosylation of the α-D-ribofuranosyl chloride **1** with the anion of the pyrimidine **2** mediated by TDA-1. Since **1** lacks a neighboring group, the glycosylation proceeds stereospecifically. In the absence of TDA-1, a two-fold excess of **2** is required for glycosylation of **1**, and the yield of **3** is only 34%.

1 + **2** $\xrightarrow[65\%]{\text{TDA-1, KOH, } CH_3CN, 25°}$ **3** $\xrightarrow{\text{1) } CF_3COOH \text{; 2) } NH_3}$ **4**

RMgX/TDA-1.[5] Grignard reagents form a 2:1 complex with TDA-1 in ether that can be obtained as a powdered solid after evaporation of the solvent. The complexes are soluble in THF but not in ether, aromatic solvents, or cyclohexane. They are stable to temperatures below 110°. The complexes react as regular Grignard reagents in most respects, but are somewhat less reactive. The slight difference is useful for selective reactions of aldehydes in the presence of ketones. The complexes also undergo 1,4-addition to enones to a greater extent than a regular Grignard reagent.

[1] W. E. Truce, E. M. Kreider, and W. W. Brand, *Org. React.*, **18**, 99 (1970).
[2] A. Greiner, *Tetrahedron Letters*, **30**, 931 (1989).
[3] J. A. Stafford and J. E. McMurry, *ibid.*, **29**, 2531 (1988).
[4] H. Rosemeyer and F. Seela, *Helv.*, **71**, 1573 (1988); *see also* F. Seela, H.-P. Muth, and U. Bindig, *Synthesis*, 670 (1988).
[5] A. Boudin, G. Cerveau, C. Chuit, R. J. P. Corriu, and C. Reye, *Tetrahedron*, **45**, 171 (1989).

Tris(trimethylsilyl)silane, $[(CH_3)_3Si]_3SiH$ (**1**).

Reduction of RX.[1] This silane[2] is as effective as Bu_3SnH for reduction of RX by a free-radical reaction and is moreover less toxic. The reductions can be initiated by UV photolysis at 40° or by dibenzoyl peroxide at 50°. Yields of RH are essentially quantitative.

It is as efficient as Bu_3SnH for initiation of intermolecular or intramolecular radical reactions with alkenes.[3]

$$C_6H_{11}I + CH_2{=}CHCN \xrightarrow[70-90^\circ]{1,\ AIBN,} C_6H_{11}CH_2CH_2CN$$

$$CH_2{=}CH(CH_2)_4Br \xrightarrow[70^\circ]{1,\ AIBN,} \text{methylcyclopentane } (93\%) + \text{cyclohexane } (2\%)$$

[1] C. Chatgilialoglu, D. Griller, and M. Lesage, *J. Org.*, **53**, 3641 (1988).
[2] Preparation: H. Bürger and W. J. Kilian, *J. Organometal. Chem.*, **18**, 299 (1969).
[3] B. Giese, B. Kopping, and C. Chatgilialoglu, *Tetrahedron Letters*, **30**, 681 (1989).

Tris(triphenylphosphine)nickel(0). This Ni(0) complex can be generated *in situ* by reduction of $Br_2Ni[P(C_6H_5)_2]$ with $Zn{-}P(C_6H_5)_3$.

ArCl → ArCN. This Ni(0) catalyst effects cyanation of aryl chlorides with KCN in HMPT or CH_3CN at 60°. Coupling of aryl chlorides to biphenyls occurs to only a slight extent.[1]

$$o\text{-}ClC_6H_4OCH_3 + KCN \xrightarrow[\text{quant.}]{Ni(O),\ HMPT,\ 60^\circ} o\text{-}NCC_6H_4OCH_3 + 2,2'\text{-dimethoxybiphenyl} \quad (97:3)$$

[1] Y. Sakakibara, F. Okuda, A. Shimobayashi, K. Kirino, M. Sakai, N. Uchino, and K. Takagi, *Bull. Chem. Soc. Japan*, **61**, 1985 (1988).

Trityl chloride (triphenylmethyl chloride).

N-Tritylamino acids. Direct reaction of trityl chloride with amino acids provides N-trityl derivatives in low yield. A superior route to these derivatives involves reaction of trimethylsilyl esters of N-trimethylsilylamino acids with trityl chloride in refluxing chloroform, which provides the N-protected amino acids in 88–92% yield. Trimethylsilyl-protected hydroxyl or thiol groups are not cleaved under these conditions.[1]

4-Polystyryltritylamino acids can be prepared in the same way from the trimethylsilyl esters of amino acids.[2] These N-protected amino acids undergo usual coupling reactions with amino acid esters, often in quantitative yield. The polymeric protective group is removed by p-TsOH, HCl, or TFA in 75–90% yield.

[1] P. Mamos, C. Sanida, and K. Barlos, *Ann.*, 1083 (1988).
[2] K. Barlos, D. Gatos, I. Kalitsis, D. Papaioannou, and P. Sotiriou, *ibid.*, 1079 (1988).

Trityl chloride–Tin(II) chloride.

Allyl ketene dithioacetals. In the presence of these two reagents, ketene dithioacetals (**1**) react with allyl methyl ethers to provide allyl-substituted ketene dithioacetals (equation I). The combination of $ClSi(CH_3)_3$ and $SnCl_2$ is an equally

(I) CH_3, H, S, S (**1**) + CH_3 ... OCH_3, CH_3 (**2**) —[TrCl–$SnCl_2$, CH_2Cl_2, 25°; 82%]→ CH_3 ... CH_3, CH_3, S, S (**3** (91:9))

effective catalyst for this substitution. A similar reaction of **1** can be effected with α,β-unsaturated ortho esters (equation II).[1]

(II) **1** + CH_3 ... OCH_3, OCH_3, OCH_3 —[TrCl, $SnCl_2$, $C_6H_5CH_3$, −23°; 83%]→ CH_3O ... O, CH_3, CH_3, S, S

[1] T. Mukaiyama, H. Sugumi, H. Uchiro, and S. Kobayashi, *Chem. Letters*, 1291 (1988).

Trityl hexachloroantimonate, $TrSbCl_6$ (1).

Reaction of α,β-unsaturated acetals with silyl nucleophiles. This trityl salt is an efficient catalyst for reaction of the dimethyl acetal (**2**) of cinnamaldehyde with successive silyl nucleophiles to form γ,δ-unsaturated β-azido ketones. The azide group can be reduced to an amino group by the Staudinger reaction.[1]

C_6H_5 ... OCH_3, OCH_3 (**2**) + $C_6H_5C(OSiR_3){=}CH_2$ —[**1**]→ [C_6H_5 ... CH_3O, O, C_6H_5] —[$N_3Si(CH_3)_3$; 85%]→

C_6H_5 ... N_3, O, C_6H_5 —[$P(C_6H_5)_3$, H_2O, THF, Δ; ~80%]→ C_6H_5 ... NH_2, O, C_6H_5

[1] T. Mukaiyama, P. Leon, and S. Kobayashi, *Chem. Letters*, 1495 (1988).

Trityl perchlorate.

Addition of silyl enol ethers to o-quinones.[1] This reaction requires mild conditions because of the sensitivity of these quinones, but can be achieved with a catalytic amount of trityl perchlorate (equation I). *o*-Quinones with a bulky group at C_4 tend to undergo 1,6-addition with silyl enol ethers.

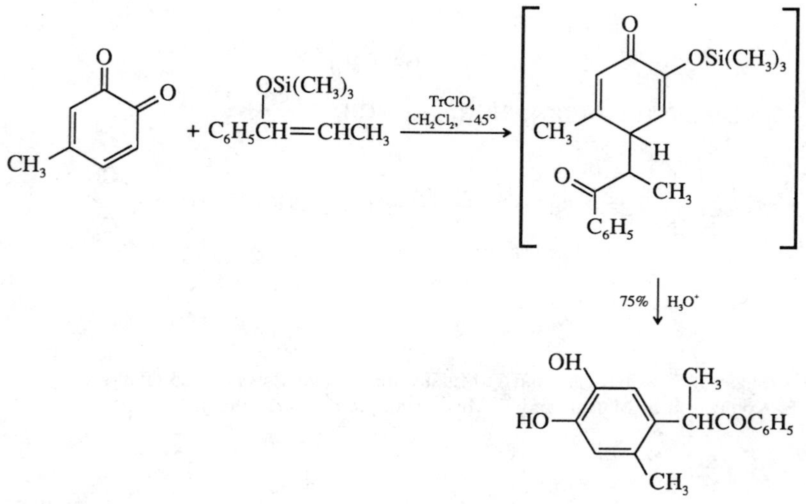

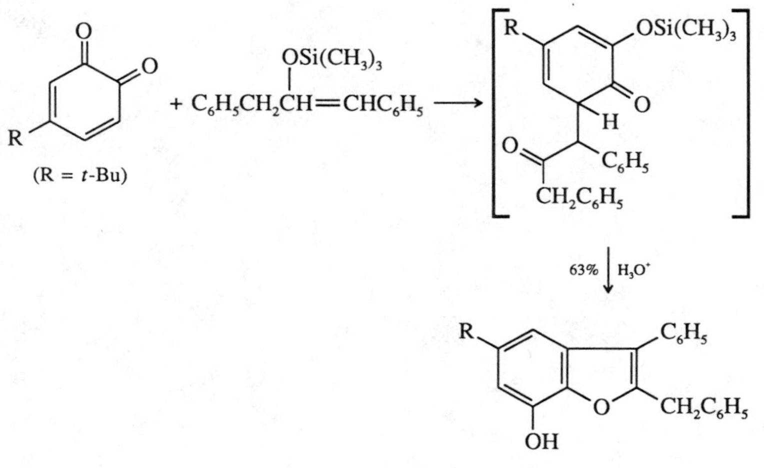

Aldol reaction between α,β-acetylenic ketones and silyl enol ethers. $TrClO_4$ is an efficient catalyst for 1,2-addition of silyl enol ethers to α,β-acetylenic ketones.[2]

$$CH_3\overset{O}{\overset{\|}{C}}-C\equiv C-C_6H_5 + C_6H_5C(OSiR_3)=CH_2 \xrightarrow[69\%]{TrClO_4,\ CH_2Cl_2} C_6H_5COCH_2C(CH_3)(OSiR_3)C\equiv CC_6H_5$$

$$CH_3\overset{O}{\overset{\|}{C}}-C\equiv C-Si(CH_3)_3 + (CH_3)_2CHCH_2-\overset{OSi(CH_3)_3}{\overset{|}{C}}=CH_2 \xrightarrow[85\%]{TrClO_4}$$

$$(CH_3)_2CHCH_2\overset{O}{\overset{\|}{C}}CH_2\underset{CH_3}{\underset{|}{\overset{OH}{\overset{|}{C}}}}-C\equiv CSi(CH_3)_3 \xrightarrow{(CF_3CO)_2O} (CH_3)_2CHCH_2\overset{O}{\overset{\|}{C}}CH=C(CH_3)C\equiv CSi(CH_3)_3$$

(*trans/cis* = 65 : 35)

[1] Y. Sagawa, S. Kobayashi, and T. Mukaiyama, *Chem. Letters*, 1105 (1988).
[2] S. Kobayashi, S. Matsui, and T. Mukaiyama, *ibid.*, 1491 (1988).

U

Ultrasound, (((.

Diels–Alder reactions.[1] Ultrasound can promote some Diels–Alder reactions with *o*-quinones, and also improve regioselectivity. Highest yields of adducts are obtained in the absence of a solvent. This reaction provides a route to some pigments of Chinese sage responsible for the biological activity of Dan Shen. An example is the synthesis of **3**, the acetonide of tanshindiol B. Thus the thermal reaction of the diene **1** with the quinone **2** gives **3** and **4** in the ratio 1:1. Sonication improves both the regioselectivity and the yield.

1 $\xrightarrow{\mathbf{2}}$ $\xrightarrow[\text{2) DDQ}]{\text{1) } SiO_2,\ O_2}$ **3** + **4**

Conditions	Yield	3 : 4
C_6H_6, Δ,	15%	1:1
10 kBar, 25°	73%	7:1
(((•, neat, 45°	76%	5:1

Review.[2] Ultrasound irradiation improves the rate and the yields of reactions with many organometallic reagents, particularly of organocopper, -lithium, and -zinc reagents (64 references).

[1] J. Lee and J. K. Snyder, *Am. Soc.*, **111**, 1522 (1989).
[2] R. F. Abdulla, *Aldrichim. Acta*, **21**, 31 (1988).

Vanadyl trichloride (Vanadium oxytrichloride), $VOCl_3$ (**1**).

Oxidative decarboxylation–deoxygenation of 3-hydroxycarboxylic acids. These acids are converted into alkenes when heated in chlorobenzene with 1 equiv. each of $VOCl_3$ and Proton Sponge® [**2**, 1,8-bis(dimethylamino)naphthalene].

[1] I. K. Meier and J. Schwartz, *Am. Soc.*, **111**, 3069 (1989).

Vinyl(fluorodimethyl)silanes, $RCH = CHSiF(CH_3)_2$ (**1**). The reagents are prepared by reaction of the corresponding vinyl(chloro)silanes with copper fluoride.

1,3-Dienes; vinylarenes.[1] These fluoro-substituted alkenylsilanes are much more reactive than trimethyl(vinyl)silanes in coupling reactions with vinyl iodides mediated by TASF and a palladium catalyst (**14**, 340–341). Oddly enough, vinyl(difluoromethyl)silanes are slightly less reactive than **1** and require longer reaction times, and a trifluorosilyl group is completely inactive in this coupling. The coupling proceeds with retention of configuration of both components, particularly

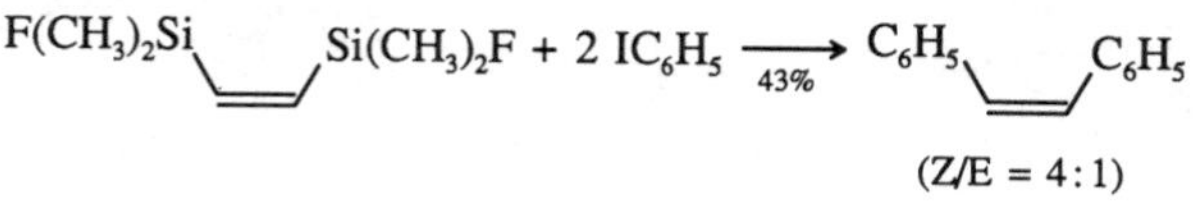

$$F(CH_3)_2Si\text{–}CH{=}CH\text{–}Si(CH_3)_2F + 2\ IC_6H_5 \xrightarrow[43\%]{} C_6H_5\text{–}CH{=}CH\text{–}C_6H_5$$

(Z/E = 4 : 1)

with $Pd[P(C_6H_5)_3]_4$ as catalyst and DMF as solvent. TASF can be replaced by Bu_4NF in coupling of the vinylsilanes with iodoarenes.

[1] K. Tamao, T. Kakui, M. Akita, T. Iwahara, R. Kanatani, J. Yoshida, and M. Kumada, *Tetrahedron*, **39**, 983 (1983); Y. Hatanaka and T. Hiyama, *J. Org.*, **54**, 268 (1989).

Vitamin B_{12} (Hydroxocobalamin hydrochloride), **1**.

Isomerization of epoxides to allylic alcohols.[1] This rearrangement has been effected with strong bases and various Lewis acids. Enantioselective rearrangement to optically active allylic alcohols can be effected with catalytic amounts of vitamin B_{12} at 25°. Thus cyclopentene oxide rearranges to (R)-2-cyclopentene-1-ol in 65% ee. The rearrangement of the *cis*-2-butene oxide to (R)-3-butene-2-ol in 26% ee is more typical.

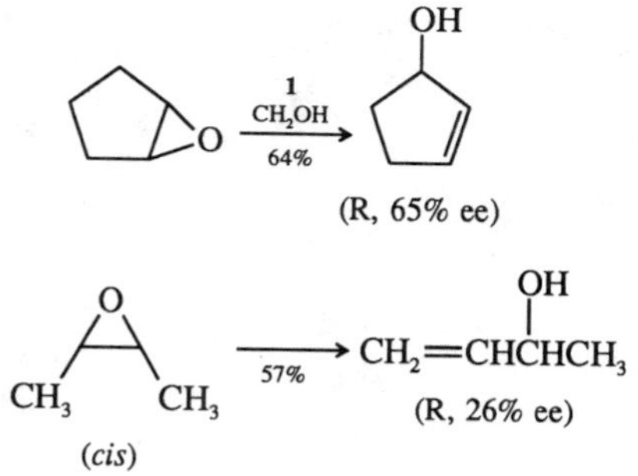

Cyclopentene oxide $\xrightarrow[64\%]{1,\ CH_3OH}$ 2-cyclopentene-1-ol (OH)

(R, 65% ee)

CH_3–(epoxide O)–CH_3 (*cis*) $\xrightarrow[57\%]{}$ $CH_2{=}CHCH(OH)CH_3$

(R, 26% ee)

[1] H. Su, L. Walder, Z.-d. Zhang, and R. Scheffold, *Helv.*, **71**, 1073 (1988).

Ytterbium(0), Yb.

Cross-coupling reactions of ArCOAr. Reaction of Yb(0) with diaryl ketones changes the reactivity of the carbonyl group from electrophilic to nucleophilic. Thus in the presence of this lanthanoid metal, diaryl ketones couple with other ketones, nitriles, and epoxides to give pinacols, α-hydroxy ketones, and 1,3-diols, respectively, via the intermediate **a**.

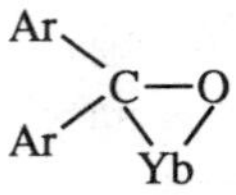

a

[1] Z. Hou, K. Takamine, O. Aoki, H. Shiraishi, Y. Fujiwara, and H. Taniguchi, *J. Org.*, **53**, 6077 (1988).

Z

Zeolites.

These natural minerals are alkali aluminum silicates and have been used as molecular sieves and cation exchangers. Recently they have been used to control some organic photochemical reactions. du Pont chemists[1] have prepared lithiated zeolites by treatment of sodium aluminum silicates such as zeolite 13X with $LiNO_3$. These silicates, designated Li-X and Li-Y, can effect rearrangement of α-alkyldeoxybenzoins to *p*-alkylbenzophenones in nearly quantitative yield. Thus photolysis

$$C_6H_5COCH(Pr)C_6H_5 \xrightarrow{h\nu} \text{2 (cyclobutanol: OH, } C_6H_5, CH_3, C_6H_5) + C_6H_5CH(Pr)\text{—}CH(Pr)C_6H_5 + C_6H_5\overset{O}{\overset{\|}{C}}\text{—}C_6H_4(CH_2)_3CH_3\text{-}p$$

	1	2	3	4
C_6H_6		54%	24%	—
+ Li–X		—	4%	95%

(2 : 3 = 2:1)

of **1** in benzene results in **2** and **3** as the major products. Photolysis of **1** in the zeolite Li-X results in a trace of **3**, with the major product being **4**, formed by selective cleavage of the bond between C—O and the CHC_6H_5 group.

[1] D. R. Corbin, D. F. Eaton, and V. Ramamurthy, *J. Org.*, **53**, 5384 (1988).

Zinc borohydride.

RCOCl → RCH₂OH.[1] This reduction can be effected by zinc borohydride catalyzed by TMEDA in ether at 0–40° in 80–95% yield. Both aromatic and aliphatic acyl halides are reduced readily without effect on a conjugated double bond. The by-product is TMEDA · $(BH_3)_2$.

[1] H. Kotsuki, Y. Ushio, N. Yoshimura, and M. Ochi, *Bull. Chem. Soc. Japan*, **61**, 2684 (1988).

Zinc–Copper couple. A highly active Zn-Cu couple can be prepared by sonication of a mixture of Zn and CuI (3:1 ratio) in an alcohol–water mixture (~65:35). A black heavy suspension is formed which should be used at once.[1]

1,4-Addition of RI to α,β-enones.[2] This couple and sonication effect addition of RI to enones. Highest yields are obtained with *i*-PrOH/H_2O (4:1) or PrOH/H_2O (65–25:35–75) as solvent for primary iodides. C_2H_5OH/H_2O (65:35) is the

$$\text{cyclohexenone} + (CH_3)_2CHI \xrightarrow[95\%]{Zn/Cu,\ \bullet)))),\ C_2H_5OH,\ H_2O} \text{3-}CH(CH_3)_2\text{-cyclohexanone}$$

solvent of choice for *tert*- or *sec*-iodides. Yields are decreased in reactions of α-methyl substituted enones and are particularly depressed by a β-methyl substituent. A radical reaction occurring on the metal surface is suggested for this addition.

[1] J. L. Luche and C. Allavena, *Tetrahedron Letters*, **29**, 5369 (1988).
[2] J. L. Luche, C. Allavena, C. Petrier, and C. Dupuy, *ibid*, **29**, 5373 (1988).

Zinc bromide.

***(E)*-α,β-Unsaturated acids.**[1] Zinc bromide is the most effective Lewis acid for promoting a reaction of C,O,O-tris(trimethylsilyl) ketene acetal (**1**) with aldehydes (but not ketones) to form α,β-unsaturated acids. The ketene acetal can be prepared as shown in equation (I).

$$\text{(I)}\quad (CH_3)_3SiCH_2COOSi(CH_3)_3 \xrightarrow[91\%]{1)\ LDA\quad 2)\ ClSi(CH_3)_3} \underset{\mathbf{1}}{(CH_3)_3SiCH{=}C[OSi(CH_3)_3]_2}$$

Example:

$$C_6H_5CHO + \mathbf{1} \xrightarrow[25^\circ]{ZnBr_2,\ THF} \left[\begin{array}{l} \qquad\quad Si(CH_3)_3 \\ C_6H_5{-}CH{-}COOSi(CH_3)_3 \\ OSi(CH_3)_3 \end{array} \right] \xrightarrow{95\%} \underset{(E)}{C_6H_5CH{=}CHCOOH}$$

[1] M. Bellassoued and M. Gaudemar, *Tetrahedron Letters*, **29**, 4551 (1988).

Zinc chloride.

Cyclopentadienes.[1] Allyl chlorides that are monosubstituted at the central allylic position, such as **1**, react with 1,3-dicarbonyl compounds such as acetylacetone (**2**) to form acetylcyclopentadienes (**3**) via an allylated dicarbonyl intermediate (**a**), which can be isolated from reactions of **1** with **2** at −78°.

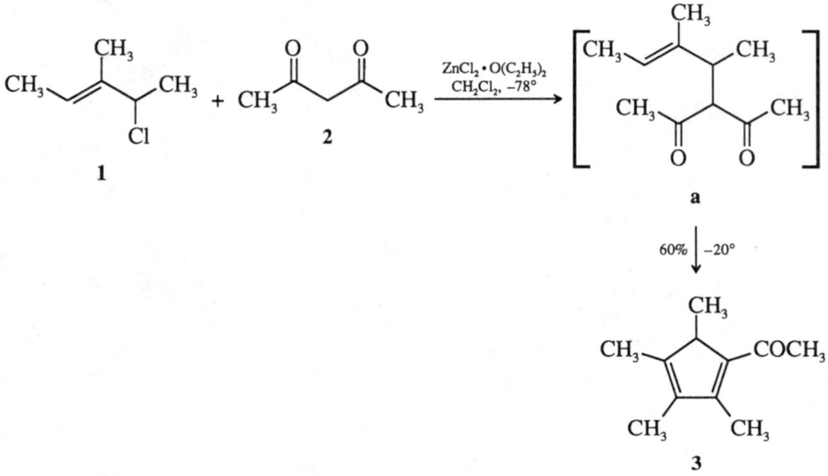

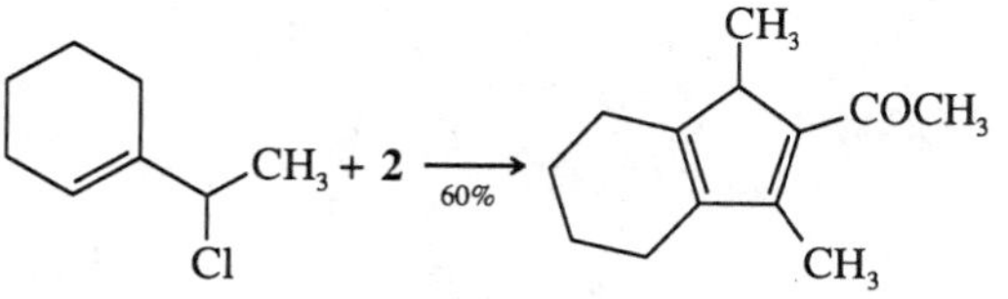

$CH_3CH=C(CH_3)CHClCH_3$ (1) + $CH_3COCH_2COCH_3$ (2) $\xrightarrow[CH_2Cl_2,\ -78^\circ]{ZnCl_2 \cdot O(C_2H_5)_2}$ [a] $\xrightarrow[60\%]{-20^\circ}$ 3

1 2 a 3

Cyclohexenyl-CH(Cl)CH$_3$ + 2 $\xrightarrow{60\%}$ product (COCH$_3$, CH$_3$)

Cyclization of quinone methides.[2] *p*-Quinone methides, particularly those substituted at the 2- and 6-positions, are stable enough to be characterized by IR and NMR, and can be generated in fairly high yield by oxidation of a phenol with Ag_2O (10 equiv.) in CH_2Cl_2. When substituted by a suitable terminator, these quinonemethides can undergo Lewis-acid-catalyzed cyclization. Suitable terminators that can survive the initial oxidation include allylsilanes and β-keto esters. In the latter case, the initial product undergoes oxidation to afford a cyclohexenone.

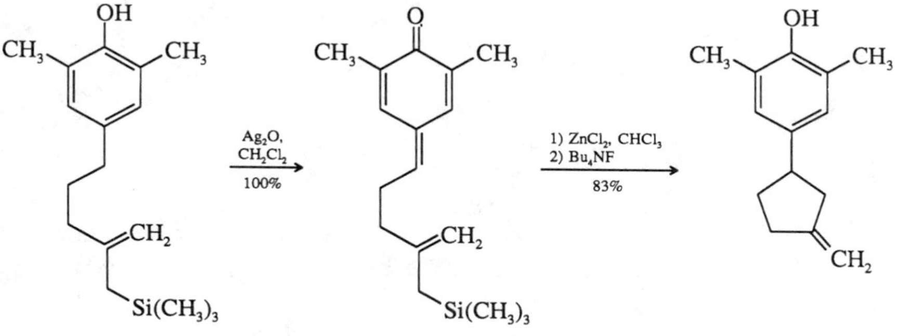

$\xrightarrow[100\%]{Ag_2O,\ CH_2Cl_2}$ $\xrightarrow[83\%]{1)\ ZnCl_2,\ CHCl_3;\ 2)\ Bu_4NF}$

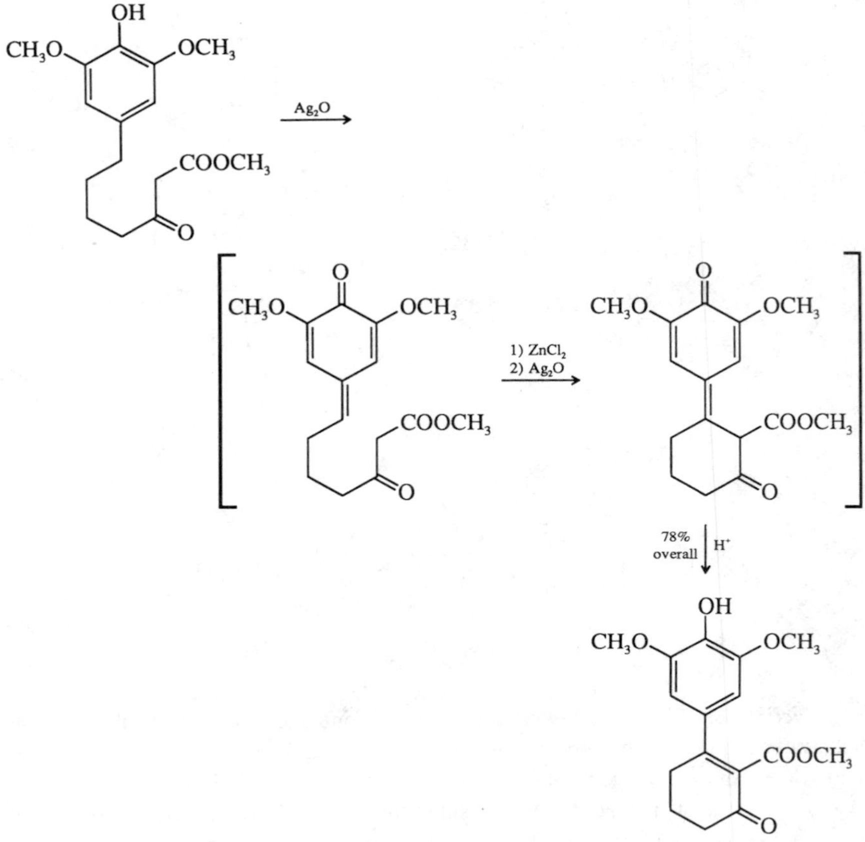

$$CH_3O\text{-}C_6H_2(OH)(OCH_3)\text{-}(CH_2)_4COCH_2COOCH_3 \xrightarrow{Ag_2O}$$

1) $ZnCl_2$
2) Ag_2O

78% overall, H^+

β-Lactams.[3] The zinc enolates of N,N-disubstituted α-amino esters react with imines to form β-lactams in 60–90% yield. The reaction can be *trans*-selective, depending on the substituent on the imine and the solvent.

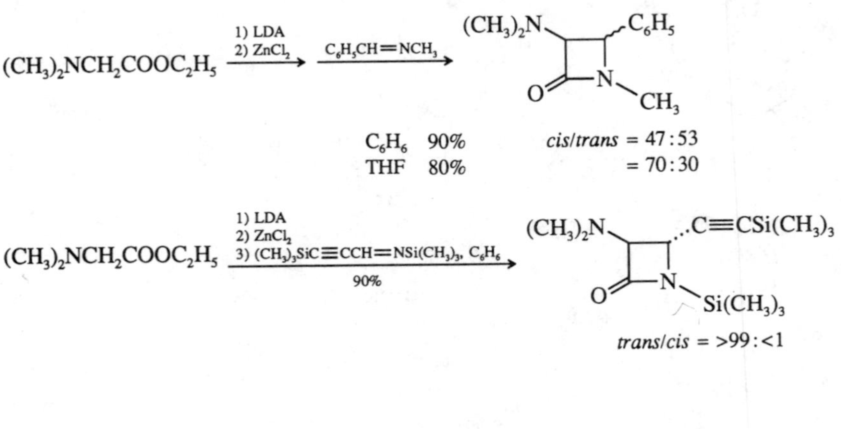

$$(CH_3)_2NCH_2COOC_2H_5 \xrightarrow[2)\ ZnCl_2]{1)\ LDA} \xrightarrow{C_6H_5CH{=}NCH_3}$$

C_6H_6	90%	*cis/trans* = 47:53
THF	80%	= 70:30

$$(CH_3)_2NCH_2COOC_2H_5 \xrightarrow[90\%]{1)\ LDA;\ 2)\ ZnCl_2;\ 3)\ (CH_3)_3SiC{\equiv}CCH{=}NSi(CH_3)_3,\ C_6H_6}$$

trans/cis = >99:<1

[1] R. Koschinsky, T.-P. Köhli, and H. Mayr, *Tetrahedron Letters*, **29**, 5641 (1988).
[2] S. R. Angle and K. D. Turnbull, *Am. Soc.*, **111**, 1136 (1989).
[3] F. H. van der Steen, H. Kleijn, J. T. B. H. Jastrzebski, and G. van Koten, *Tetrahedron Letters*, **30**, 765 (1989).

Zirconium sulfate.

Monoacetalization of glyoxal.[1] The reaction of glyoxal with CH_3OH or C_2H_5OH in large excess and catalyzed by zirconium sulfate results in monoacetalization, followed slowly by diacetalization. The derived water should be eliminated without decreasing the concentration of the alcohol (Soxhlet).

[1] F. H. Sangsari, F. Chastrette, and M. Chastrette, *Syn. Comm.*, **18**, 1343 (1988).

AUTHOR INDEX

SUBJECT INDEX